Marine Structures

Marine Structures

Special Issue Editor
Erkan Oterkus

MDPI • Basel • Beijing • Wuhan • Barcelona • Belgrade

Special Issue Editor
Erkan Oterkus
University of Strathclyde
UK

Editorial Office
MDPI
St. Alban-Anlage 66
4052 Basel, Switzerland

This is a reprint of articles from the Special Issue published online in the open access journal *Journal of Marine Science and Engineering* (ISSN 2077-1312) from 2018 to 2019 (available at: https://www.mdpi.com/journal/jmse/special_issues/jz_marine_structures).

For citation purposes, cite each article independently as indicated on the article page online and as indicated below:

LastName, A.A.; LastName, B.B.; LastName, C.C. Article Title. *Journal Name* **Year**, *Article Number*, Page Range.

ISBN 978-3-03928-182-4 (Pbk)
ISBN 978-3-03928-183-1 (PDF)

Contents

About the Special Issue Editor

Erkan Oterkus is a professor in the department of Naval Architecture, Ocean and Marine Engineering at the University of Strathclyde. He is also the director of PeriDynamics Research Centre (PDRC). He received his PhD from University of Arizona, USA and was a researcher at NASA Langley Research Center, USA before joining the University of Strathclyde. His research is mainly focused on the computational mechanics of materials and structures using state-of-the-art techniques including peridynamics and the inverse finite element method. Some of his recent research focuses on multiscale modelling of stress corrosion cracking, the underwater shock response of marine composite structures, the failure analysis of electronic packages, the collision and grounding of ships and real-time monitoring of ship structures. He is the co-author of numerous publications, including the first of book on peridynamics, and journal and conference papers. Dr. Oterkus has been a visiting professor at Stanford University (USA), University of Padova (Italy), Otto von Guericke University (Germany) and Nihon University (Japan). Dr. Oterkus is an associate editor of the *Journal of Peridynamics* and *Nonlocal Modeling* (Springer) and *Sustainable Marine Structures* (NASS). He is also a subject editor of the *Journal of the Faculty of Engineering and Architecture* of Gazi University. In addition, Dr. Oterkus is Special Issue Editor for Computational *Materials Science* (Elsevier), *Journal of Mechanics* (Cambridge), *Journal of Marine Science and Engineering* (MDPI) and *AIMS Materials Science*. Dr. Oterkus is a member of the editorial boards of *International Journal of Naval Architecture and Ocean Engineering* (Elsevier), *Journal of Marine Science and Engineering* (MDPI), *Composite Materials*, and *Annals of Limnology and Oceanography*.

Journal of
Marine Science and Engineering

Editorial

Marine Structures

Erkan Oterkus

Department of Naval Architecture, Ocean and Marine Engineering, University of Strathclyde, Glasgow G4 0LZ, UK; erkan.oterkus@strath.ac.uk

Received: 1 October 2019; Accepted: 1 October 2019; Published: 3 October 2019

Structural mechanics is an important field of engineering. The main goal of structural mechanics is to ensure that structures are safe and durable so that catastrophic situations can be prevented which could otherwise cause environmental pollution, financial loss, and the loss of lives. The usage of the structure and the conditions that the structure is subjected to require special treatment for the analysis. Specifically, marine structures are subjected to harsh environmental conditions due to marine environment which can cause several different damage mechanisms including fatigue and corrosion. This special issue on 'marine structures' considers a wide range of areas related to marine structures and provides a compilation of numerical and experimental studies related to research.

Qiu et al. [1] performed nonlinear finite element analysis in order to investigate the degradation of an offshore protective device by considering the effect of material corrosion. Gao and Oterkus [2] utilised a new computational technique, peridynamics, to study the state of damage to marine composites subjected to shock loading by performing fully coupled thermomechanical analysis. Plodpradit et al. [3] presented modal and coupled analyses of tripod-supported offshore wind turbines using X-SEA and FAST software packages. Hemmati and Oterkus [4] demonstrated a new model for offshore wind turbine systems equipped with a semi-active time-variant tuned mass damper by considering the nonlinear soil–pile interaction phenomenon and time-variant damage conditions. Ding et al. [5] studied towing operation methods, specifically for an offshore integrated meteorological mast to be used for offshore wind farms. Yu et al. [6] performed a failure analysis of topside facilities on oil/gas platforms in the Bohai Sea. Gao et al. [7] determined the cross-flow vortex-induced vibration (VIV) responses and hydrodynamic forces of a long flexible and low mass ratio pipe by performing laboratory tests. Liu and Yang [8] studied the characteristics of acoustic wave transmission in a metamaterial-type seawater piping system. Karambas and Samaras [9] developed an integrated coastal engineering numerical model which is capable of stimulating linear wave propagation, wave-induced circulation, sediment transport, and bed morphology evolution. Ameryoun et al. [10] performed stochastic modelling of bio-colonisation for the computation of stochastic hydrodynamic loading on jacket-type offshore structures. Finally, Chang [11] numerically investigated wave generation and vortex formation under the action of viscous fluid flows over a depression.

As a final remark, I hope the readers who are interested in marine structures will find this special issue useful for their studies.

Acknowledgments: I would like to thank the authors for their contributions to this special issue.

Conflicts of Interest: The authors declare no conflict of interest.

References

1. Qiu, A.; Han, X.; Qin, H.; Lin, W.; Tang, Y. Anti-Collision Assessment and Prediction Considering Material Corrosion on an Offshore Protective Device. *J. Mar. Sci. Eng.* **2017**, *5*, 37. [CrossRef]
2. Gao, Y.; Oterkus, S. Peridynamic Analysis of Marine Composites under Shock Loads by Considering Thermomechanical Coupling Effects. *J. Mar. Sci. Eng.* **2018**, *6*, 38.

3.   Plodpradit, P.; Dinh, V.N.; Kim, K.-D. Tripod-Supported Offshore Wind Turbines: Modal and Coupled Analysis and a Parametric Study Using X-SEA and FAST. *J. Mar. Sci. Eng.* **2019**, *7*, 181. [CrossRef]

4.   Hemmati, A.; Oterkus, E. Semi-Active Structural Control of Offshore Wind Turbines Considering Damage Development. *J. Mar. Sci. Eng.* **2018**, *6*, 102. [CrossRef]

5.   Ding, H.; Hu, R.; Le, C.; Zhang, P. Towing Operation Methods of Offshore Integrated Meteorological Mast for Offshore Wind Farms. *J. Mar. Sci. Eng.* **2019**, *7*, 100. [CrossRef]

6.   Yu, S.; Zhang, D.; Yue, Q. Failure Analysis of Topside Facilities on Oil/Gas Platforms in the Bohai Sea. *J. Mar. Sci. Eng.* **2019**, *7*, 86. [CrossRef]

7.   Gao, X.; Xu, Z.; Xu, W.; He, M. Cross-Flow Vortex-Induced Vibration (VIV) Responses and Hydrodynamic Forces of a Long Flexible and Low Mass Ratio Pipe. *J. Mar. Sci. Eng.* **2019**, *7*, 179. [CrossRef]

8.   Liu, B.; Yang, L. Transmission of Low-Frequency Acoustic Waves in Seawater Piping Systems with Periodical and Adjustable Helmholtz Resonator. *J. Mar. Sci. Eng.* **2017**, *5*, 56.

9.   Karambas, T.V.; Samaras, A.G. An Integrated Numerical Model for the Design of Coastal Protection Structures. *J. Mar. Sci. Eng.* **2017**, *5*, 50. [CrossRef]

10.   Ameryoun, H.; Schoefs, F.; Barillé, L.; Thomas, Y. Stochastic Modeling of Forces on Jacket-Type Offshore Structures Colonized by Marine Growth. *J. Mar. Sci. Eng.* **2019**, *7*, 158. [CrossRef]

11.   Chang, C.-H. Numerical Analyses of Wave Generation and Vortex Formation under the Action of Viscous Fluid Flows over a Depression. *J. Mar. Sci. Eng.* **2019**, *7*, 141. [CrossRef]

Article

Anti-Collision Assessment and Prediction Considering Material Corrosion on an Offshore Protective Device

Ang Qiu [1,2], Xiangxi Han [3], Hongyu Qin [2], Wei Lin [4,*] and Youhong Tang [2,*]

[1] Department of Marine Structure, Guangdong Shipping Science Research Institute, Guangzhou 510110, China; qiuang88@gmail.com

[2] College of Science and Engineering, Flinders University, Adelaide, SA 5042, Australia; hongyu.qin@flinders.edu.au

[3] College of Mechanical and Marine Engineering, Qinzhou University, Qinzhou 535000, China; hanxiangxi@qzhu.edu.cn

[4] Department of Naval Architecture and Ocean Engineering, School of Civil Engineering and Transportation, South China University of Technology, Guangzhou 510648, China

* Correspondence: wlin@scut.edu.cn (W.L.); youhong.tang@flinders.edu.au (Y.T.); Tel.: +86-20-8711-1030-3512 (W.L.); +61-8-820-12138 (Y.T.)

Received: 13 July 2017; Accepted: 9 August 2017; Published: 15 August 2017

Abstract: Corrosion deterioration of steel can heavily degrade the performance of marine and offshore structures. A typical steel protective device, which has worked for a dozen years in a river estuary, is selected as the research object. Its current corrosion response is measured on site and its further corrosive response is predicted based on measurement data and the structure's current state. Nonlinear finite element method is utilized to analyze the degradation of the protective device's anti-collision performance. Meanwhile the rubber buffer effect has been investigated for its anti-collision on the protective device. A prediction method is proposed that can accurately forecast degradation of the anti-collision performance of a protective device as time progresses.

Keywords: corrosion detection; protective device; anti-collision assessment; performance prediction; steel

---

1. Introduction

Steel is one of the most popular materials used in offshore and marine structures such as ships, ocean platforms, and submarine pipelines. The corrosion deterioration of steel can lead to heavy losses [1]: in 2001 the annual worldwide cost of corrosion was around 3% of the world's GDP [2], and Khan [3] summarized that 1034 incidents (44.7% of the total incidents) were caused by steel corrosion and the corresponding 'mechanical failure', including a series of serious incidents in 1988 and 1992 in the U.S. [1,4]. Many researchers have studied the mechanism of steel corrosion in marine and offshore environments. Jyoti et al. [5] summarized previous research work and reported that temperature [6], pH value [7], salinity [8], dissolved oxygen [9], and water velocity [10] were the main physical factors that affected the corrosion responses of steel. Besides the investigation of corrosion factors, other studies have focused on mechanism optimization. For example, Li et al. [11] adjusted element contents within steel, and found that tin-containing steel had better corrosion resistance than steel itself.

The assessment of steel under a state of corrosion is another area of current research interest. Calabrese et al. [12] studied the corrosion degradation effects of aluminum/steel clinched joints and pointed out that thicker aluminum foil could endow superior durability properties in large-scale applications. Chen et al. [13] used finite element analysis (FEA) to investigate the failure processes of high-strength pipelines with single and multiple corrosions and produced an assessment procedure to predict the failure pressure of a pipeline with multiple corrosions. Jin et al. [14] used FEA in considering the shear buckling behavior of a web panel with pits and through-thickness corrosion damage, and found that through-thickness corrosion damage significantly decreased the critical buckling load.

Bridge protective devices are among the many large-scale steel structures used in marine environments. They are energy-absorptive structures designed to withstand ship impact and depend on internal steel plastic damage [15]. Many studies of ship–bridge collision responses [16,17] and device usage [18,19] have been reported. For example, Qiu et al. [20] analyzed an original protective device scheme and proposed structural and material optimization for the protective device [21–23].

Because of the many influencing factors, it is difficult to predict realistic corrosion responses of engineering structures based only on existing research conclusions and results, especially given the coupling effect of an antirust coat. Degradation of the anti-collision performance of protective devices influenced by corrosion has been insufficiently analyzed. In this study, a protective device that had worked for 12 years in a river estuary was selected as a research object, and its current corrosion response was detected on-site based on the corresponding standards [24,25]. Further corrosive response was predicted based on measurement data and the current state of the structure. The current degradation of the protective device's anti-collision performance and the device's further corrosion states were analyzed and predicted separately. An anti-collision performance prediction method for the protective device is proposed.

2. Model Description

Yamen Bridge is located in the Pearl River estuary, China. The estuary experiences severe engineering environments such as a high tidal range, high swell velocity, and large-tonnage ships shuttling back and forth over it. Thus, a protective device had been designed to protect the bridge piers from ship collisions. The main geometrical parameters of the studied device are listed in Table 1. This protective device was a large-scale shell structure containing 14 watertight cabins and 4 non-watertight cabins assembled by deck plates, bottom plates, longitudinal bulkheads, transverse bulkheads, and trunk bulkheads. Each plate was enhanced by stiffeners with rib distance of 650 mm to ensure structural stability. Non-watertight cabins were used to adjust the protective device's floating state when it was assembled around bridge piers. After assembly of the device, seawater could flow into the non-watertight cabins. This maintained the same water level within and outside the protective device and ensured that the protective device was unaffected by tidal levels. Seventy drum-type buffer rubbers were assembled on the inside bulkhead of the protective device, designed to absorb impact forces that the bridge pier received during ship collision.

Table 1. Key geometrical parameters of the protective device.

Length (L)	Breadth (B)	Depth (H)	Draught (d)
47.95 m	30.20 m	7.50 m	various

Front collision is selected as the study case here, with the design value of the ultimate impact force being 28 MN. Cabins in the front direction would perform the main anti-collision functions. Figure 1 shows a sketch of the protective device.

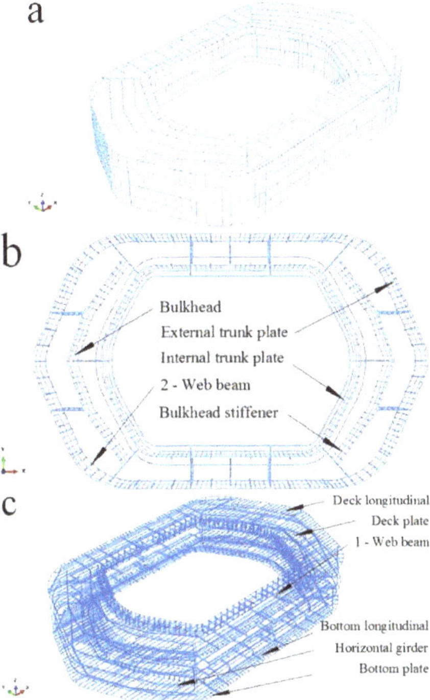

Figure 1. Sketch of protective device (**a**) isometric view, (**b**) top view and (**c**) perspective view with a code of each cabin.

3. Corrosion Measurement and Prediction

The protective device had been manufactured and placed in the river estuary 12 years earlier and corrosion had certainly occurred within it. External views of the protective device were obtained first through a diver's visual inspection; subsequently, each steel component's residual thickness was measured by an ultrasonic thickness gauge. The whole structure of the protective device was examined to check whether pitting corrosion, bulking, weld crack, steel damage, or corrosion punching had occurred. Visual inspection and hammer testing were utilized in the tidal and spray zones, and diver inspection was utilized to detect the corrosion state of the immersed zone. Typical corrosion states of the protective device are shown in Figure 2 and the detection results are listed in Table 2. As shown in Table 2, no apparent pitting corrosion, weld crack, or steel plate punching was observed, implying that no severe local damage had occurred.

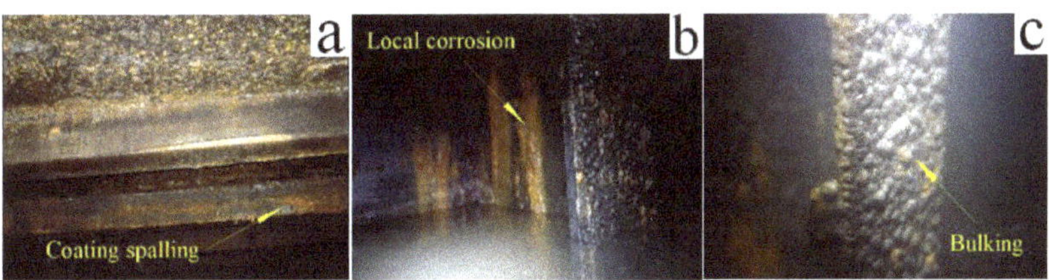

Figure 2. Typical images of corrosion states within the protective device of (**a**) the spray zone; (**b**) the tidal zone and (**c**) the full immersion zone.

Table 2. Visual detection results of corrosion states within the protective device.

Corrosion Extent	Spray Zone	Tidal Zone	Full Immersion Zone
Coating spalling	●	●	●
Local corrosion	●	●	●
Bulking	●	●	●
Aquatic adhesion		●	●

Originally, undercoat and finishing had been painted on the protective device for anticorrosion purposes and a special undercoating containing Zn [26] was used in the tidal zone, as steel in this zone oxidized more severe than in other zones [5]. After years of immersion, however, rusting had occurred almost everywhere (see Figure 2), the coating retained little anticorrosion effect, meanwhile, corrosion and structural loss had occurred in almost every part of the protective device. Components in the different zones (spray, tidal, and full immersion) were measured separately as they were situated in different environmental conditions (differences in temperature, dissolved oxygen, salinity, pH value, redox potential, etc.). As the protective device was a non-watertight structure, seawater perfused into the undersea cabins and thus seawater corrosion occurred on both sides of the steel components. In total, 500 points were selected to measure the residual thickness of the steel components and data were classified according to the design thickness of the steel component, as summarized in Table 3. Comparisons of the different corrosion zones showed that the protective device experienced the most severe corrosion in the full immersion zone rather than in the tidal zone and its anticorrosion performance was best in the spray zone.

Table 3. Corrosion state classified by design thickness of steel component.

Design Thickness (mm)	Zone	Measuring Points	Average Value (mm)	Average Corrosion Rate [1] (mm/a) [2]	Minimal Value (mm)	Maximum Corrosion Rate [1] (mm/a) [2]
6.00	Spray zone	12	5.60	0.03	5.30	0.06
	Tidal zone	12	5.50	0.04	5.30	0.06
	Full immersion zone	12	5.36	0.05	4.30	0.14
8.00	Spray zone	113	7.50	0.04	6.90	0.09
	Tidal zone	69	7.32	0.06	5.10	0.24
	Full immersion zone	101	7.30	0.06	5.10	0.24
10.00	Spray zone	39	9.00	0.08	7.30	0.23
	Tidal zone	47	8.70	0.11	6.70	0.28
	Full immersion zone	95	8.60	0.12	6.50	0.29

Note: [1] Corrosion rate [25] was calculated based on measurement data and service years, not considering the influence of the coating's guarantee period; [2] mm/a is mm per annual.

Pitting corrosion is regarded as one of the most hazardous forms of corrosion for traditional marine and offshore structures as it is a localized accelerated dissolution of metal that can weaken structural stiffness and lead to early catastrophic failure (oil leakage, product loss, environmental pollution, loss of life, for example) [4]. Pitting corrosion does not greatly affect the function of a protective device, as pitting corrosion (even consequent piercing) merely causes local damage relative to the large-scale structure of the whole device and its influence on anti-collision performance is negligible [27]. The design principle of the protective device is to absorb a large amount of the kinetic energy from the hull's impact through plastic damage to the metal within itself. The amount of energy absorbed through plastic damage to the metal correlates with the mass loss and consequent thickness reduction of the metal. The average residual metal thicknesses of typical steel components of the protective device are listed in Table 4. In the subsequent FEA, each steel plate or enhanced component is modeled with the average thickness detected by a thickness gauge as reported above.

According to the simulation results presented in the next section, the protective device at the current time could meet anti-collision requirements. As long as the protective device remained in active service around the bridge pier, everyday corrosion progressed. Further analysis to predict whether the protective device could retain its anti-collision performance after a long period of service is an important issue. The protective device was located in a marine environment and, when manufactured in a shipyard, it was designed with only 10 years' service life. Most of its components were painted by hand-brushing using common marine coatings whose normal durability was 10 years. Thus, further corrosive responses need to be predicted if we want to study the subsequent corrosion response of the device and judge its ultimate service life. Steel components within the protective device were classified according to design thickness and the corrosion zone in which they were located, and then the further corrosive response of each classified group was predicted. The average residual thicknesses of typical components in their corrosion environment (5-year prediction) are listed in Table 5.

Table 4. Average measured residual thickness of typical plates and stiffeners in the protective device.

Component Name	Design Thickness (mm)	Measured Thickness (mm)		
		Spray Zone	Tidal Zone	Full Immersion Zone
		Plates		
Deck plate	8.00	7.50	—	—
Tween deck plate	8.00	—	7.32	—
Bottom plate	8.00	—	—	7.30
External trunk plate	8.00	7.50	7.32	7.30
Internal trunk plate	8.00	7.50	7.32	7.30
Bulkhead	10.00	9.00	8.70	8.60
		Stiffeners		
Deck longitudinal	$L140 \times 90 \times 8$	$L140 \times 90 \times 7.5$	—	—
Tween deck longitudinal	$L100 \times 75 \times 8$	—	$L100 \times 75 \times 7.32$	—
Bottom longitudinal	$L150 \times 100 \times 10$	—	—	$L150 \times 100 \times 8.59$
Bulkhead stiffener	$L80 \times 50 \times 6$	$L80 \times 50 \times 5.6$	$L80 \times 50 \times 5.5$	$L80 \times 50 \times 5.3$
1-Web beam	$\perp\dfrac{8 \times 350}{12 \times 150}$	$\perp\dfrac{7.5 \times 350}{11.2 \times 150}$	$\perp\dfrac{7.32 \times 350}{11.0 \times 150}$	$\perp\dfrac{7.3 \times 350}{10.9 \times 150}$
2-Web beam	$\perp\dfrac{8 \times 225}{12 \times 125}$	$\perp\dfrac{7.5 \times 225}{11.2 \times 125}$	$\perp\dfrac{7.32 \times 225}{11.0 \times 125}$	$\perp\dfrac{7.3 \times 225}{10.9 \times 125}$
Horizontal girder	$\perp\dfrac{8 \times 350}{12 \times 200}$	$\perp\dfrac{7.5 \times 350}{11.2 \times 200}$	$\perp\dfrac{7.32 \times 350}{11.0 \times 200}$	$\perp\dfrac{7.3 \times 350}{10.9 \times 200}$

Note: $L100 \times 75 \times 8$ means L-section stiffener with length 100 mm, width 75 mm and thickness 8 mm; $\perp\dfrac{8 \times 350}{12 \times 150}$ means T-section stiffener with web length 350 mm, web thickness 8 mm, panel length 150 mm and panel thickness 12 mm.

Table 5. Average 5-year predicted residual thickness of typical plates and stiffeners in the protective device.

Component Name	Design Thickness (mm)	Measured Thickness (mm)		
		Spray Zone	Tidal Zone	Full Immersion Zone
		Plates		
Deck plate	8.00	6.50	—	—
Tween deck plate	8.00	—	5.96	—
Bottom plate	8.00	—	—	5.90
External trunk plate	8.00	6.50	5.96	5.90
Internal trunk plate	8.00	6.50	5.96	5.90
Bulkhead	10.00	7.00	6.10	5.80
		Stiffeners		
Deck longitudinal	$L140 \times 90 \times 8$	$L140 \times 90 \times 6.5$	—	—
Tween deck longitudinal	$L100 \times 75 \times 8$	—	$L100 \times 75 \times 5.96$	—
Bottom longitudinal	$L150 \times 100 \times 10$	—	—	$L150 \times 100 \times 5.8$
Bulkhead stiffener	$L80 \times 50 \times 6$	$L80 \times 50 \times 4.8$	$L80 \times 50 \times 4.5$	$L80 \times 50 \times 4.08$
1-Web beam	$\perp\dfrac{8 \times 350}{12 \times 150}$	$\perp\dfrac{6.5 \times 350}{9.8 \times 150}$	$\perp\dfrac{5.96 \times 350}{8.9 \times 150}$	$\perp\dfrac{5.9 \times 350}{8.8 \times 150}$
2-Web beam	$\perp\dfrac{8 \times 225}{12 \times 125}$	$\perp\dfrac{6.5 \times 225}{9.8 \times 125}$	$\perp\dfrac{5.96 \times 225}{8.9 \times 125}$	$\perp\dfrac{5.9 \times 225}{8.8 \times 125}$
Horizontal girder	$\perp\dfrac{8 \times 350}{12 \times 200}$	$\perp\dfrac{6.5 \times 350}{9.8 \times 200}$	$\perp\dfrac{5.96 \times 350}{8.9 \times 200}$	$\perp\dfrac{5.9 \times 350}{8.8 \times 200}$

4. FE Simulation

4.1. FE Model of the Protective Device

A detailed FE model of the protective device was built, with each plate and stiffener (Tables 4 and 5) modeled with shell elements. The majority of shell elements were S4R (4-node double-curved thick shell with reduced integration); some small components were modeled with S3R (3-node triangular thick shell) elements. Although the macroscopic structure size of the protective device was very large, the mesh size was only between 130 and 260 mm, and in total 186,126 elements (connected by 244,102 nodes) were used. The main parameters of the protective device are listed in Table 1 and its FE model is shown in Figure 3.

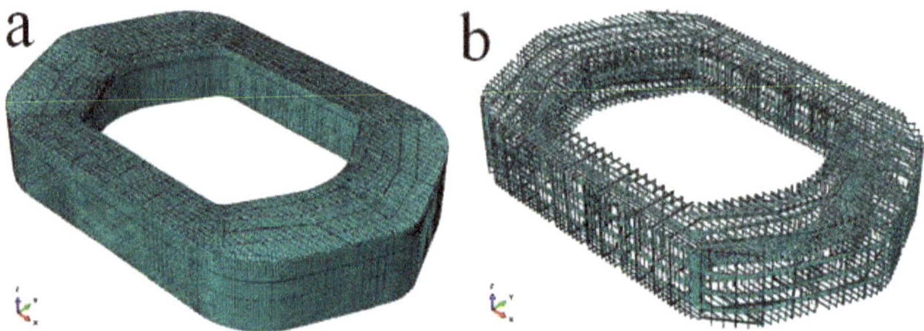

Figure 3. FE model of the protective device (**a**) Integration sketch and (**b**) Inner frame sketch.

4.2. FE Model of a Striking Ship

Both ocean-going and inland vessels with deadweight tonnage (DWT) normally around 5000 tons commonly pass through a river estuary, so a 5000 DWT striking ship model was proposed, built with shell elements. As the ship's response was not the research object here, the ship's FE model was simplified to minimize calculation time. Only the ship's bow was modeled as an elastic body; other parts were modeled as rigid bodies as they were relatively distant from the protective device and direct collision with them should not occur. Enhanced stiffeners and bulkheads within the ship's bow were deleted and their mass was compensated through increasing the bulkhead thickness. Furthermore, a centerline bulkhead was added in the ship's bow to compensate for the loss of section modulus. The main parameters of the striking ship are listed in Table 6.

Table 6. Key parameters of the ship.

Principal Dimensions				Mass Matrix (Considering Attach Water)			
Overall length (L)	112 m	Load draught (d)	7 m	M11	12,276 t	I11	447,841 t·m^2
Modeled breadth (B)	17 m	Block coefficient (C_b)	0.8	M22	23,100 t	I22	17,248,000 t·m^2
Molded depth (H)	9.2 m	Displacement	11,000 t	M33	23,100 t	I33	17,248,000 t·m^2

4.3. Boundary Conditions

Front collision was selected as the case involving the maximum impact force received by a bridge pier [20]. The ship's stroking velocity was defined as 5 m/s and the high tide level was selected, considering that navigation is relatively busy at this tide level. The full immersion zone was below the low tide level, the tidal zone was between the low and high tide levels, and the spray zone was above the high tide level. Detailed parameters of the boundary conditions are listed in Table 7.

Table 7. Boundary conditions of the ship's collision.

Ship Draught (m)	Protective Device Draught (m)	Stroking Velocity (m/s)	Impact Direction	Impact Location
7.00	6.66	5	Transverse bridge direction	Transverse vertex

The attached water mass proposed was based on relevant theory [28] and was uniformly distributed onto the under-water part of the protective device in order to solve the fluid-solid coupling problem without increasing calculation time. Different corrosion states of the protective device were defined and the anti-collision effect of the buffer rubber was also considered. In total, 5 work cases were studied. The buffer rubber's anti-collision effect was analyzed in cases 1 and 2 (without and with buffer rubber equipment), and the influence of the protective device's corrosion was analyzed based on cases 3–5 (case 3 corresponds to the original design state, case 4 to the current corrosion state, and case 5 to the predicted future corrosion state).

4.4. Material Properties

This protective device was made of common marine steel Q235 with the chemical composition was well documented in the literature and the steel's mechanical parameters concerning metal plasticity are listed in Table 8. Ductile and shear failure criteria were utilized to determine the damage response of the marine steel. The mechanical properties of the rubber (simulated by a viscous elastic spring) are listed in Table 9. The friction coefficient between steel components was assigned as 0.10 [29].

Table 8. Mechanical properties of marine steel Q235.

Basic Parameters			Yield Stress (MPa)	Plastic Strain
Elasticity Modulus E (GPa)	Poisson Ratio	Density (kg·m^{-3})		
			235	0
			245	0.01
			251	0.02
			255	0.03
			262	0.06
210	0.3	7800	267	0.10
			271	0.15
			276	0.25
			279	0.40
			289	2.00

Table 9. Stiffness and damping factors of the buffer rubber.

Standard Code	Stiffness Factor	Damping
SC500	250,000	250,000

5. Results and Discussion

5.1. Anti-Collision Effect of Buffer Rubber

As the protective device had been used and immersed in a corrosive environment for 12 years, the mechanical performance of the buffer rubber had significantly decreased. The current mechanical performance of the buffer rubber was not measured as it could not simply be represented by thickness reduction. Two models (cases 1 and 2) were built: one was equipped with buffer rubber and the other not, to represent typical/extreme cases, and the protective device's anti-collision response is analyzed here.

The simulation results are listed in Table 10 and the corresponding reaction force histories are shown in Figure 4. The maximum value of the impact force received by the protective device is 26.79 MN in case 1, compared with 26.12 MN in case 2, an increase of 2.56%. The maximum ship stroke in case 1 is 8.69 m, compared with 8.71 m in case 2, a decrease of 0.23%. According to the discussion above, the protective device equipped with buffer rubber could decrease the maximum impact force and prolong the ship stroke, but the difference is not significant, as shown in Figure 4. However, it is worth mentioning that buffer rubber can obviously decrease the impact force at the beginning of impact and allow the protective device to activate its function more quickly, as shown in Figure 5. At the beginning stage of impact (within 0.15 s), the maximum impact force in case 1 is 15.75 MN (0.14 s), whereas in case 2 it is 13.43 MN (0.06 s), a decrease of 14.73%. Therefore, if the ship impact is too weak to damage the protective device's steel components, the anti-collision effect of the buffer rubber is efficient and sufficient, but if the collision is too violent and must be resisted through structural damage within the protective device, the buffer rubber plays almost no role.

As the performance of the protective device was demonstrated by the ultimate collision state and the ship collision cases discussed here were all violent, the buffer rubber was insignificant with regard to the impact response in the ultimate collision state. In the following cases, the protective device models are all equipped with buffer rubber and the corrosive state of the steel components is the only parameter investigated.

Table 10. Simulation results of bridge pier with and without buffer rubber equipment.

Case No.	Feature	Maximum Impact Force Received by Bridge Pier (MN)	Maximum Internal Energy of Protective Device (kJ)	Maximum Ship Stroke (m)	Impact Duration (s)
1	No buffer rubber	26.79	148,371	8.69	3.23
2	Buffer rubber present	26.12	147,883	8.71	3.23

Note: Impact duration is defined as the period during which a ship obtains its minimum kinetic energy.

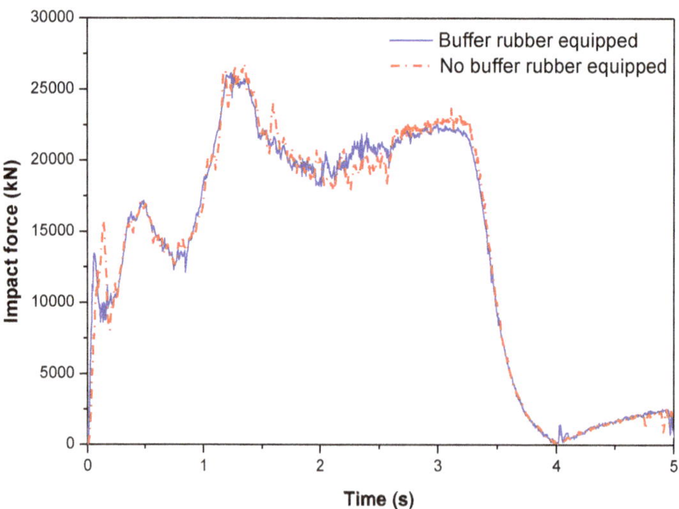

Figure 4. Reaction force histories of bridge pier with and without buffer rubber equipment.

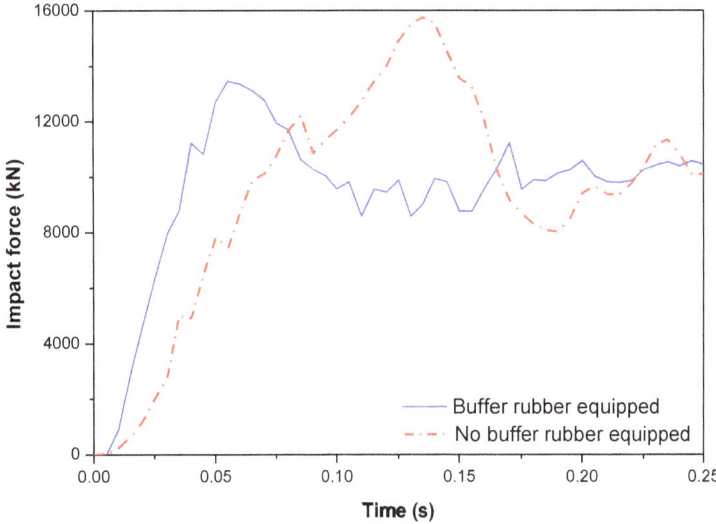

Figure 5. Initial stage of reaction force histories of the bridge pier with and without buffer rubber.

5.2. Corrosion Effects on the Protective Device

Models of the protective device in the original design, with current corrosion, and with predicted further corrosion were built separately. Simulation results are listed in Table 11. The lowest maximum impact force experienced by the bridge piers occurs in case 3 (25.28 MN) and the highest maximum impact force increases as corrosion continues, with values in cases 4 and 5 of 26.62 MN and 28.98 MN respectively, increases of 5.3% and 14.6% respectively from the original design state. It is worth noting that the maximum impact force in case 5 exceeds the ultimate design value (28 MN), which means that the protective device cannot meet design requirements in this predicted further corrosion state. Reaction force histories are shown in Figure 6, in which the shape of the reaction force history with the original design is somewhat rectangular, revealing that the range of variation of the impact force is greater with the original design than in the corrosion states. The ship collision is the most severe in the predicted further corrosion state: not only is the peak value of the reaction force (28.98 MN) higher, but also the variation of the reaction force and the corresponding acceleration velocity is more turbulent, with higher acceleration velocity than in the other two cases.

Table 11. Simulation results of bridge pier under different corrosion states.

Feature	Maximum Impact Force Received by Bridge Pier (MN)	Maximum Internal Energy of Device (kJ)	Maximum Ship Stroke (m)	Impact Duration (s)	Mass of Protective Device (t)
Original design	25.28	147,563	8.51	3.29	463.64
Current corrosion state	26.62	147,883	8.69	3.23	413.85
Predicted further corrosion state	28.98	146,698	8.81	3.17	327.06

Note: Impact duration is defined as the period in which ship obtains its minimum kinetic energy.

The buffering effect of the protective device tended to be less when the plate thickness decreased under the action of corrosion, and the extent of corresponding structural deformation was more severe once the ship collision occurred. Deformation responses of the protective device during impact are shown in Figure 7. In the beginning stage of impact, the protective device in the further corrosion state deforms most easily. During the entire impact process, the extent of deformation and damage of the protective device is the greatest in the further corrosion state.

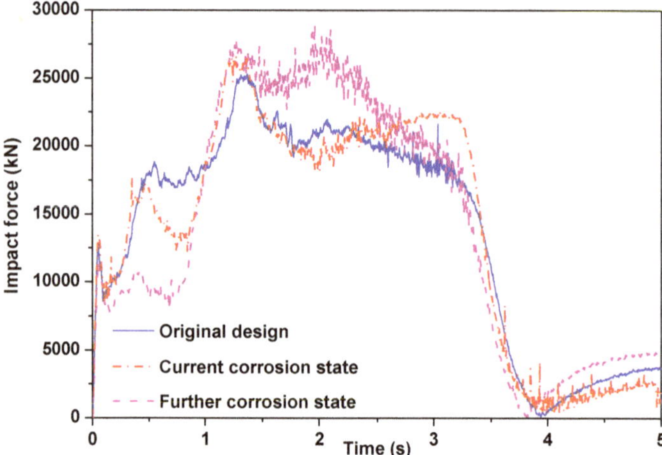

Figure 6. Reaction force histories of the bridge pier under different corrosion states.

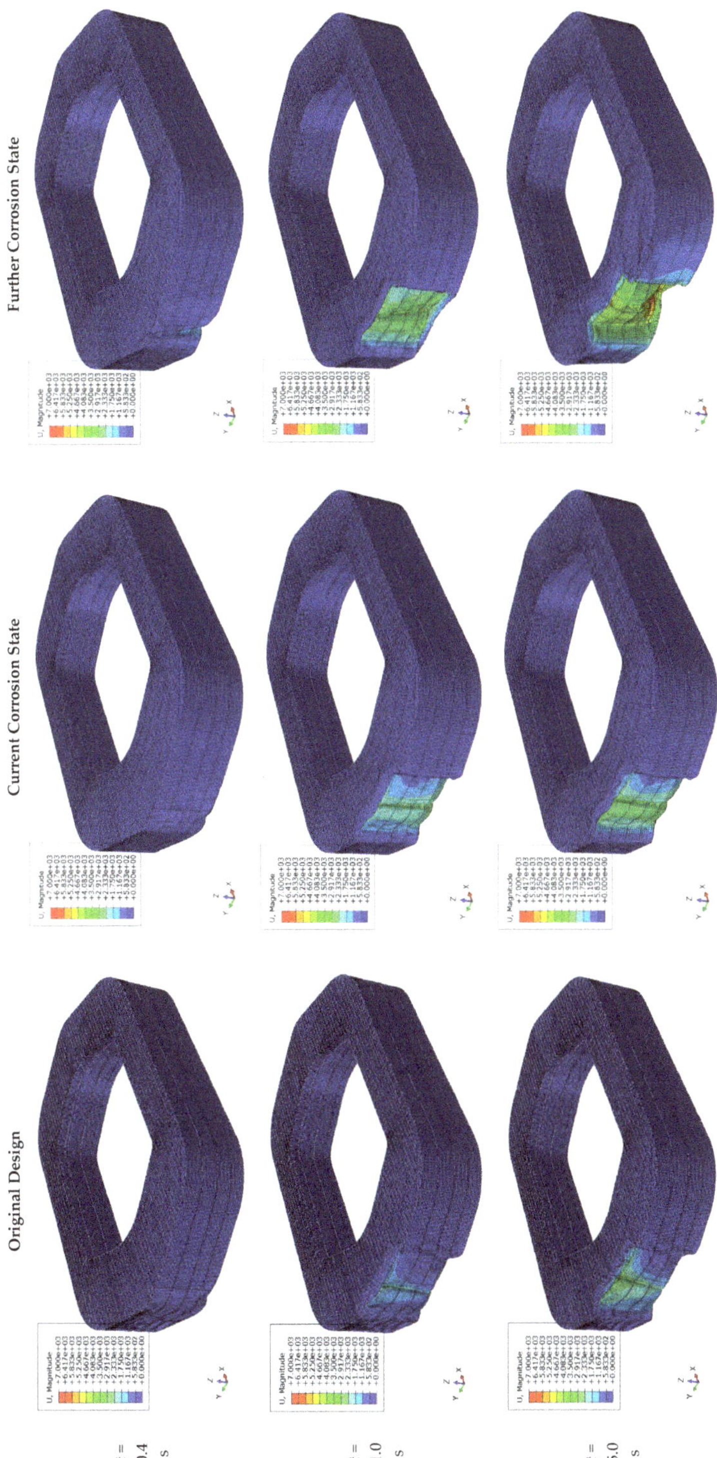

Figure 7. Deformation distribution in protective device during impact for original design, current corrosion state, and further corrosion state.

5.3. Prediction of Performance Degradation

The protective device was a large-scale shell structure assembled with different steel components located in different corrosion zones. As the corrosion characteristics in the spray zone, the tidal zone, and the full immersion zone differed, and the size of steel components could also influence the corrosion process, it was difficult to accurately quantify the overall corrosion response of the protective device. Moreover, ship collision is a complex impact process that involves problems of material nonlinearity (metal plasticity and rubber hyperelasticity), geometric nonlinearity (large deformation), and boundary nonlinearity (boundary variation). All these factors increase the difficulty in predicting the anti-collision performance of the protective device after further corrosion. Based on the simulation results discussed here, a polynomial equation is proposed to judge the variation trend of the maximum impact force. Then a new case is calculated to judge the validity of the equation through comparison with the design value. The anti-collision performance of the protective devices is further discussed from the perspectives of ship stroke and residual mass of the device.

The maximum impact forces under different corrosion states were calculated and are listed in Table 11. According to these data plotted in Figure 8, a prediction equation (Equation (1)) was derived by a polynomial fitting method to show the evolution trend:

$$\text{Impact force} = 0.0212 \cdot \text{year}^2 - 0.1427 \cdot \text{year} + 25.28 \ (\text{MN}) \tag{1}$$

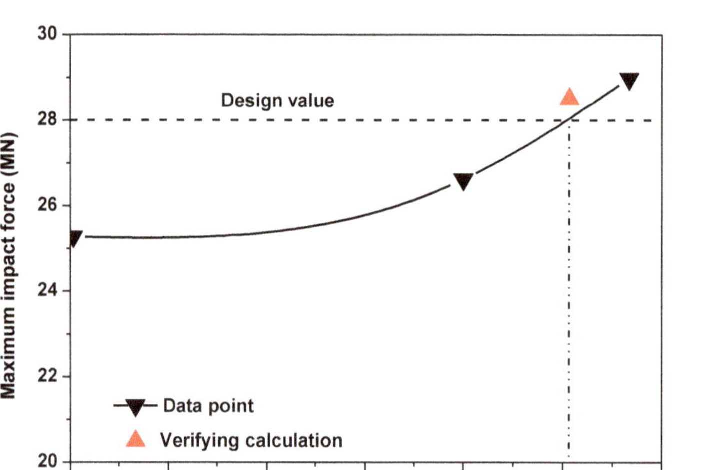

Figure 8. Prediction of maximum impact force and verification considering corrosion effect.

With substitution of the design value (28 MN) into Equation (1), the calculated service life was 15.18 years. A new FE model was built up considering the corrosion effect after 15.18 years. The upper triangle (with red color) in Figure 8 represents the maximum reaction force calculated with the new FE model, and it is evident that it matches the evolution trend well. The evolution trend of anti-collision performance follows a quadratic nonlinear relationship and is unlike the corrosion effect of the steel component, which is determined by a unique spot in the marine environment and is an approximately linear relationship [5]. To further investigate the corrosion response in the protective device, other impact factors were analyzed and verified by this simulation case, from which the corresponding data points and fitted curves are shown in Figure 9. Each verification point falls exactly on the respective fitting curve, which means that, under the corrosion effect, both the quadratic curves plotted are valid and they fit the variation trend of each factor well. Therefore, the variation trend of device's anti-collision response follows a quadratic nonlinear relationship under the corrosion effect, which is unlike the linearly changing corrosion effect, because too many influencing factors are involved. However, prediction of this variation trend depends merely on three calculation values referring to different corrosion durations, respectively.

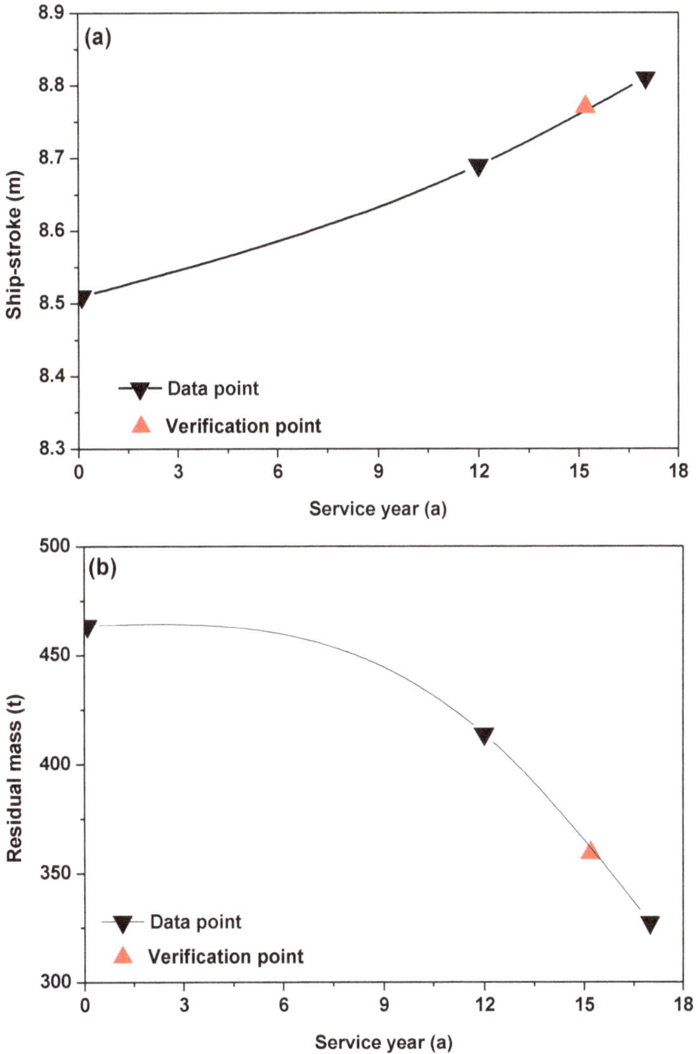

Figure 9. Prediction and verification of key impact factors (**a**) ship stroke prediction and (**b**) residual mass prediction considering corrosion effects.

6. Conclusions

When the Yamen Bridge protective device located in the Pearl River estuary had been in active service for 12 years, on-site measurement was utilized to detect its current corrosion response. Further, the evolution of its corrosion was predicted based on measurement data and corresponding theories. The anti-collision performance of the protective device was analyzed taking into account the corrosion effect, and a performance prediction method was proposed. If a ship's impact is too weak to damage the protective device's steel components, the anti-collision effect of the buffer rubber is considerable, but if the ship's impact is so great that it must be resisted through structural damage within the protective device, the buffer rubber plays almost no role. The anti-collision response of the protective device under the corrosion effect followed a quadratic nonlinear variation trend that was accurately predicted by a quadratic curve determined by three calculation values with reference to different corrosion durations.

Acknowledgments: This study was partially supported by "Haiou Talent Plan" of Qinzhou City, China. We would like to thank Chengbi Zhao from South China University of Technology for his valuable comments in this study.

Author Contributions: Y.T., W.L. and A.Q. conceived and designed the analysis; A.Q. performed the experiments; A.Q. and X.H. analyzed the data; H.Q. contributed analysis tools; A.Q. and W.L. wrote the paper.

Conflicts of Interest: The authors declare no conflict of interest.

References

1. Maureen, W.; Lisa, V.A.A.; Lorenzo, V.W. *Corrosion-Related Accidents in Petroleum Refineries: Lessons Learned from Accidents in EU and OECD Countries*; Publications Office of the European Union: Luxembourg, 2013.
2. Velazquez, J.; Van Der Weide, J.; Hernandez, E.; Hernandez, H.H. Statistical modeling of pitting corrosion: Extrapolation of the maximum pit depth growth. *Int. J. Electrochem. Sci.* **2014**, *9*, 4129–4143.
3. Khan, F.; Howard, R. Statistical approach to inspection planning and integrity assessment. *Insight-Non-Destr. Test. Cond. Monit.* **2007**, *49*, 26–36. [CrossRef]
4. Roberge, P.R. *Corrosion Engineering: Principles and Practice*; McGraw-Hill Professional: New York, NY, USA, 2008.
5. Bhandari, J.; Khan, F.; Abbassi, R.; Garaniya, V.; Ojeda, R. Modelling of pitting corrosion in marine and offshore steel structures—A technical review. *J. Loss Prevent. Proc.* **2015**, *37*, 39–62. [CrossRef]
6. Ramana, K.V.S.; Anita, T.; Mandal, S.; Kaliappan, S.; Shaikh, H.; Sivaprasad, P.V.; Dayal, R.K.; Khatak, H.S. Effect of different environmental parameters on pitting behavior of AISI type 316L stainless steel: Experimental studies and neural network modeling. *Mater. Des.* **2009**, *30*, 3770–3775. [CrossRef]
7. Younis, A.A.; EI-Sabbah, M.M.B.; Holze, R. The effect of chloride concentration and pH on pitting corrosion of AA7075 aluminum alloy coated with phenyltrimethoxysilane. *J. Solid State Electrochem.* **2012**, *16*, 1033–1040. [CrossRef]
8. Zakowski, K.; Narozny, M.; Szocinski, M.; Darowicki, K. Influence of water salinity on corrosion risk-the case of the southern Baltic Sea coast. *Environ. Monit. Assess.* **2014**, *186*, 4871–4879. [CrossRef] [PubMed]
9. Zamaletdinov, I.I. Pitting on passive metals. *Prot. Met.* **2007**, *43*, 470–475. [CrossRef]
10. Melchers, R.E. Pitting corrosion of mild steel in marine immersion environment-Part 2: Variability of maximum pit depth. *Corrosion* **2004**, *60*, 937–944. [CrossRef]
11. Li, H.; Yu, H.; Zhou, T.; Yin, B.; Yin, S.; Zhang, Y. Effect of tin on the corrosion behavior of sea-water corrosion-resisting steel. *Mater. Des.* **2015**, *84*, 1–9. [CrossRef]
12. Calabrese, L.; Proverbio, E.; Galtieri, G.; Borsellino, C. Effect of corrosion degradation on failure mechanisms of aluminium/steel clinched joints. *Mater. Des.* **2015**, *87*, 473–481. [CrossRef]
13. Chen, Y.; Zhang, H.; Zhang, J.; Li, X.; Zhou, J. Failure analysis of high strength pipeline with single and multiple corrosions. *Mater. Des.* **2015**, *67*, 552–557. [CrossRef]
14. Ahn, J.H.; Cheung, J.H.; Lee, W.H.; Oh, H.; Kim, I.T. Shear buckling experiments of web panel with pitting and through-thickness corrosion damage. *J. Constr. Steel Res.* **2015**, *115*, 290–302. [CrossRef]
15. Lu, G.; Yu, T. *Energy Absorption of Structures and Materials*; Woodhead Publishing: Cambridge, UK, 2003.
16. Macduff, T. The Probability of Vessel Collisions. *Ocean Ind.* **1974**, *9*, 144–148.
17. Sha, Y.; Hao, H. Nonlinear finite element analysis of barge collision with a single bridge pier. *Eng. Struct.* **2012**, *41*, 63–76. [CrossRef]
18. Laura, P.A.A.; Nava, L.C. An economic device for protecting bridge piers against ship collisions. *Ocean Eng.* **1981**, *8*, 331–333. [CrossRef]
19. Wang, L.; Yang, L.; Huang, D.; Zhang, Z.; Chen, G. An impact dynamics analysis on a new crashworthy device against ship-bridge collision. *Int. J. Impact Eng.* **2008**, *35*, 895–904. [CrossRef]

20. Qiu, A.; Lin, W.; Ma, Y.; Zhao, C.; Tang, Y. Novel material and structural design for large-scale marine protective devices. *Mater. Des.* **2015**, *68*, 29–41. [CrossRef]
21. Qiu, A.; Fu, K.; Lin, W.; Zhao, C.; Tang, Y. Modelling low-speed drop-weight impact on composite laminates. *Mater. Des.* **2014**, *60*, 520–531. [CrossRef]
22. Qiu, A.; Tang, Y.; Zhao, C.; Lin, W. Numerical investigation of transverse tensile behaviors of marine composites under different strain rates. *Adv. Mater. Res.* **2013**, *774–776*, 944–948. [CrossRef]
23. Qiu, A.; Zhao, C.; Tang, Y.; Lin, W. Rapid predicting the impact behaviors of marine composite laminates. *Mater. Sci. Forum* **2015**, *813*, 19–27. [CrossRef]
24. ASTM International. *Standard Practice for Exposing and Evaluating Metals and Alloys in Surface Seawater*; ASTM G52-00; ASTM International: West Conshohocken, PA, USA, 2011.
25. Standards China. Technical Specification for Corrosion Protection of Steel Structures for Sea Port Construction. Chinese Standard JTS-153-3-2007. 2007. Available online: http://www.chinasybook.com (accessed on 14 August 2017).
26. Li, J.; Bao, G. Main girder protective shell construction of Yamen Bridge. *J. Chongqing Jiaotong Univ.* **2003**, *22*, 11–13.
27. Lu, G.; Wang, X. On the quasi-static piercing of square metal tubes. *Int. J. Mech. Sci.* **2002**, *44*, 1101–1115. [CrossRef]
28. Zhan, D.; Zhang, L.; Zhao, C.; Wu, J.; Zhang, S. Numerical simulation and visualization of immersed tube tunnel maneuvering and immersing. *J. Wuhan Univ. Technol.* **2001**, *25*, 16–20.
29. Johannes, V.I.; Green, M.A.; Brockley, C.A. The role of the rate of application of the tangential force in determining the static friction coefficient. *Wear* **1973**, *24*, 381–385. [CrossRef]

Journal of
Marine Science and Engineering

Article

Peridynamic Analysis of Marine Composites under Shock Loads by Considering Thermomechanical Coupling Effects

Yan Gao and Selda Oterkus *

Department of Naval Architecture, Ocean and Marine Engineering, University of Strathclyde,
Glasgow G4 0LZ, UK; yan.gao@strath.ac.uk
* Correspondence: selda.oterkus@strath.ac.uk; Tel.: +44-141-548-4979

Received: 15 February 2018; Accepted: 31 March 2018; Published: 6 April 2018

Abstract: Nowadays, composite materials have been increasingly used in marine structures because of their high performance properties. During their service time, they may be exposed to extreme loading conditions such as underwater explosions. Temperature changes induced by pure mechanical shock loadings cannot to be neglected especially when smart composite materials are employed for condition monitoring of critical systems in a marine structure. Considering this fact, both the thermal loading effect on deformation and the deformation effect on temperature need to be taken into consideration. Consequently, an analysis conducted in a fully coupled thermomechanical manner is necessary. Peridynamics is a newly proposed non-local theory which can predict failures without extra assumptions. Therefore, a fully coupled thermomechanical peridynamic model is developed for laminated composites materials. In this study, numerical analysis of a 13 ply laminated composite subjected to an underwater explosion is conducted by using the developed model. The pressure shocks generated by the underwater explosion are applied on the top surface of the laminate for uniform and non-uniform load distributions. The damage is predicted and compared with existing experimental results. The simulation results obtained from uncoupled case are also provided for comparison. Thus the coupling term effects on crack propagation paths are investigated. Furthermore, the corresponding temperature distributions are also investigated.

Keywords: peridynamics; thermomechanical; composites; shock loads

---

1. Introduction

Laminated composite materials have many outstanding mechanical, physical, and chemical properties. For example, they are an easily fabricated and cost-effective alternative to some other monolithic materials [1]. Therefore, in recent years, composites have become common materials in marine industries. One application is for the construction of military vessels [2]. Composite materials can provide low-radar signatures for stealth operations. In addition, the low electro-magnetic signature these materials provide can reduce the possibility of detonating magnetic sea mines [3]. However, due to the special working conditions for military vessels, the composite materials may be subjected to some severe environments, such as mechanical shock loads, large temperature variations, and exposures [4]. Hence, the damage level of composites induced by such extreme loading conditions becomes a critical factor with regards to the safety issue in the designation of the vessels. As a result, the failure analyses of composite materials under shock loadings draws a lot of interest, and has been investigated for years.

It is a challenging task to predict damage in composites. Composites can be defined as two or more materials combined to form a single material [5]. For fiber-reinforced laminates, there are mainly four modes of failure: matrix cracking occurs parallel to the fiber; delamination; fiber breakage in tension, as well as fiber bulking in compression; and penetration due to impact [6]. Therefore, the inhomogeneous nature of composites must be taken into consideration in the analysis, in order to predict the corresponding failure modes. Furthermore, the stacking sequence and thickness also have an important effect on the failure initiation and evolution [7]. In addition to the complexity of the composite material properties, shock loadings, which result in high strain rates, also give rise to additional complexity in the analysis. Large safety factors are typically used in composite structure design to make sure no damage will occur, resulting a conservative solution or over-design [8]. Therefore, a good understanding of the responses of composite materials under shock loadings (i.e., explosions) is necessary for the balance between safety and economy issues.

There are three major methods to investigate the responses of composite materials under explosions: the experimental method, the analytical method, and the numerical simulation method. As to the experiment method, there are two kinds of experimental tests, according to the scale, i.e., a full-scale test and a laboratory-scale test. The full-scale explosive tests can provide important information on survivability, damage tolerance, and failure modes [9]. They are necessary to validate the results of analytical and numerical simulations [10]. In 1989, a 3 m × 3 m composite plate was tested under an underwater blast, to be investigated in full scale [11]. However, the full-scale tests are performed infrequently, due to high costs. For this reason, the explosive test in the laboratory scale is adopted for research. A divergent shock tube was designed to investigate the responses of a clamped test plate under shock loadings [12]. Thus, plane wave fronts and wave parameters were easily controlled and repeated. LeBlanc and Shukla used a tube filled with water to reproduce the underwater explosive loads [13]. Wadley [14] developed another test method to investigate the compressive responses of multilayered lattices during underwater shock loadings. Analytical methods are generally adopted in the initial design state of composite structures, which give relatively faster solutions compared to the other two methods. Rabczuk et al. [15] proposed a simplified method to investigate the effects of fluid–structure interaction in composite structures subjected to dynamic underwater loads. Hoo Fatt and Palla [16] derived analytical solutions for transient response and damage initiation of a composite panel subjected to blast loading. However, analytical solutions are mainly limited to special and simple cases. In contrast, numerical simulation methods can be applied on various types of loadings, complicated geometries of structures, and complex boundary conditions. Kazancı [17] conducted a review of the available numerical achievements regarding the simulation of composite plates under a blast load. The finite element method (FEM) [18], smooth particle hydrodynamics (SPH) [19], and the finite strip method (FSM) [20] have all been applied to model composite materials.

Unlike the FEM, peridynamics (PD) is a new, non-local theory, and utilizes a mesh-free approach [21]. In FEM, partial differential equations are used to predict the motions of a body, which creates non-physical singular stress and strain at discontinuities. As a result, remedies such as the cohesive zone element (CZE) [22] and the extended finite element method (XFEM) [23] are proposed to improve the shortcomings of the FEM. However, additional assumptions are still needed to predict the crack propagation path in these methods. In contrast, PD converts the equation of motion from its traditional partial differential form into an integral form, which remains valid even at discontinuities [24]. Consequently, PD is well-suited for problems involving discontinuities. As to the composite materials, PD has been successfully applied on the failure analyses of composites [7,25–29]. However, the composites are mainly the focus on the mechanical field only. When explosion loads are applied to the test plate, the plate experiences high strain rate stages. Therefore, the coupling effect of deformation on temperature cannot be neglected, which may have an effect on the crack propagation path with the induced temperature changes. Therefore, a fully coupled thermomechanical composite model is necessary for the simulation of thermal and mechanical responses of composites under shock loadings. Here, a fully coupled approach means both the temperature effects on deformation and the deformation effects on temperature are included in the simulation [30]. Oterkus et al. [31] proposed a PD thermal model to simulate the heat conduction in isotropic materials. Then, they generalized the model to include the coupling effects on both deformation and temperature [32]. Furthermore, the model is developed for composite lamina by considering its directional dependency properties [33]. Based on previous work, the fully coupled thermomechanical laminate model formulated by PD is developed in this paper, including heat conduction and coupling effects in the thickness direction of the composite laminate.

The paper is organized as follows. Firstly, the PD theory is introduced, and some basic concepts are explained. Secondly, a mechanical and thermal model for composite materials is formulated by bond-based PD. Then, the responses of a 13-ply composite plate subjected to an underwater explosion load are studied, by considering the fully coupled thermomechanical effects. The crack propagation evolutions are predicted and compared with uncoupled cases. The predicted temperature distributions are also provided.

2. Peridynamic (PD) Theory

2.1. Basic Concepts in PD Theory

The PD theory which is proposed by Silling and Askari [34] falls into the category of non-local theory. The material points, $\mathbf{x}$, can interact other material points, i.e., $\mathbf{x}'$, in a neighbourhood $H_{\mathbf{x}}$. The maximum interaction distance is called horizon and denoted by δ. $H_{\mathbf{x}}$ is called the family of point $\mathbf{x}$. As shown in Figure 1, the initial relative position vector is denoted as $\xi = \mathbf{x}' - \mathbf{x}$, in the deformed configuration, the positions of material points $\mathbf{x}$ and $\mathbf{x}'$ are represented by $\mathbf{y}$ and $\mathbf{y}'$, respectively. Hence, the displacements of the points $\mathbf{x}$ and family member $\mathbf{x}'$ are $\mathbf{u}(\mathbf{x}) = \mathbf{y} - \mathbf{x}$ and $\mathbf{u}(\mathbf{x}') = \mathbf{y}' - \mathbf{x}'$, respectively. Consequently, the relative displacement between $\mathbf{x}$ and $\mathbf{x}'$ can be defined as

$$\eta = \mathbf{u}(\mathbf{x}') - \mathbf{u}(\mathbf{x}) = (\mathbf{y}' - \mathbf{y}) - (\mathbf{x}' - \mathbf{x}) \tag{1}$$

The stretch between two points can be defined as

$$s = \frac{|\mathbf{y}' - \mathbf{y}| - |\mathbf{x}' - \mathbf{x}|}{|\mathbf{x}' - \mathbf{x}|} = \frac{|\eta + \xi| - |\xi|}{|\xi|} \tag{2}$$

The equation of motion in bond based PD theory is [34]

$$\rho(\mathbf{x})\ddot{\mathbf{u}}(\mathbf{x},\, t) = \int_{H_{\mathbf{x}}} \mathbf{f}(\xi,\, \eta,\, t)dV + \mathbf{b}(\mathbf{x},\, t) \tag{3}$$

where $\rho(\mathbf{x})$ represents density, V represents volume, $\ddot{\mathbf{u}}(\mathbf{x},\, t)$ represents the acceleration, $\mathbf{b}(\mathbf{x}, t)$ represents the body force and $\mathbf{f}(\xi,\, \eta,\, t)$ represents the pairwise PD force. The pairwise PD force can be defined as [32]

$$\mathbf{f}(\xi,\, \eta,\, t) = c(s - \alpha T)\frac{\xi + \eta}{|\xi + \eta|} \tag{4}$$

In which α is the linear thermal expansion coefficient of the material, T is the average temperature of point $\mathbf{x}$ and $\mathbf{x}'$ with respect to reference temperature, c is the PD constant. It should be noted that in bond based PD, the pairwise PD forces $\mathbf{f}$ and $\mathbf{f}'$ are forced to be equal in magnitude and parallel in direction. Hence, the Poisson's ratio is forced to be $1/3$ in two dimensional (2D) analysis and $1/4$ in three dimensional (3D) analysis [21].

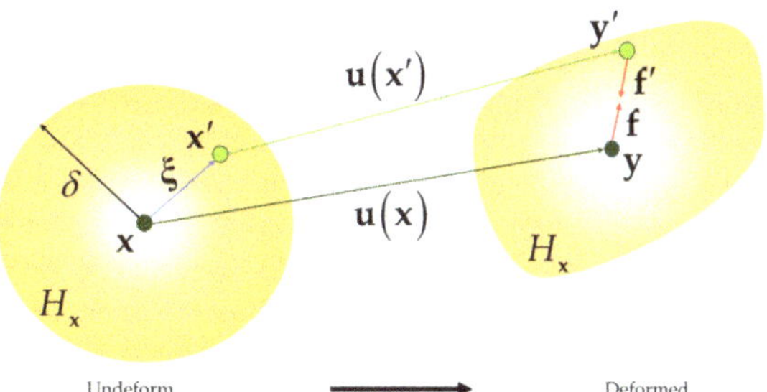

Figure 1. Illustration of PD forces

2.2. PD Mechanical Laminate Model

The PD mechanical model developed by Oterkus and Madenci [21,25] for composite laminates is adopted in this paper. As illustrated in Figure 2, each ply in a laminate is modelled by one-layer PD nodes (shown in blue, red, and yellow colours for different plies). The multi-layer laminate is modelled by assembling the single layer models according to the stacking sequence. For a resin-rich laminate, the properties in the thickness direction are treated as its matrix material properties. Due to the directionally-dependent properties of the laminate, four kinds of PD bonds are defined in the model: in-plane fibre bonds, in-plane matrix bonds, interlayer normal bonds, and interlayer shear bonds. The grid size is represented by Δx and the fibre direction is denoted by Φ.

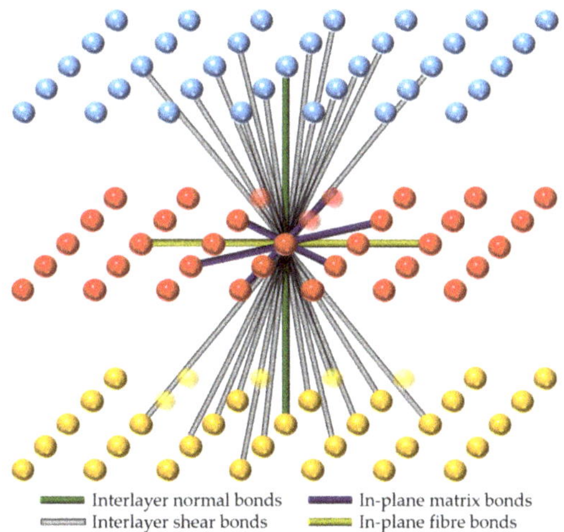

Figure 2. Illustration of PD laminate model for $\delta = 2\Delta x$ and fibre direction, $\Phi = 0$.

The discretized form of the PD equation of motion for a material point x_k^n in the n layer of a laminate can be written as

$$
\begin{aligned}
\rho(x_k^n)\ddot{u}(x_k^n, t) = & \sum_{j=1}^{N} \left(\mu_f c_f + c_m\right)\left(s_{kj}^n - \alpha_\vartheta T_{kj}^n\right)\frac{y_j^n - y_k^n}{|y_j^n - y_k^n|}V_j^n \\
& + \sum_{m=n+1,\,n-1} c_n\left(s_k^{nm} - \alpha_m T_k^{nm}\right)\frac{y_k^m - y_k^n}{|y_k^m - y_k^n|}V_k^m + \sum_{m=n+1,\,n-1}\sum_{j=1}^{N_s} c_s \varphi(\Delta x)^2 \frac{y_j^m - y_k^n}{|y_j^m - y_k^n|}V_j^m \\
& + b(x_k^n, t)
\end{aligned}
\tag{5}
$$

The first three terms on the right hand side of Equation (5) represent the PD forces developed by in-plane bonds (including fibre bonds and matrix bonds), interlayer normal bonds, and interlayer shear bonds in sequence. If the bond direction is parallel to the fibre direction, μ_f is equal to 1, otherwise it is 0. c_f, c_m, c_n, and c_s are PD material constants associated with in-plane fiber bonds, in-plane matrix bonds, interlayer normal bonds, and interlayer shear bonds, respectively. The definitions for PD material constants are listed as [21,25,27]

$$c_f = \frac{2E_1(E_1 - E_2)}{\left(E_1 - \frac{1}{9}E_2\right)\left(\sum\limits_{j=1}^{N_F} \left|\xi_{jk}\right| V_j\right)} \tag{6}$$

$$c_m = \frac{8E_1 E_2}{\left(E_1 - \frac{1}{9}E_2\right)\pi h \delta^3} \tag{7}$$

$$c_n = \frac{E_m}{hV} \tag{8}$$

$$c_s = \frac{2G_m}{\pi h}\frac{1}{(\delta^2 + h^2 \ln(h^2/\delta_s^2))} \tag{9}$$

In the above equations, E_1 and E_2 represent the elastic moduli of a single ply in fiber and transverse directions, respectively. E_m represents the elastic modulus of matrix material and G_m represents the shear modulus of matrix material, h represents the thickness of one ply and Δx represents the spacing between material points on the plane of a ply. It is assumed that a material point interacts with other points in adjacent plies through interlayer normal bonds and interlayer shear bonds. Therefore, the horizon of interlayer normal bond is taken as equal to thickness of one ply, h. δ_s is the horizon of interlayer shear bond determined as $\delta_s = \sqrt{\delta^2 + h^2}$. In Equation (6), N_f represents the total number of family members those connect to the material point with fibre bonds. In Equation (8), the value of V can be calculated as the average volume of material points connected through interlayer normal bonds.

As to the thermal expansion coefficients appeared in Equation (5), the formulation developed in [27] is utilized in the current model as

$$\alpha_\vartheta = \alpha_x \cos^2(\vartheta) + \alpha_y \sin^2(\vartheta) + \alpha_x \alpha_y \cos(\vartheta)\sin(\vartheta) \tag{10}$$

where α_ϑ is the thermal expansion coefficient in the bond direction, ϑ. α_x, α_y and α_{xy} are thermal expansion coefficients of the located ply with respect to the global coordinate system. α_m represents thermal expansion coefficient of matrix material.

In Equation (5) φ represents the shear angle of the diagonal shear bonds, the effect of temperature is included as

$$\varphi = \left[\left(s_{kj}^{nm} - \alpha_m T_{kj}^{nm}\right) - \left(s_{jk}^{nm} - \alpha_m T_{jk}^{nm}\right)\right]\left|\xi_{kj}^{nm}\right|/(2h) \tag{11}$$

where s_{kj}^{nm} is the stretch between nodes $\mathbf{x}_k^n$ and $\mathbf{x}_j^m$, and T_{kj}^{nm} is the temperature difference between nodes $\mathbf{x}_k^n$ and $\mathbf{x}_j^m$, with respect to reference temperature. It should be noted that because of the adoption of bond based PD theory, the four material constants existing in a laminate—i.e., E_1, E_2, v_{12}, and G_{12}—reduce to two constants: E_1 and E_2. The major Poisson's ratio v_{12} is limited to $1/3$, and the major shear modulus is $G_{12} = \frac{v_{12}E_2}{1-v_{12}v_{21}}$ with $v_{12}/E_1 = v_{21}/E_2$.

2.3. PD Thermal Laminate Model

PD thermal model was developed in [32] and extended into composite lamina by Oterkus and Madenci [33] as

$$\rho c_v \dot{T}(\mathbf{x}_k^n, t) = \sum_{j=1}^{N} \left[\left(\mu_f \kappa_f + \kappa_m \right) \frac{\Theta\left(\mathbf{x}_j^n, t\right) - \Theta\left(\mathbf{x}_k^n, t\right)}{\left|\xi_{kj}^n\right|} - \Theta_0 \left(\mu_f \beta_f + \beta_m \right) \dot{e}_{kj}^n \right] V_j^n \tag{12}$$

where c_v is the specific heat capacity, $T(\mathbf{x}_k^n, t)$ is the temperature change, $T(\mathbf{x}_k^n, t) = \Theta(\mathbf{x}_k^n, t) - \Theta_0$, $q_b(\mathbf{x}_k^n, t)$ is the volumetric heat source, and $\dot{e}$ is the time rate of the change of stretch, which is defined as $\dot{e} = \frac{\eta + \xi}{|\eta + \xi|} \cdot \dot{\eta}$.

In Equation (12) κ_f and κ_m represent the micro-conductivities associated with in-plane fibre bonds and in-plane matrix bonds. The expressions for these micro-conductivities in PD concept are given as [33]

$$\kappa_f = \frac{2(k_1 - k_2)}{\sum\limits_{j=1}^{N_f} \left|\xi_{kj}^n\right| V_j^n} \tag{13}$$

$$\kappa_m = \frac{6k_2}{\pi h \delta^3} \tag{14}$$

where k_1 and k_2 represent the thermal conductivities for fibre and transverse directions in a ply. PD thermal modulus β depends on the PD material bond constant [32,33,35]. β_f and β_m are associated with in-plane fibre bonds and in-plane matrix bonds which can be expressed as

$$\beta_f = \frac{1}{2} c_f \alpha_\theta \tag{15}$$

$$\beta_m = \frac{1}{2} c_m \alpha_\theta \tag{16}$$

where c_f and c_m represent the PD material bond constant provided in Equations (6) and (7).

In this study, the heat conduction equation given for a lamina is modified to represent a composite laminate by taking into account the interaction between the plies as

$$
\begin{aligned}
\rho c_v \dot{T}(\mathbf{x}_k^n, t) = \; & \sum_{j=1}^{N} \left[\left(\mu_f \kappa_f + \kappa_m \right) \frac{\Theta\left(\mathbf{x}_j^n, t\right) - \Theta\left(\mathbf{x}_k^n, t\right)}{\left|\xi_{kj}^n\right|} - \Theta_0 \left(\mu_f \beta_f + \beta_m \right) \dot{e}_{kj}^n \right] V_j^n \\
& + \sum_{m=n+1, n-1} \left[\kappa_n \frac{\Theta\left(\mathbf{x}_k^m, t\right) - \Theta\left(\mathbf{x}_k^n, t\right)}{\left|\xi_k^{nm}\right|} - \Theta_0 \beta_n \dot{e}_k^{nm} \right] V_k^m \\
& + \sum_{j=1}^{N_s} \left[\kappa_n \frac{\Theta\left(\mathbf{x}_j^m, t\right) - \Theta\left(\mathbf{x}_k^n, t\right)}{\left|\xi_{kj}^{nm}\right|} - \Theta_0 \beta_s \dot{e}_{kj}^{nm} \right] V_j^m + \rho q_b(\mathbf{x}_k^n, t)
\end{aligned}
\tag{17}
$$

where κ_n represents the micro-conductivity for both interlayer normal and shear bonds as

$$\kappa_n = \frac{k_m}{2\pi h^3 (\delta_s - h)} \tag{18}$$

where k_m is the thermal conductivity of the matrix material. PD thermal modulus for interlayer normal bonds and interlayer shear bonds can be expressed as

$$\beta_n = \frac{1}{2} c_n \alpha_m \tag{19}$$

$$\beta_s = \frac{1}{2} c_s \alpha_m \tag{20}$$

2.4. Failure Criteria

In classical mechanics, singular stress and strain occur at crack tips, which are non-physical. Therefore, additional assumptions are needed to simulate crack propagation paths. On the contrary, in PD theory, the equation of motion is converted into its non-local form. As a result, failure can be simulated without any additional assumptions [36]. The form of equations in PD theory makes it suitable for failure analysis. The approach to simulate failure in PD theory is simply to break a bond once its stretch is beyond the critical stretch value, s_0. The PD forces of broken bonds become zero permanently. Because of the existing of four kinds of PD bonds in PD laminate model, different critical stretch values are defined for different types of PD bonds. The definitions of these critical stretch values are listed as [7,25,34]

$$s_{ft} = \frac{\sigma_{1t}}{E_1} \tag{21}$$

$$s_{fc} = \frac{\sigma_{1c}}{E_1} \tag{22}$$

$$s_m = \sqrt{\frac{5G_{IC}}{9K_m\delta}} \tag{23}$$

$$s_n = \sqrt{\frac{2G_{IC}}{hE_m}} \tag{24}$$

$$s_s = \sqrt{\frac{G_{IIC}}{hG_m}} \tag{25}$$

In the above equations, s_{ft} and s_{fc} are critical stretch values of fibre bonds in tension and compression states, respectively. s_m, s_n, and s_s are related with matrix bonds, interlayer normal bonds, and interlayer shear bonds. G_{IC} and G_{IIC} are the critical energy release rates for first and second failure modes, respectively. K_m is the bulk modulus of the matrix material. σ_{1t} and σ_{1c} are longitudinal tension and compression strength properties of a single ply. By applying the above failure criteria, it can be observed that the fibre bonds can fail both in tension and compression. The matrix bonds, interlayer normal bonds, and interlayer shear bonds are only allowed to fail in tension.

A history dependent function, $\mu(\xi, t)$, is introduced to indicate the status of a bond, i.e., being 1 for intact bond and being zero for broken bond. The definitions of parameter, $\mu(\xi, t)$ for different kinds of PD bonds can be defined as

$$\mu_{ff} = \begin{cases} 1, & (s - \alpha_\vartheta T) < s_{ft} \text{ and } (s - \alpha_\vartheta T) > s_{fc} \\ 0, & (s - \alpha_\vartheta T) \geq s_{ft} \text{ or } (s - \alpha_\vartheta T) \leq s_{fc} \end{cases} \tag{26}$$

$$\mu_m = \begin{cases} 1, & (s - \alpha_\vartheta T) < s_m \\ 0, & (s - \alpha_\vartheta T) \geq s_m \end{cases} \tag{27}$$

$$\mu_n = \begin{cases} 1, & (s - \alpha_m T) < s_n \\ 0, & (s - \alpha_m T) \geq s_n \end{cases} \tag{28}$$

$$\mu_s = \begin{cases} 1, & \varphi < s_s \\ 0, & \varphi \geq s_s \end{cases} \tag{29}$$

where μ_{ff}, μ_m, μ_n, and μ_s are related with fibre, matrix, interlayer normal, and interlayer shear bonds, respectively. Subsequently, a local damage parameter, i.e., the ratio of the number of broken bonds to the number of total bonds, is introduced to represent the damage level of a point, shown as [21]

$$\varphi(\mathbf{x}, t) = 1 - \frac{\int_H \mu(\xi, t)dV}{\int_H dV} \tag{30}$$

3. Numerical Implementation

In this study, the heat conduction equation and the equation of motion are solved simultaneously for each time increment by using explicit time integration.

3.1. Problem Description

The bond based PD laminate model is implemented in FORTRAN program to predict the responses of a 13 ply laminate subjected to shock loading which was previously considered by Diyaroglu et al. [37]. Note that, in this study, the temperature changes due to mechanical deformations and their effects on damage evolution are taken into account by solving fully coupled thermomechanical equations whereas thermal effects are ignored in [37]. The composite material properties is provided in Table 1.

Table 1. Material properties of composite [8].

Mechanical Properties		Thermal Properties	
E_1 (GPa)	39.3	α_1 (μm/m/K)	8.6
E_2 (GPa)	9.7	α_2 (μm/m/K)	22.1
G_{12} (GPa)	3.32	k_1 (W/mK)	10.4
Poisson's ratio ν_{12}	0.33	k_2 (W/mK)	0.89
ρ $\left(\text{kg/m}^3\right)$	1850	c_v (J/(kg $\cdot$ K))	879
E_m (GPa)	3.792	α_m (μm/m/K)	63
G_m (GPa)	1.422	k_m (W/mK)	0.34
Poisson's ratio ν_m	0.33	Θ_0(K)	285

Because of the adoption of bond based PD, the major shear modulus changed to be 3.32 GPa according to the constraint on material constants. As illustrated in Figure 3, the 13 ply test plate is in a circle shape with outer radius, $R_{out} = 132.715$ mm and inner radius, $R_{in} = 114.3$ mm. The thickness of each ply in the laminate is same as $h = 0.254$ mm. The region between the inner circle and outer circle is constrained in top and bottom plies, and is left free for other plies. The constraint is implemented by applying six bolts with a radius of $r = 4$ mm. Thus the fixed end allows the specimen to absorb the full energy of the applied load. The stacking sequence is $\left[0/90/0/90/0/90/\overline{0}\right]$ (shown in Figure 3).

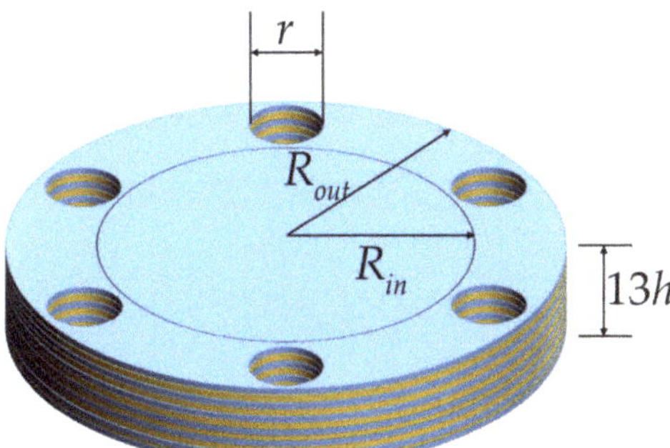

Figure 3. Geometry dimension illustration of the test laminate. (Blue colour represents 0° and yellow colour represents 90° plies).

The PD discretization of one ply is presented in Figure 4. The grid size is $\Delta x = 2.6543 \times 10^{-3}$ m. The horizon size is chosen as $\delta = 3.015\Delta x$. The material points located within the bolt regions are deleted in order to represent the actual shape of the test plate. Based on such discretization, the critical stretch value related with bonds failures can be calculated [7]. The critical energy release rate for matrix failure is $G_{IC} = 11.85 \times 10^{-3}$ MPa, thus s_m is calculated as $s_m = 1.47 \times 10^{-2}$. The tension and compression strength properties are $\sigma_{1t} = 965$ MPa and $\sigma_{1c} = -883$ MPa. Therefore, the critical stretch value for fibre failure in tension is $s_{ft} = 2.46 \times 10^{-2}$ and in compression is $s_{fc} = -2.25 \times 10^{-2}$. As to the interlayer bonds, the critical stretch values are calculated as $s_n = 7.015 \times 10^{-2}$ with $G_{IC} = 2.73 \times 10^{-3}$ MPa and $s_s = 0.14$ with $G_{IIC} = 7.11 \times 10^{-3}$ MPa. The time step size for an explicit time integration is $\Delta t = 7.69 \times 10^{-8}$ s. The total simulation time is set as 0.3641×10^{-3} s.

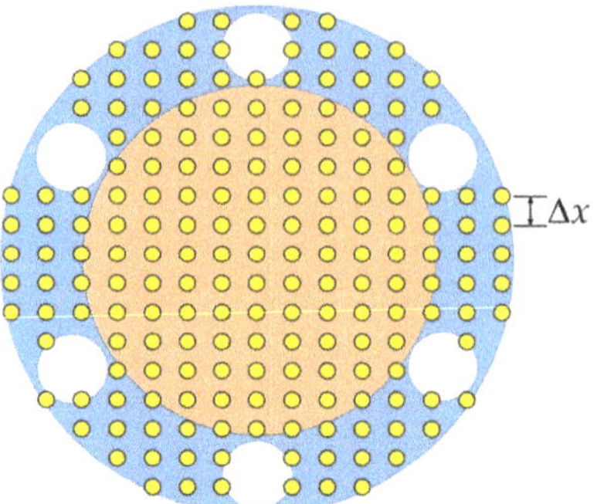

Figure 4. Illustration of PD discretization for one ply (blue colour represents the fixed boundary region and orange colour represents the inner part).

Several dynamic loadings generated by explosions are modelled by using different time-dependent pressure functions. The pressure shock applied in the experiment conducted by LeBlanc and Shukla [8] is utilized here. The charge which is equivalent to 1.32 g TNT is located at 5.25 m away from the test plate. The pressure wave is cause by rapid expansion of explosive gases. The speed of these gases can be approximated as the speed of sound in water [38]. The pressure linearly increases until it reached its peak value, $P_{\max}$, followed by the exponential decay, expressed in Equation (31) and shown in Figure 5. Here $P_{\max}$ is set to be 9.65 MPa.

$$P(t) = \begin{cases} P_{\max} \times \left(t/4 \times 10^{-5}\right) & t < 0.04 \times 10^{-3}\,\text{s} \\ P_{\max} & 0.04 \times 10^{-3}\,s < t < 0.08 \times 10^{-3}\,\text{s} \\ P_{\max}e^{-1000(t-0.08)/0.2} & 0.04 \times 10^{-3}\,s < t < 1 \times 10^{-3}\,\text{s} \end{cases} \tag{31}$$

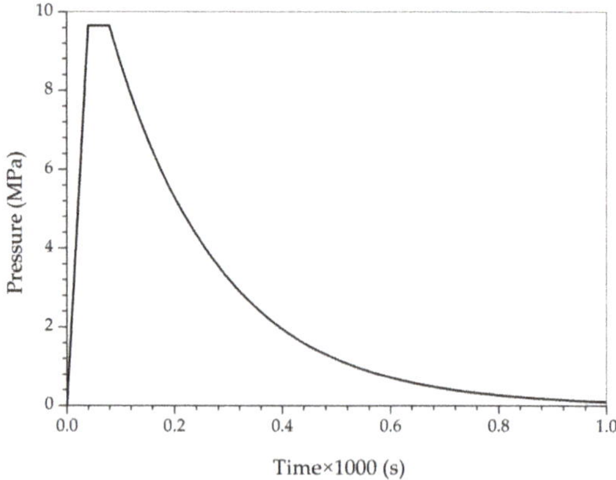

Figure 5. Pressure load distribution for the test plate.

Generally, there are two approaches for modelling the shock load depending on the distance (stand-off distance) between the charge source and the object of interest [17]. The explosion load is assumed to be uniform if the stand-off distance is long enough, which is termed as far-field explosion. On the contrary, the near-field explosion adopts non-uniform load distribution. There are also two approaches to simulate the non-uniform pressure shock loads, i.e., decoupling the load and the structural response and coupling the load and response. In this paper, a non-uniform pressure load simulated and decoupled approach is utilized, i.e., the pressure shock load is in a form of $P(r,t) = P_1(r)P_2(t)$. A non-uniform distribution of shock loading over the plate is simulated by adopting the pressure distribution derived by Turkmen and Mecitoglu [39] as

$$P(r) = -0.0005r^4 + 0.01r^3 - 0.0586r^2 - 0.001r + 1 \tag{32}$$

where r represents the distance from the collective node to the centre of the test plate. The test plate adopted here is slightly larger than the one in [39]. Consequently, the distribution profile is extended by 0.83 cm correspondingly, as illustrated in Figure 6. Finally, the explosion load is defined as

$$P(r,t) = \begin{cases} P(t)\left(-0.0005(r-0.83)^4 + 0.01(r-0.83)^3 - 0.0586(r-0.83)^2 - 0.001(r-0.83) + 1\right) & r > 0.83\,\text{cm} \\ P(t)\,r < 0.83\,\text{cm} \end{cases} \tag{33}$$

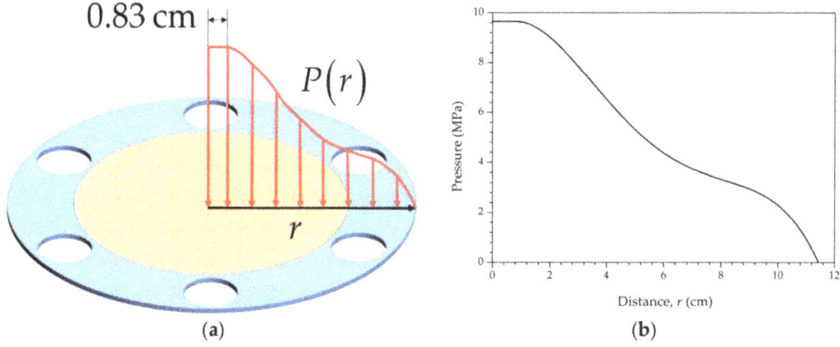

Figure 6. (**a**) Illustration of non-uniform pressure distribution over the top ply and (**b**) pressure profile.

3.2. Numerical Results

3.2.1. Subjected to Uniform Pressure Loading

First, the test laminate is subjected to uniform pressure load, $P(t)$ without allowing failure. The regions between the inner circle and outer circle are fixed in three dimensions for all plies. During the simulation, the central points in each ply experience the same vertical (z) displacement evolutions. Therefore, the vertical displacement evolution of the central point on the top ply is plotted in Figure 7a. It can be observed that the test plate firstly deforms in the negative z direction, then it will recover to some extent with velocity in positive z direction. The largest deformation occurs at approximately 3700 time steps, corresponding to 0.28453×10^{-3} s. The vertical displacement distribution over the top ply at 0.28453×10^{-3} s is shown in Figure 7b.

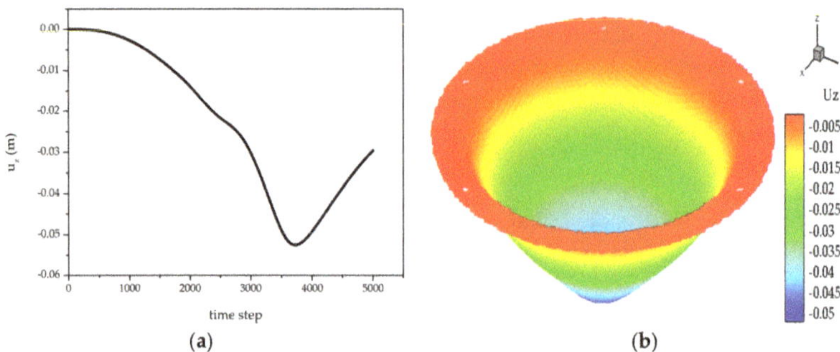

Figure 7. (**a**) Variation of the displacement in z direction of the central point as a function of time; (**b**) Vertical displacement distribution for the top ply at 0.28453×10^{-3} s.

Fully coupled thermomechanical simulation under the uniform pressure load $P(t)$, i.e., far field explosion, is also investigated for further comparison. The crack propagations and temperature change distributions at 0.1538×10^{-3} s are provided in Figure 8 for top ply, Figure 9 for middle ply, and Figure 10 for bottom ply. It can be inferred from the matrix damage plots that all the plies in the laminate experience the tear failure near the constraint boundary condition. Furthermore, the damage region in the bottom ply is larger than the top ply, indicting a combination of tension failure mode and tear failure mode. As to the temperature distribution, the temperature increases near cracks are observed for all plies, which are more obvious in the top ply provided in Figure 8b. Temperature drop is also observed in tension state, which is obvious in the bottom ply provided in Figure 10b.

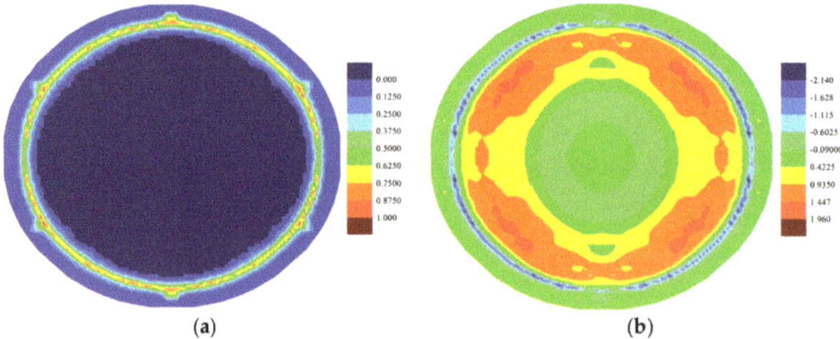

Figure 8. (**a**) Matrix damage and (**b**) temperature change distribution (K) of top ply at 0.1538×10^{-3} s.

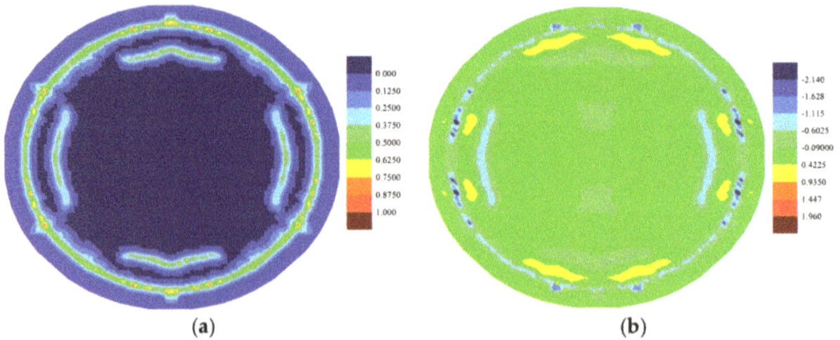

Figure 9. (**a**) Matrix damage and (**b**) temperature change distribution (K) of middle (7th) ply at 0.1538×10^{-3} s.

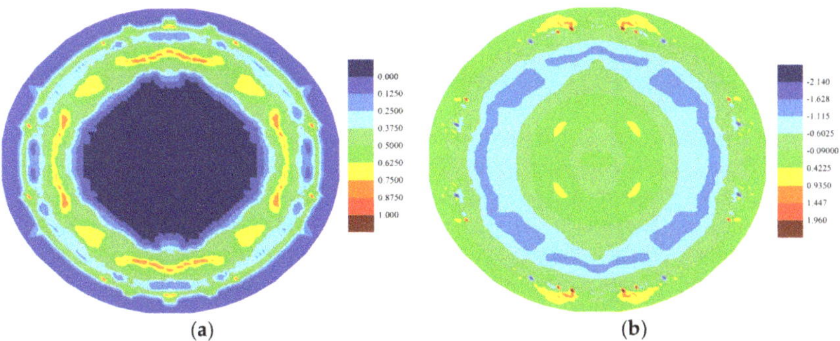

(a)   (b)

Figure 10. (**a**) Matrix damage and (**b**) temperature change distribution (K) of bottom ply at 0.1538×10^{-3} s.

3.2.2. Subjected to Uniform Non-Uniform Pressure Load

In this section, the test laminate is subjected to non-uniform pressure load $P(r,t)$, i.e., near field explosion. The matrix damage and temperature distribution in deformed shape are provided in Figure 11. Matrix damage predictions at 0.28453×10^{-3} s and 0.3461×10^{-3} s obtained from coupled and uncoupled cases are are shown in Figures 12–17.

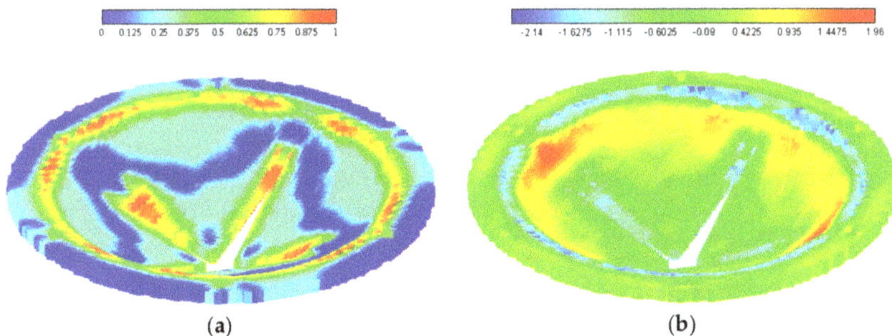

(a)   (b)

Figure 11. (**a**) Matrix damage and (**b**) temperature change distribution (K) of the laminate at 0.3461×10^{-3} s.

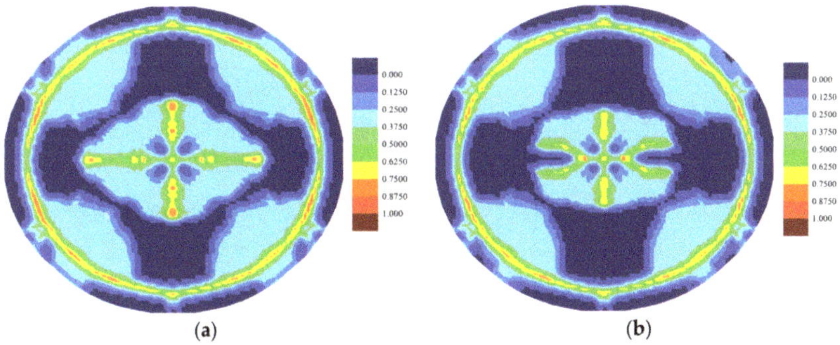

(a)   (b)

Figure 12. Matrix damage comparison of top ply for (**a**) coupled case and (**b**) uncoupled case at 0.28453×10^{-3} s.

For the fully coupled simulation case, by comparing the damage of the plies at different times, it is obvious that the damaged zone gets larger as time progresses. The damage patterns are different for each ply when compared at the same time. The cracks mainly occur near the clamped boundary region for the top ply, indicating a tear failure mode. On the other hand, the central part experiences the largest level of damage for the bottom ply, indicating a tension failure mode. Consequently, the different force states give rise to the different level of damages. However, for all plies, the crack propagations present a cross-shaped pattern. It can be explained that the fiber direction of each ply is either zero or 90 degrees. The matrix damage occurs parallel to the fiber direction. For a ply with fiber direction being zero, the matrix crack will occur along the horizontal direction. However, the fiber directions for its adjacent plies are 90 degrees. Hence, the matrix crack will also occur in the vertical direction due to the contribution of the interlayer bonds. Consequently, the final cracks are in cross shapes. The damages present highest levels near the central vertical lines for all plies. This phenomenon is also observed in the experiment [13], as shown in Figure 18. As it can be seen in Figures 8–10, there are damages around the bolt holes and these damages were also observed in experiments [13] as it can be seen in Figure 18.

As shown in Figures 12–17, different damage patterns are observed for coupled and uncoupled cases. As the time progresses, temperature change increases and the differences in damage plots become more obvious. Considering the small temperature changes induced by the applied pressure shock, the coupling term effect on damage is significant. It can be inferred that the difference in damage due to coupling effect will become more significant with larger strain rates. Temperature decreases where there is local tension and as a result local compression is created due to temperature drop which reduces the extent of damage observed by the uncoupled cases (Figures 12–17).

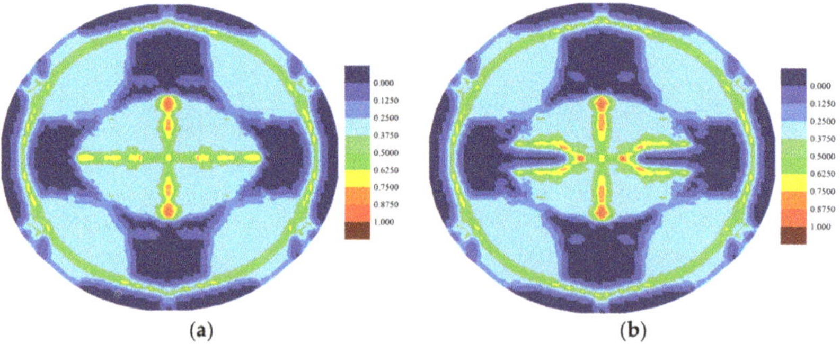

(a)   (b)

Figure 13. Matrix damage comparison of middle (7th) ply for (**a**) coupled case and (**b**) uncoupled case at 0.28453×10^{-3} s.

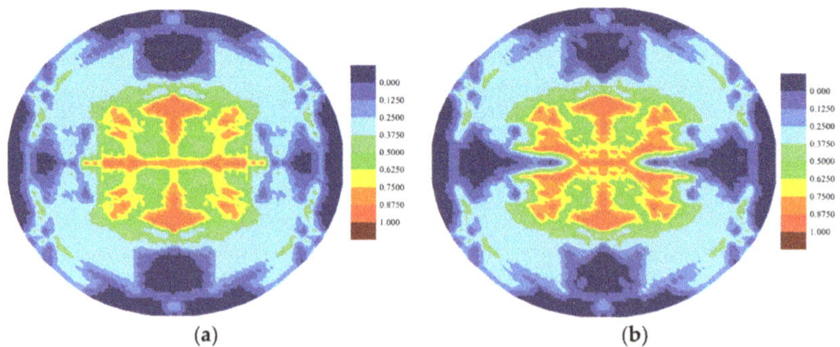

(a)   (b)

Figure 14. Matrix damage comparison of bottom ply for (**a**) coupled case and (**b**) uncoupled case at 0.28453×10^{-3} s.

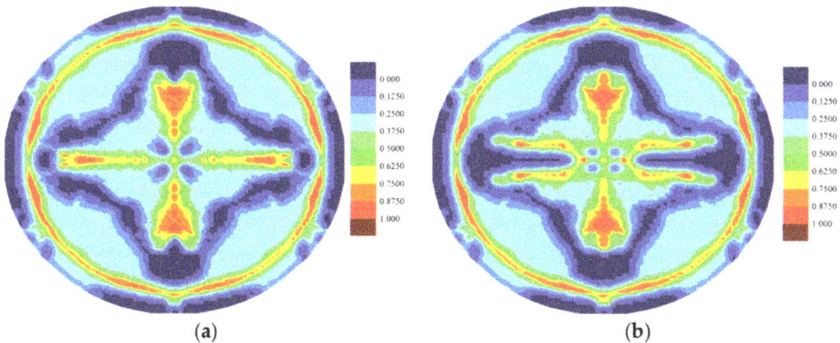

Figure 15. Matrix damage comparison of top ply for (**a**) coupled case and (**b**) uncoupled case at 0.3461×10^{-3} s.

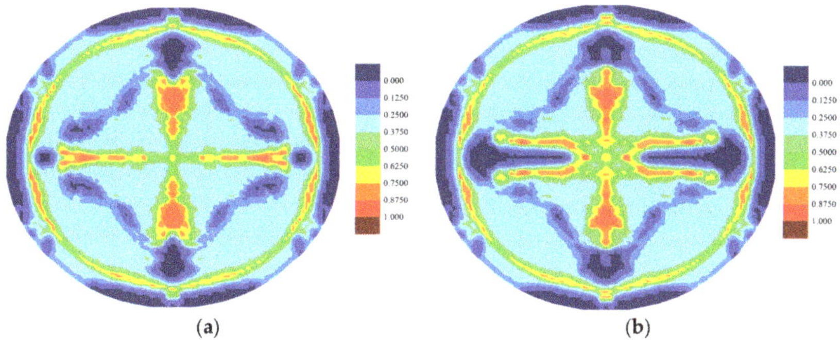

Figure 16. Matrix damage comparison of middle (7th) ply for (**a**) coupled case and (**b**) uncoupled case at 0.3461×10^{-3} s.

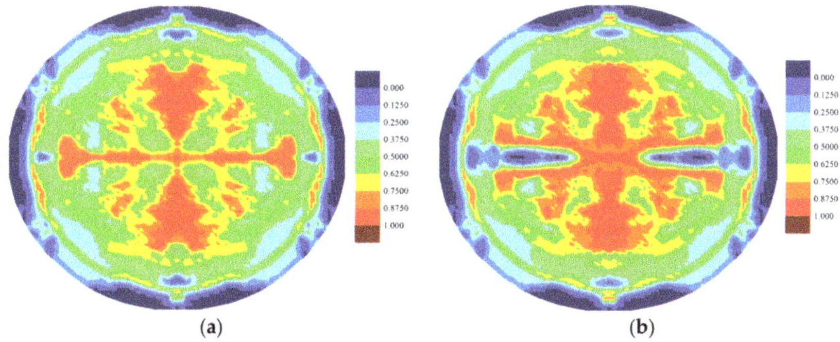

Figure 17. Matrix damage comparison of bottom ply for (**a**) coupled case and (**b**) uncoupled case at 0.3461×10^{-3} s.

Figure 18. Material damage during test [13].

The extent of damage in interlayer shear bonds was also investigated, and only slight differences were observed in the top few plies between the coupled and uncoupled cases (Figure 19). The middle ply experienced the most severe damage, as shown in Figure 20. Thus, it can be inferred that the interlayer shear bond damages occur mainly in the middle plies of the test laminate. Hence, it can be concluded that there is delamination failure in the middle plies.

Temperature changes induced by the applied pressure shock loading are presented for different plies in Figures 21–23. It is observed that as the loading increased, the temperature changes of PD nodes increased. For all plies, the temperature change profiles all have similar patterns as the corresponding crack damage patterns. As shown in Figures 21–23, there is a temperature rise where there is local compression, and there is a temperature drop where there is local tension, as explained in [21]. In the top ply, most of the region was under compression and a temperature rise was observed; on the other hand, the bottom ply was mostly under tension, and a consequent temperature drop was observed, as shown in Figure 23. In the cracked surfaces, temperature drops were observed because of the local tension; however, temperature rise was observed near the crack tips. Thus, the crack propagation paths do have effects on the temperature distributions.

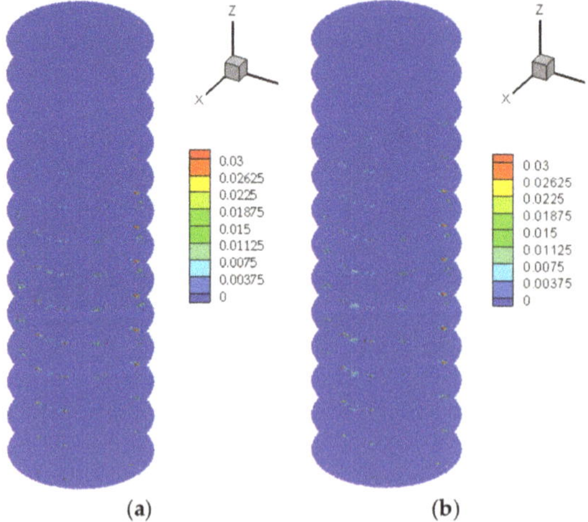

(**a**)   (**b**)

Figure 19. Interlayer shear damage comparison for (**a**) coupled case and (**b**) uncoupled case at 0.3461×10^{-3} s.

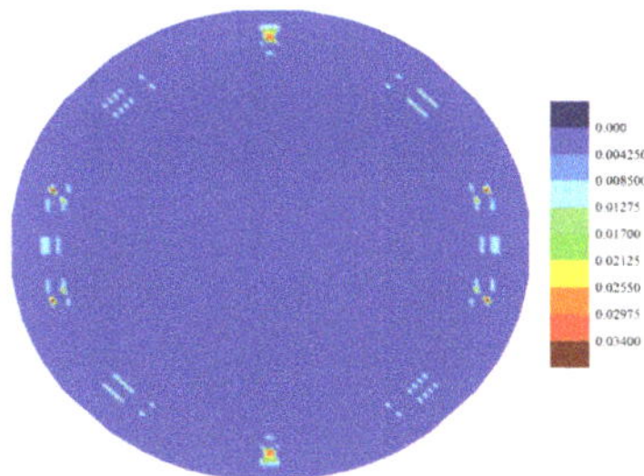

Figure 20. Interlayer shear damage of middle ply in coupled case at 0.3461×10^{-3} s.

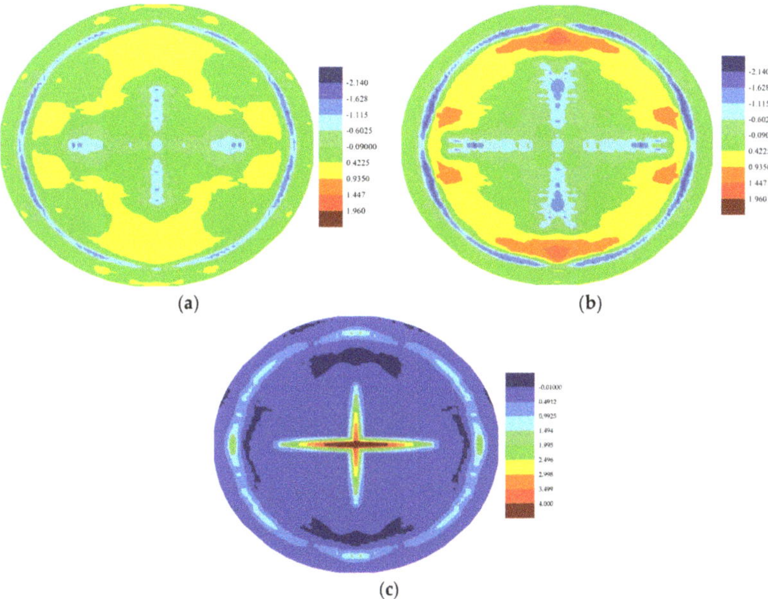

Figure 21. (**a**) Distribution of temperature change (K) of top ply at 0.28453×10^{-3} s; (**b**) Distribution of temperature change (K) of top ply at 0.3461×10^{-3} s; (**c**) Maximum stretch distribution of top ply at 0.3461×10^{-3} s.

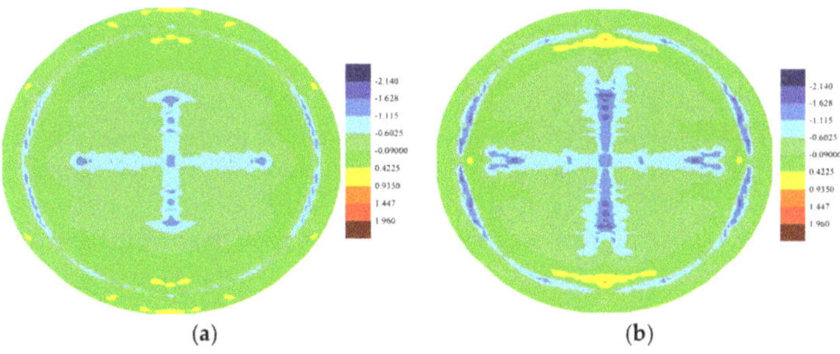

Figure 22. *Cont.*

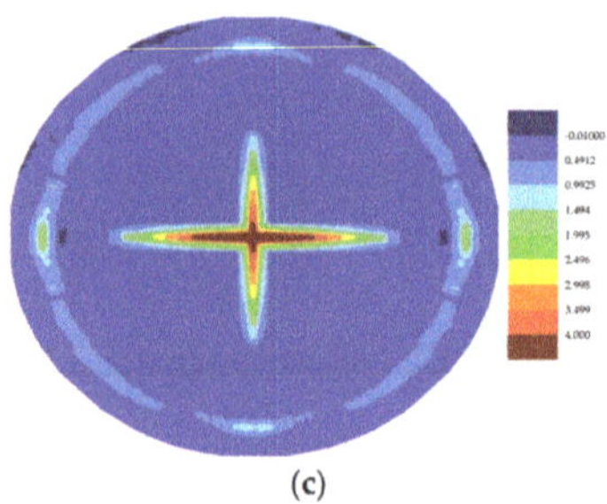

(c)

Figure 22. (**a**) Distribution of temperature change (K) of middle ply at 0.28453×10^{-3} s; (**b**) Distribution of temperature change (K) of middle ply at 0.3461×10^{-3} s; (**c**) Maximum stretch distribution of middle ply at 0.3461×10^{-3} s.

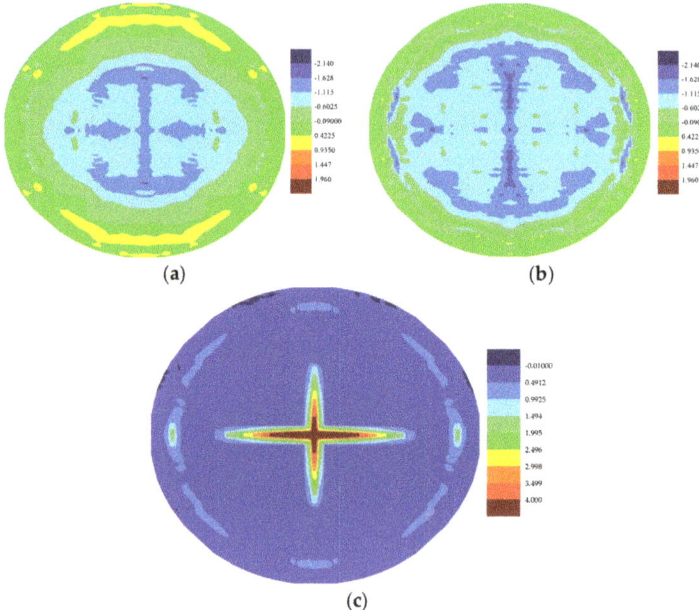

(a)

(b)

(c)

Figure 23. (**a**) Distribution of temperature change (K) of bottom ply at 0.28453×10^{-3} s; (**b**) Distribution of temperature change (K) of bottom ply at 0.3461×10^{-3} s; (**c**) Maximum stretch distribution of bottom ply at 0.3461×10^{-3} s.

4. Conclusions

In this paper, a bond-based PD laminate model was applied to predict the responses of a 13-ply composite under a pressure shock loading. Both the deformation effect on the temperature field and the temperature effect on deformation were taken into consideration in the generalized PD model. Hence, the simulation was conducted by considering fully coupled thermomechanical effects. Firstly, the matrix damages of the laminate were predicted at different simulation times. It was observed that each ply within the laminate experienced a different level of matrix crack. Then, the breakages of the interlayer shear bonds were investigated. The delamination failure occurred mainly in the middle part of the laminate. The PD-predicted damage pattern agrees well with the experimental result. Furthermore, the temperature change evolution induced by the applied mechanical pressure shock was also simulated. The temperature distributions presented similar patterns with the crack propagations. Finally, the crack propagation patterns were compared for coupled and uncoupled cases. Results showed that the coupling term has an effect on crack propagation pattern. The developed model can be used for predicting more realistic crack patterns. In conclusion, the PD theory is able to predict both thermal and mechanical responses of marine composites under shock loadings.

Acknowledgments: The authors gratefully acknowledge financial support from China Scholarship Council (CSC No. 201506230126) and University of Strathclyde.

Author Contributions: Yan Gao and Selda Oterkus conceived and designed the numerical model; Yan Gao performed the numerical simulation; Yan Gao and Selda Oterkus analyzed the data; Yan Gao and Selda Oterkus wrote the paper.

Conflicts of Interest: The authors declare no conflict of interest.

References

1. Mouritz, A.P. The damage to stitched GRP laminates by underwater explosion shock loading. *Compos. Sci. Technol.* **1995**, *55*, 365–374. [CrossRef]
2. Hellbratt, S.-E. Experiences from Design and Production of the 72 m CFRP-Sandwich Corvette Visby. In Proceedings of the 6th International Conference on Sandwich Construction; CRC Press: Fort Lauderdale, FL, USA, 2003; pp. 15–24.
3. Li, H.C.H.; Herszberg, I.; Davis, C.E.; Mouritz, A.P.; Galea, S.C. Health monitoring of marine composite structural joints using fibre optic sensors. *Compos. Struct.* **2006**, *75*, 321–327. [CrossRef]
4. Baley, C.; Davies, P.; Grohens, Y.; Dolto, G. Application of interlaminar tests to marine composites. A literature review. *Appl. Compos. Mater.* **2004**, *11*, 99–126. [CrossRef]
5. Ghobadi, A. Common type of damages in composites and their inspections. *World J. Mech.* **2017**, *7*, 24–33. [CrossRef]
6. Richardson, M.O.W.; Wisheart, M.J. Review of low-velocity impact properties of composite materials. *Compos. Part A Appl. Sci. Manuf.* **1996**, *27*, 1123–1131. [CrossRef]
7. Diyaroglu, C. Peridynamics and Its Applications in Marine Structures. Ph.D. Thesis, University of Strathclyde, Glasgow, UK, 2016.
8. LeBlanc, J.; Shukla, A. Dynamic response of curved composite panels to underwater explosive loading: Experimental and computational comparisons. *Compos. Struct.* **2011**, *93*, 3072–3081. [CrossRef]
9. Tagarielli, V.L.; Schiffer, A. The response to underwater blast. In *Dynamic Deformation, Damage and Fracture in Composite Materials and Structures*; Woodhead Publishing: Sawston, UK, 2016; pp. 279–307.
10. Arora, H.; Rolfe, E.; Kelly, M.; Dear, J.P. Chapter 7—Full-scale air and underwater-blast loading of composite sandwich panels. In *Explosion Blast Response of Composites*; Woodhead Publishing: Sawston, UK, 2017; pp. 161–199.
11. Hall, D.J. Examination of the effects of underwater blasts on sandwich composite structures. *Compos. Struct.* **1989**, *11*, 101–120. [CrossRef]
12. Latourte, F.; Gregoire, D.; Zenkert, D.; Wei, X.D.; Espinosa, H.D. Failure mechanisms in composite panels subjected to underwater impulsive loads. *J. Mech. Phys. Solids* **2011**, *59*, 1623–1646. [CrossRef]
13. LeBlanc, J.; Shukla, A. Dynamic response and damage evolution in composite materials subjected to underwater explosive loading: An experimental and computational study. *Compos. Struct.* **2010**, *92*, 2421–2430. [CrossRef]
14. Wadley, H.; Dharmasena, K.; Chen, Y.C.; Dudt, P.; Knight, D.; Charette, R.; Kiddy, K. Compressive response of multilayered pyramidal lattices during underwater shock loading. *Int. J. Impact Eng.* **2008**, *35*, 1102–1114. [CrossRef]
15. Rabczuk, T.; Samaniego, E.; Belytschko, T. Simplified model for predicting impulsive loads on submerged structures to account for fluid-structure interaction. *Int. J. Impact Eng.* **2007**, *34*, 163–177. [CrossRef]
16. Hoo Fatt, M.S.; Palla, L. Analytical modeling of composite sandwich panels under blast loads. *J. Sandw. Struct. Mater.* **2009**, *11*, 357–380. [CrossRef]
17. Kazancı, Z. A review on the response of blast loaded laminated composite plates. *Prog. Aerosp. Sci.* **2016**, *81*, 49–59. [CrossRef]

18. Carrera, E.; Demasi, L.; Manganello, M. Assessment of plate elements on bending and vibrations of composite structures. *Mech. Adv. Mater. Struct.* **2002**, *9*, 333–357. [CrossRef]
19. Liew, K.M.; Zhao, X.; Ferreira, A.J.M. A review of meshless methods for laminated and functionally graded plates and shells. *Compos. Struct.* **2011**, *93*, 2031–2041. [CrossRef]
20. Dawe, D.J. Use of the finite strip method in predicting the behaviour of composite laminated structures. *Compos. Struct.* **2002**, *57*, 11–36. [CrossRef]
21. Madenci, E.; Oterkus, E. *Peridynamic Theory and Its Applications*; Springer: Berlin, Germany, 2014.
22. Hillerborg, A.; Modéer, M.; Petersson, P.E. Analysis of crack formation and crack growth in concrete by means of fracture mechanics and finite elements. *Cem. Concr. Res.* **1976**, *6*, 773–781. [CrossRef]
23. Belytschko, T.; Black, T. Elastic crack growth in finite elements with minimal remeshing. *Int. J. Numer. Methods Eng.* **1999**, *45*, 601–620. [CrossRef]
24. Silling, S.A.; Epton, M.; Weckner, O.; Xu, J.; Askari, E. Peridynamic states and constitutive modeling. *J. Elast.* **2007**, *88*, 151–184. [CrossRef]
25. Oterkus, E.; Madenci, E. Peridynamic analysis of fiber-reinforced composite materials. *J. Mech. Mater. Struct.* **2012**, *7*, 45–84. [CrossRef]
26. Kilic, B.; Agwai, A.; Madenci, E. Peridynamic theory for progressive damage prediction in center-cracked composite laminates. *Compos. Struct.* **2009**, *90*, 141–151. [CrossRef]
27. Oterkus, E. Peridynamic Theory for Modeling Three-Dimensional Damage Growth in Metallic and Composite Structures. Ph.D. Thesis, University of Arizona, Tucson, AZ, USA, 2010.
28. Oterkus, E.; Madenci, E. Peridynamic theory for damage initiation and growth in composite laminate. *Key Eng. Mater.* **2011**, *488–489*, 355–358. [CrossRef]
29. Hu, Y.L.; Madenci, E. Bond-based peridynamic modeling of composite laminates with arbitrary fiber orientation and stacking sequence. *Compos. Struct.* **2016**, *153*, 139–175. [CrossRef]
30. Biot, M.A. Thermoelasticity and irreversible thermodynamics. *J. Appl. Phys.* **1956**, *27*, 240–253. [CrossRef]
31. Oterkus, S.; Madenci, E.; Agwai, A. Peridynamic thermal diffusion. *J. Comput. Phys.* **2014**, *265*, 71–96. [CrossRef]
32. Oterkus, S.; Madenci, E.; Agwai, A. Fully coupled peridynamic thermomechanics. *J. Mech. Phys. Solids* **2014**, *64*, 1–23. [CrossRef]
33. Oterkus, S.; Madenci, E. Fully Coupled Thermomechanical Analysis of Fiber Reinforced Composites Using Peridynamics. In Proceedings of the 55th AIAA/ASME/ASCE/AHS/SC Structures, Structural Dynamics, and Materials Conference—SciTech Forum and Exposition, National Harbor, MD, USA, 13–17 January 2014.
34. Silling, S.A.; Askari, E. A meshfree method based on the peridynamic model of solid mechanics. *Comput. Struct.* **2005**, *83*, 1526–1535. [CrossRef]
35. Oterkus, S. Peridynamics for the Solution of Multiphysics Problems. Ph.D. Thesis, University of Arizona, Tucson, AZ, USA, 2015.
36. Silling, S.; Weckner, O.; Askari, E.; Bobaru, F. Crack nucleation in a peridynamic solid. *Int. J. Fract.* **2010**, *162*, 219–227. [CrossRef]
37. Diyaroglu, C.; Oterkus, E.; Madenci, E.; Rabczuk, T.; Siddiq, A. Peridynamic modeling of composite laminates under explosive loading. *Compos. Struct.* **2016**, *144*, 14–23. [CrossRef]
38. Cimpoeru, S.J.; Ritzel, D.V.; Brett, J.M. Chapter 1—Physics of explosive loading of structures. In *Explosion Blast Response of Composites*; Woodhead Publishing: Sawston, UK, 2017; pp. 1–22.
39. Turkmen, H.S.; Mecitoglu, Z. Nonlinear structural response of laminated composite plates subjected to blast loading. *AIAA J.* **1999**, *37*, 1639–1647. [CrossRef]

Article

Tripod-Supported Offshore Wind Turbines: Modal and Coupled Analysis and a Parametric Study Using X-SEA and FAST

Pasin Plodpradit [1], Van Nguyen Dinh [2] and Ki-Du Kim [1,*]

[1] Department of Civil and Environmental Engineering, Konkuk University, 120 Neungdong-ro, Seoul 05029, Korea; P_tuay@hotmail.com
[2] MaREI Centre for Marine and Renewable Energy, ERI, University College Cork, P43C573 Cork, Ireland; nguyen.dinh@ucc.ie or nguyendhhh@yahoo.com
* Correspondence: kimkd@konkuk.ac.kr

Received: 16 May 2019; Accepted: 31 May 2019; Published: 9 June 2019

Abstract: This paper presents theoretical aspects and an extensive numerical study of the coupled analysis of tripod support structures for offshore wind turbines (OWTs) by using X-SEA and FAST v8 programs. In a number of site conditions such as extreme and longer period waves, fast installation, and lighter foundations, tripod structures are more advantageous than monopile and jacket structures. In the implemented dynamic coupled analysis, the sub-structural module in FAST was replaced by the X-SEA offshore substructure analysis component. The time-histories of the reaction forces and the turbine loads were then calculated. The results obtained from X-SEA and from FAST were in good agreement. The pile-soil-structure interaction (PSSI) was included for reliable evaluation of OWT structural systems. The superelement concept was introduced to reduce the computational time. Modal, coupled and uncoupled analyses of the NREL 5MW OWT-tripod support structure including PSSI were carried out and the discussions on the natural frequencies, mode shapes and resulted displacements are presented. Compared to the uncoupled models, the physical interaction between the tower and the support structure in the coupled models resulted in smaller responses. Compared to the fixed support structures, i.e., when PSSI is not included, the piled-support structure has lower natural frequencies and larger responses attributed to its actual flexibility. The models using pile superelements are computationally efficient and give results that are identical to the common finite element models.

Keywords: offshore wind turbine; tripod support structures; coupled analysis; uncoupled analysis; soil-pile-structure interaction; superelement

---

1. Introduction

Among the common types of fixed-bottom substructures, shown in Figure 1, monopile structures are most suitable for regions with shallow water depths (less than 30 m) [1]. Tripod and jacket structures can be constructed in transitional water depths (between 30 and 50 m). An alternative design that makes use of the various advantages of both monopile and jacket structures is known as the tripod support structure. The main part consists of a mono pile tubular section and the lower part consists of braces and three legs. Compared to a standard lattice structure, the tripod support structure is considered to be a relatively lightweight three-legged steel jacket. The central column beneath the tower and turbine transfers the forces from the tower into the three inclined members. In order to anchor the tripod to the seabed, piles are usually installed at each leg position. Suction caisson [2,3] and suction buckets [4] can be used to support the tripod structures, which have good stability and overall stiffness.

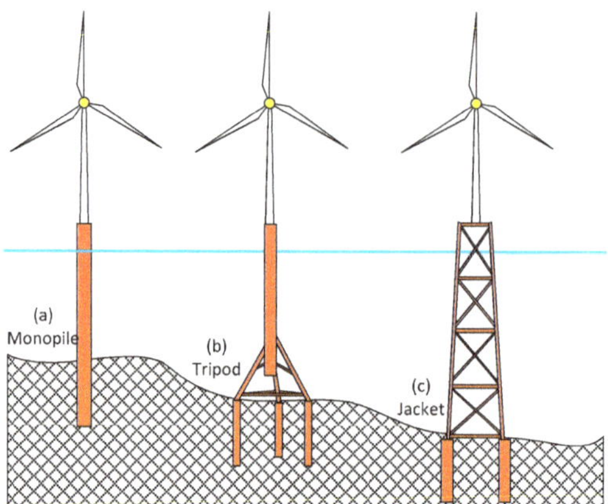

Figure 1. Common types of support structures for offshore wind turbines.

Although the jacket foundation concept is preferable to the tripod concept in terms of scour, ship collision, complexity of joints and deflection at tower top [5], the stiffer tripods are less resonant with waves and provide more opportunities to tune the natural frequency [6]. This becomes more profound as the turbines get higher; the natural frequencies of the tower-support structure system decrease and may match the high energy part of the wave spectrum [6]. Tripod support structures are therefore a good option for transitional water regions that have extreme wave conditions or long wave periods.

A tripod suction caisson foundation has been adopted by Korea's second Herald of Meteorological and Oceanographic Special Research Unit (HeMOSU-2) [2,3] after a jacket pile substructure was used for the HeMOSU-1 [7]. The construction cost of the tripod suction caisson foundation is only half that of the jacket piles for the same seabed geology, according to the cost analysis for HeMOSU-2 [2]. Additionally, installing HeMOSU-2 with a tripod suction caisson foundation took only 6 h to complete, while two months were required for the installation of HeMOSU-1 [2,3]. In terms of construction and installation, suction caissons appear to be an excellent solution.

Compared to monopile, suction bucket and gravity-based structures, the piled tripod is the lightest support structure because of its light foundation piles [6]. If the manufacturing process of the tripods, which requires more space than the monopiles, can be optimized, the use of piled tripods is more beneficial. Besides that, the types and sizes of conventional offshore structures, whose installation is currently routine are similar to those of tripod piles. Monopiles supporting 6 MW or larger wind turbines must have much larger diameters than the current piles, which causes practical problems including the lack of sufficiently heavy hammers [6]. Similar to jacket foundations, the global moments in tripods are dissolved into pairs of forces that are transferred as axial loads to the soil. Thus, tripods are especially advantageous in weak soils compared to monopiles, which transfer the lateral and moment loads by bending to the soil [5]. In a comparison study on the structural properties of monopile and tripod support structures for offshore wind-turbines [8], tripods exhibited higher stiffness, greater stress-control capacity, and a longer lifetime than monopiles. If the challenges related to tripods, such as the complex joints required to connect the three legs to the upper monopile and their susceptibility to fatigue damage can be overcome, tripod foundations could be a better alternative to monopiles in transitional water depths of around 30 m.

As discussed, tripod support structures are more advantageous than monopile and jacket structures in a number of site and installation conditions. In terms of geotechnical aspects, the group effect [4] and combined horizontal-moment bearing capacities of tripod bucket foundations in clay [9] and in sand [10] have been investigated by using advanced three-dimensional (3D) finite element analysis. Parametric studies were carried out by varying the spacing between each bucket foundation, embedded depths and loading directions [4]. In terms of structural aspects, static and modal analyses have been used to study the structural properties of tripod support structures for offshore wind-turbines [8]. A finite element model of a tripod substructure was constructed for global optimization of the best design considering uncertainties [11]. In a recent design of tripod foundation, the superstructure of the turbine was first simulated under wind and wave dynamic loading using FAST (Fatigue, Aerodynamic, Structures and Turbulence)—a CAE tool developed by the National Renewable Energy Laboratory (NREL), USA, to obtain time histories of internal actions for the pylons [12] using 3D finite element analysis software. These iterative analyses were utilized to obtain the required pile lengths and cross sections for tripod options [12]. However, the structural analyses and design of tripod support structures under simultaneous action of various environmental and operational conditions are less reported in the literature.

As soil models are generally complicated and are very expensive in terms of computing time [13], the simulation codes for offshore wind turbines often exclude the detailed modeling of soil-structure interaction [14]. Simpler approaches have therefore been adopted in the literature or the assumption of a support structure clamped at the seabed has been used [13]. A coupled, linear approach with six directions and soil-structure interaction matrices was introduced to modify the FAST simulation code [13]. The dynamic soil properties obtained by comparing nonlinear spring models of soils and experimental results were then incorporated [14]. Moreover, an offshore wind turbine (OWT) support structure must withstand the environmental and operational loading without failure. The behavior of the surface layer of soft, poorly consolidated marine clays and the stiffer clay or sand strata under this loading, and their influence on the responses of the foundation-turbine system are also important considerations. An analysis approach that considers the simultaneous interaction among the turbine, the tower, the support structure and the soil layers, the so-called coupled analysis, is required to ensure the safety of all structural components and the serviceability of the OWT system [15]. However, lengthy computation time for the coupled analysis of complex systems is an issue [16].

This paper therefore addresses the lack of coupled analysis of tripod support structures for offshore wind turbines under the simultaneous actions of various environmental and operational conditions, and develops efficient measures to reduce the computation times of the coupled analysis. In order to reduce the excessive computation times for the coupled analysis of a complex offshore support structure—a wind turbine—the superelement modeling technique was combined with the modal truncation augmentation concept by using the Craig–Bampton (C-B) method [17]. The procedure for dynamic coupled analysis proposed in this study was implemented in the X-SEA program, a 3-D finite element analysis software developed by the authors for the analysis and design of fixed and floating offshore structures for the oil/gas and offshore wind energy industry [18,19]. The procedure was validated by FAST v8 [20]. Theoretical aspects of the coupled analysis approach using X-SEA were initially discussed in the study of jacket structures [21] and are extensively described in this paper for tripod support structures. The soil-pile-structure interaction formulation was included in the implementation. However, the differences in soil conditions among the supports [22], the seismic loading [23,24], and the combination of seismic and aerodynamic loading [25] will be considered in a future study. The displacements resulting from the coupled analysis of the tripod structure are compared with that from the uncoupled analysis. A parametric study of an NREL 5MW OWT supported by a tripod structure using coupled analysis with soil-pile interactions is also carried out.

2. Development of Coupled Analysis for OWT and Tripod Support Structures

For simulating the coupled dynamic response of onshore, offshore fixed-bottom and floating wind turbines, a Glue-Code joins hydrodynamic, structure dynamic, SeveroDyn or electronic, aerodynamic, and structure modules, which are used in the FAST program [20] as depicted in Figure 2. This is an open source program and enables coupled nonlinear simulation in the time domain and the analysis of a range of wind turbine configurations. However, its hydrodynamic and structural element concepts were limited to the engineering field. Hence, the 3D finite element analysis software X-SEA was developed in Konkuk University, Seoul, Korea to solve several types of offshore wind structures [26].

The solution options of X-SEA range from simple static to highly advanced nonlinear dynamic analysis. In the case of uncoupled analysis, X-SEA receives the components of forces (F_x, F_y and F_z) and moments (M_x, M_y and M_z) at the top of the tower from FAST v8 and applies the loads to the support structure as shown in Figure 3a. The coupled analysis, which interchanges modules between the two programs is achieved through a modular interface and coupler as illustrated in Figure 3b.

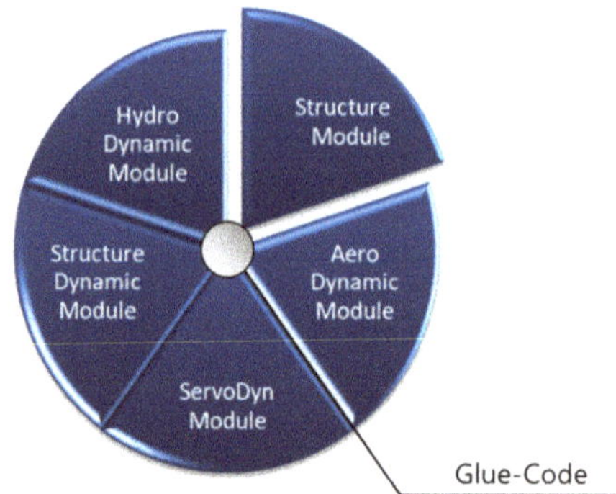

Figure 2. The concept of the interchange substructure module of the FAST V.8 program.

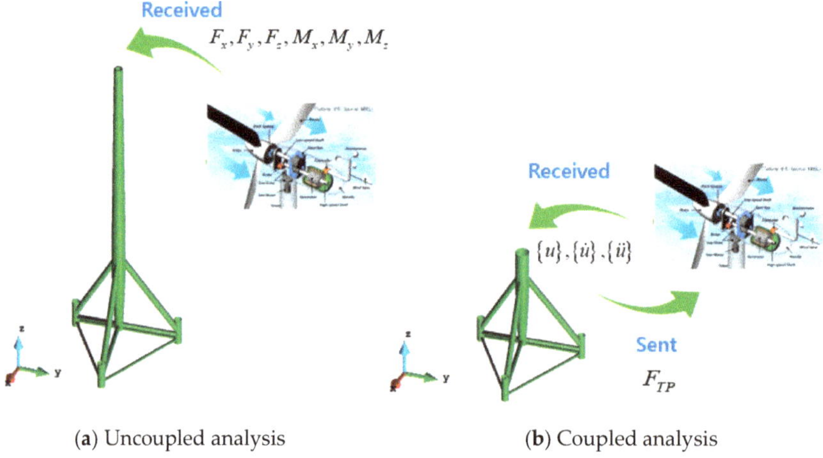

(**a**) Uncoupled analysis   (**b**) Coupled analysis

Figure 3. The concepts of uncoupled and coupled analysis.

The coupling of the substructure module in FAST V.8 program is loosened and replaced by that of X-SEA as illustrated in Figure 4. At the exchange position, the eighteen components of motion represented in the displacement $\{u\}$, velocity $\{\dot{u}\}$ and acceleration $\{\ddot{u}\}$ vectors are the input to the present program from the structural dynamic module in the FAST program. The six components of action F_{TP} from the X-SEA are input to the structural dynamic module in the FAST program.

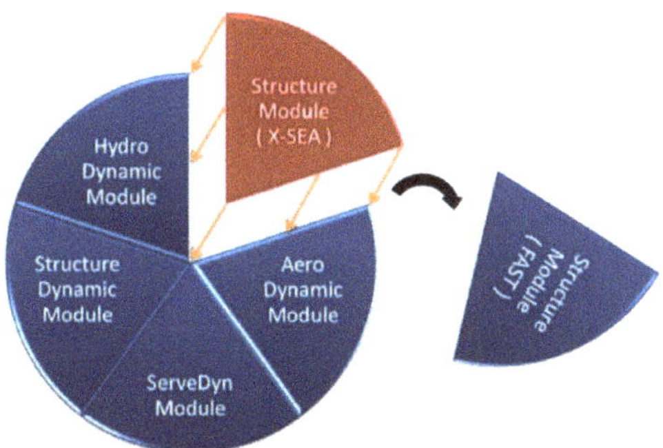

Figure 4. Basic layout of structure module within X-SEA.

2.1. Wind Turbine Dynamics

In FAST code, the representations of the modal and multibody system dynamics are combined [20]. Linear modal representation with a small deflection assumption, two flapwise bending modes and one edgewise bending mode per blade are used to model the wind turbine blades. The tower can be modelled by two fore-aft and two side-to-side bending modes. The X-SEA uses the wind turbine dynamics modules in FAST v8. When the aerodynamic module, AeroDyn is coupled to FAST, the wind profile and structural motions are inputted at each coupling time step. In the aero-elastic calculation, the aerodynamic loads on the blade and tower nodes are computed by AeroDyn and returned back to FAST.

2.2. Structural Dynamics of Substructures

The platform has full six degrees-of-freedom (DOF) with flexible body motion based on the X-SEA structural frame elements. X-SEA is the software used for integrated finite element structural analysis, which provides the nonlinear dynamic analysis and design of offshore steel and concrete structures, including oil and gas platforms and offshore wind farms [18,19]. The current version of X-SEA V3.04 is the result of extensive research and development of the finite element program XFINAS [27], originally developed at Imperial College, London. The solution options of X-SEA range from simple static to highly advanced dynamic analysis using Morison equation and diffraction theory [19,21]. The Element Library in X-SEA consists of various element types essential to the modelling of offshore structures and foundations such as shell, solid, truss, cable, spring and tendon elements [21]. By using the pre/post processor software, GiD, a user-friendly graphical interface of FAST was developed for the calculation of wind turbine loading.

2.3. Hydrodynamics in X-SEA

The dynamic response of fixed offshore structures can be carried out by using X-SEA hydrodynamic modules, which includes Airy wave, Stokes 5th, Cnoidal wave, Solitary wave, and Stream functions. The Morison equation is used to compute hydrodynamic forces with contributions including hydrodynamic added mass and damping, and incident wave excitations [28]. For a single pile, the motion equation in terms of mass (m), damping (c) and stiffness (k) that is limited by the assumptions mentioned above is

$$(m + \widetilde{m})\ddot{w} + (c + \widetilde{c})\dot{w} + kw = \frac{1}{2}C_D \rho A |v| v + C_M \rho \Delta \frac{\partial v}{\partial t} \tag{1}$$

where the parameter ρ is a water density, C_D and C_M are drag and inertia coefficients, respectively, and v is the velocity of the water particle acting on the structural node and normal to the structure. The term A is the cross-sectional area of the element, and Δ denotes the volume of the displaced fluid. The terms $\ddot{w}$, $\dot{w}$ and w are the displacement, velocity and acceleration, respectively, of the structure in its local coordinate, which are normal to or in its longitudinal axis. When the motion of the structure is considered, the inertia force is reduced by a factor proportional to the structural acceleration, and the drag force is reduced by the relative velocity and given in the form:

$$\widetilde{m} = \rho(C_M - 1)\Delta$$
$$\widetilde{c} = C_D \rho A \bar{v}$$

(2)

in which, $\bar{v}$ is time-dependent cylinder velocity [26]. The terms in Equation (1) obtained in structure local coordinates are then transformed to the global coordinates depicted in Figure 3. When the diameter (D) of the cross section in the structure are considerably large in comparison with the wavelength (L), i.e., $D/L \geq 0.2$, the Morison theory is considered to be inapplicable [28]. Therefore, a diffraction theory implemented in X-SEA [28] can be considered.

2.4. Coupled Dynamic Analysis of Offshore Wind Turbine and Support Structures

In the simulation and analysis of OWT structures, there are two approaches: uncoupled and coupled methods [24,29]. The uncoupled method is a single way of simulation by transmitting forces and moments from the turbine to the tower or support structure. The interactions among the turbine, the tower and the support structure at the time of analysis were not considered in the uncoupled approach [15]. In order to account for these interactions, the coupled analysis approach is therefore used in this paper. That can be done by following the concept of exchanging displacement (u), velocity ($\dot{u}$), acceleration ($\ddot{u}$) and reaction forces (F_{TP}) at the interface node as depicted in Figure 3b. The equation of motion of the support structure can be written as

$$[M]\{\ddot{U}\} + [C]\{\dot{U}\} + [K]\{U\} = \{F(t)\}$$

(3)

The Craig-Bampton (C-B) reduction [17] was introduced by following the FAST program [30]. This is used to reduce the number of degrees of freedom, which easily grow to thousands for the typical frame elements. The nodes were classified into the boundary nodes "S" and the interior nodes "N". The derivation of the systematic reduction is presented as follows where the reduced equation of motion is:

$$\begin{bmatrix} M_{SS} & M_{SN} \\ M_{NS} & M_{NN} \end{bmatrix} \begin{Bmatrix} \ddot{U}_S \\ \ddot{U}_N \end{Bmatrix} + \begin{bmatrix} C_{SS} & C_{SN} \\ C_{NS} & C_{NN} \end{bmatrix} \begin{Bmatrix} \dot{U}_S \\ \dot{U}_N \end{Bmatrix} + \begin{bmatrix} K_{SS} & K_{SN} \\ K_{NS} & K_{NN} \end{bmatrix} \begin{Bmatrix} U_S \\ U_N \end{Bmatrix} = \begin{Bmatrix} F_S + F_{Sg} \\ F_N + F_{Ng} \end{Bmatrix}$$

(4)

The applied forces include the external forces (F_S, F_N) through the interface node of the substructure and the gravitation forces (F_{Sg}, F_{Ng}), which are considered as static forces lumped at each node. The forces at the boundary nodes are separate into the hydrodynamic forces, $F_{Hydrodynamic}$ and those transferred from the structural dynamic module, $F_{Structural\_Dynamic\_Module}$.

$$F_S = F_{Hydrodynamic} + F_{Structural\_Dynamic\_Module}$$

(5)

The fundamental assumption of the C-B reduction method is that the nodal displacements can be simply approximated by the interior generalized vector q_N as:

$$\begin{Bmatrix} U_S \\ U_N \end{Bmatrix} = \begin{bmatrix} I & 0 \\ \Phi_S & \Phi_N \end{bmatrix} \begin{Bmatrix} U_S \\ q_N \end{Bmatrix}$$

(6)

where I is the identity matrix, Φ_S is the physical displacement of the interior nodes for static analysis, and Φ_N is the internal eigenmode. These can be obtained by:

$$\Phi_S = -K_{NN}^{-1} K_{NS}$$

(7)

$$I = \Phi_N^T M_{NN} \Phi_N \tag{8}$$

By reducing the number of generalized displacements to "m", Φ_m is chosen to denote the truncated set of Φ_N and Ω_m is the diagonal matrix containing the corresponding frequencies. The nodal displacements can be written as:

$$\left\{ \begin{array}{c} U_S \\ U_N \end{array} \right\} = \left[\begin{array}{cc} I & 0 \\ \Phi_S & \Phi_m \end{array} \right] \left\{ \begin{array}{c} U_S \\ q_m \end{array} \right\} \tag{9}$$

The equations of motion finally become:

$$\left[\begin{array}{cc} M_{BB} & M_{Bm} \\ M_{mB} & I \end{array} \right] \left\{ \begin{array}{c} \ddot{U}_S \\ \ddot{U}_N \end{array} \right\} + \left[\begin{array}{cc} C_{SS} + C_{SN}\Phi_S + \Phi_S^T C_{NS} + \Phi_S^T C_{NN}\Phi_S & C_{SN}\Phi_m + \Phi_S^T C_{NN}\Phi_R \\ \Phi_m^T C_{NS} + \Phi_m^T C_{NN}\Phi_S & \Phi_m^T C_{NN}\Phi_S \end{array} \right] \left\{ \begin{array}{c} \dot{U}_S \\ \dot{U}_N \end{array} \right\}$$
$$+ \left[\begin{array}{cc} K_{SS} & 0 \\ 0 & \Omega_m^2 \end{array} \right] \left\{ \begin{array}{c} U_S \\ U_N \end{array} \right\} = \left\{ \begin{array}{c} (F_S + F_{Sg}) + \Phi_m^T(F_S + F_{Sg}) \\ \Phi_m^T(F_N + F_{Ng}) \end{array} \right\} \tag{10}$$

The matrix partition can be calculated as follows:

$$\begin{aligned} M_{BB} &= M_{SS} + M_{SN}\Phi_S + \Phi_S^T M_{NS} + \Phi_S^T M_{NN}\Phi_S \\ M_{mB} &= \Phi_m^T M_{NS} + \Phi_m^T M_{NN}\Phi_S \\ M_{Bm} &= M_{mb}^T \\ K_{BB} &= K_{SS} + K_{SN}\Phi_S \end{aligned} \tag{11}$$

The fixed boundary condition applied at the bottom of the support structure can be written as:

$$U_S = \left[\begin{array}{c} \overline{U}_S \\ 0 \end{array} \right] \tag{12}$$

Finally, the interface nodes are treated as rigidly connected to the transition pieces as follows:

$$\overline{U}_S = T U_{TP} \tag{13}$$

where U_{TP} is the displacement and rotation of the interface node or transition piece. The matrix T is the global coordinate due to the interface node and is defined by following the local coordinate system of motion as:

$$t_i = \left[\begin{array}{cccccc} 1 & 0 & 0 & 0 & Z_i - Z_{TP} & -(Y_i - Y_{TP}) \\ 0 & 1 & 0 & -(Z_i - Z_{TP}) & 0 & X_i - X_{TP} \\ 0 & 0 & 1 & Y_i - Y_{TP} & -(X_i - X_{TP}) & 0 \\ 0 & 0 & 0 & 1 & 0 & 0 \\ 0 & 0 & 0 & 0 & 1 & 0 \\ 0 & 0 & 0 & 0 & 0 & 1 \end{array} \right] \tag{14}$$

By following the equation of motion in the present program, the velocity, acceleration, and displacement received from FAST program at a time step can be written as:

$$[M]_i\{\ddot{U}_{TP}\}_i + [C]_i\{\dot{U}_{TP}\}_i + [K]_i\{U_{TP}\}_i = \{F(t)\}_i \tag{15}$$

In Equation (15), i is the interface node number applied to the structure in the present program. In terms of the interface node, the boundary node is classified as "B". After applying the fixed constraints at those nodes, Equation (10) can be written as:

$$\left[\begin{array}{cc} \widetilde{M}_{BB} & \widetilde{M}_{Bm} \\ \widetilde{M}_{mB} & I \end{array} \right] \left\{ \begin{array}{c} \ddot{U}_{TP} \\ \ddot{q}_m \end{array} \right\} + \left[\begin{array}{cc} C_{SS} + C_{SN}\Phi_S + \Phi_S^T C_{NS} + \Phi_S^T C_{NN}\Phi_S & C_{SN}\Phi_m + \Phi_S^T C_{NN}\Phi_R \\ \Phi_m^T C_{NS} + \Phi_m^T C_{NN}\Phi_S & \Phi_m^T C_{NN}\Phi_S \end{array} \right] \left\{ \begin{array}{c} \dot{U}_S \\ \dot{q}_m \end{array} \right\}$$
$$+ \left[\begin{array}{cc} \widetilde{K}_{SS} & 0 \\ 0 & \Omega_m^2 \end{array} \right] \left\{ \begin{array}{c} U_{TP} \\ q_m \end{array} \right\} = \left\{ \begin{array}{c} F_{TP} \\ \widetilde{F}_m \end{array} \right\} \tag{16}$$

where the terms in Equation (16) are defined as:

$$\begin{aligned}
\widetilde{M}_{BB} &= T_I^T \overline{M}_{BB} T_I \\
\widetilde{M}_{mB} &= T_I^T \overline{M}_{mB} \\
\widetilde{M}_{Bm} &= \widetilde{M}_{mb}^T \\
\widetilde{K}_{BB} &= T_I^T \overline{K}_{BB} T_I \\
\overline{F}_{TP} &= F_{TP} + T_I^T \overline{F}_{HDR} + T_I^T \overline{F}_{Sg} + T_I^T \overline{\Phi}_S^T (F_N + F_{Ng}) \\
\overline{F}_m &= \Phi_m^T (F_N + F_{Ng})
\end{aligned} \tag{17}$$

Finally, the force and moment at the interface node can be written as:

$$F_{TP} = T_I^T F_{Structural\_Dynamic\_Module} \tag{18}$$

2.5. Soil-Pile-Structure Interaction Analysis

The computer simulation of a pile foundation accounts for the stiffness of the pile and the lateral and horizontal behavior of the soil. The nonlinear behavior for pile-soil interaction, as shown in Figure 5, is considered based on the geotechnical data for the lateral load deflection ($P - Y$), axial load transfer and pile displacement ($T - Z$), and tip-load displacement ($Q - Z$) curves in order to obtain a rigorous solution to the pile-soil-structure interaction. The variation in soil stiffness should be considered [22] if the soil conditions at the supports are different from each other. In addition, the superelement concept [17] is introduced to reduce the computational time where the basic idea is to use the condensation of the stiffness matrix to reduce the number of degrees of freedom and time consumption in the finite element analysis.

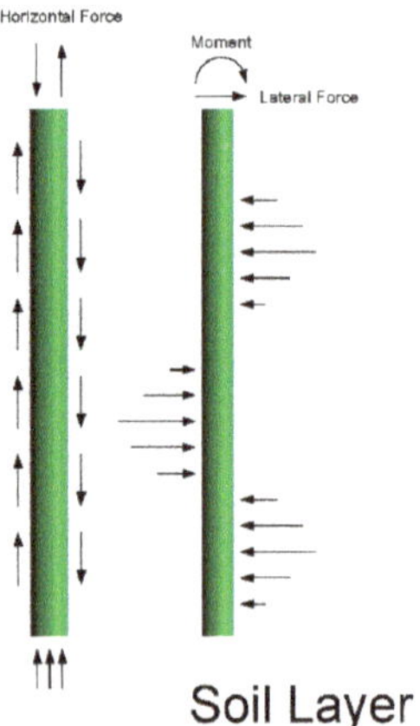

Figure 5. Definition of pile-soil interaction.

2.6. A Scheme for Coupled Analysis Including the Soil-Pile-Structure Interaction

The proposed method required a four-step pre-processing procedure as shown in Figure 6. In order to define the reaction forces with turbine effects, the coupled approach described in Sections 2.1–2.5 is used. The X-SEA program was simulating substructure with a pile model as an individual pile analysis to calculate the stiffness, which considers the condensation of stiffness for pile superelement. This procedure implies that the pile-soil-structure interaction behavior has to be evaluated several times to calculate the appropriate stiffness matrix.

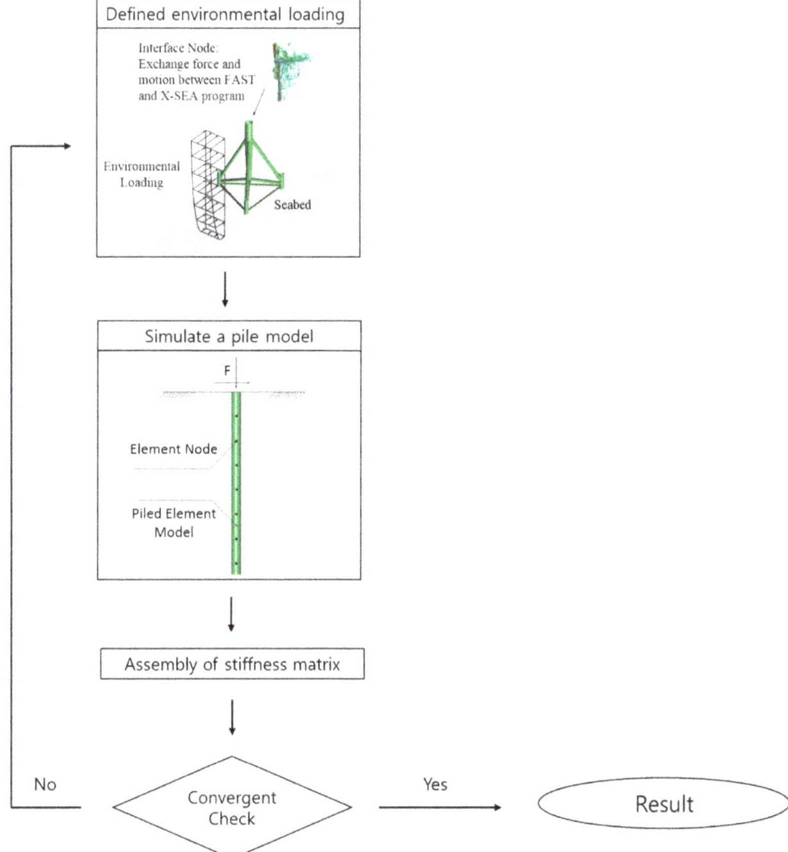

Figure 6. Scheme for coupled analysis of turbine effects including pile-soil-structure interactions.

3. Numerical Example

3.1. Verification of Tripod Structures Supporting NREL 5MW Offshore Wind Turbines

The tripod support structure of a wind turbine, which was researched in NREL, was sitting on the seabed with a fixed boundary condition of 55 m long driven into a 45 m water depth, and extended 10 m above the mean sea level. This structure was modelled in the X-SEA program with external forces acquired from the hydrodynamic module in the FAST program. The water density is 1027 kg/m^3, the significant wave height is 8 m, the wave period is 10 s and steady wind velocity is 8 m/s at a reference height of 90 m above the mean sea level. Based on the longest natural periods, the structure and the wave period, the analysis period was selected as the first 60 s, which should correspond to the start-up to power production and consists of both transient and steady behaviour. The support structure was modeled by using 158 nodes and 163 elements as illustrated in Figure 7. The geometry and material properties of the tripod structure are given in Table 1. The verification example focused on comparing six components of dynamic reaction forces and top-tower forces that resulted from FAST and from X-SEA coupled with the FAST program, as theoretically presented in Section 2 and Figures 3 and 4. The comparison was aimed at confirming that the X-SEA program could produce the same correct results as FAST.

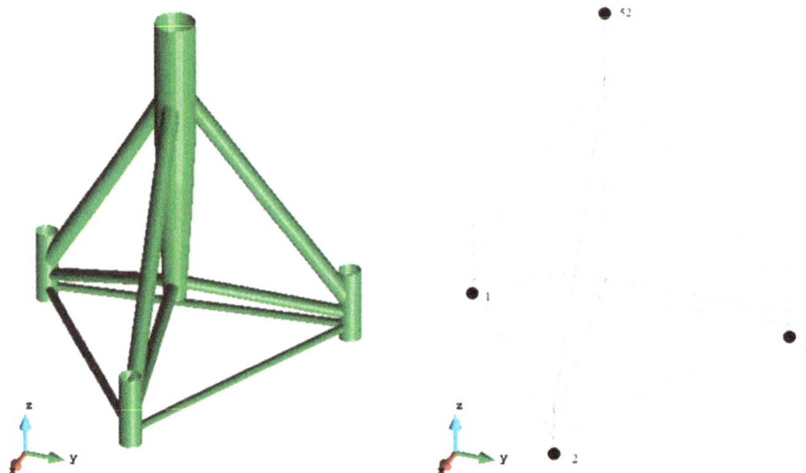

Figure 7. Model of the NREL 5MW tripod support structure.

Table 1. Geometry and material properties of the tripod support structure.

Outer diameter of diagonal brace (m)	2.475–1.200
Wall thickness of diagonal brace (m)	0.035–0.025
Outer diameter of main tubular (m)	5.412–1.875
Wall thickness of main tubular (m)	0.05–0.035
Young's Modulus (N/m^2)	2.1×10^{11}
Density (kg/m^3)	7850

Figures 8 and 9 show that the six components of the reaction forces (R_x, R_y, R_z) and moments (R_{mx}, R_{my}, R_{mz}) resulting from X-SEA are in good agreement with the corresponding results from FAST. These responses, except for R_x, appear to be transient and significantly attenuated in the first 30 s and then become steady. Such initial transient behaviour is due to the small amount of damping in the structure and the method by which the hydrodynamic load and aerodynamic loading are initialized at the start of the simulation. Therefore, the transient responses are quickly damped out.

The marginal differences between the X-SEA and FAST results in Figures 8 and 9 can be attributed to the fact that the ratio of X-SEA/FAST obtained from the natural frequency of the six mode shapes are 0.99942, 0.99943, 0.99954, 0.99960, 0.99960 and 0.99944, respectively, as illustrated in Figure 10. The six dynamic components of the turbine load that were obtained by applying the X-SEA and FAST programs on the frame elements are identical, as shown in Figures 11 and 12. It can therefore be concluded that the present study program performs normally.

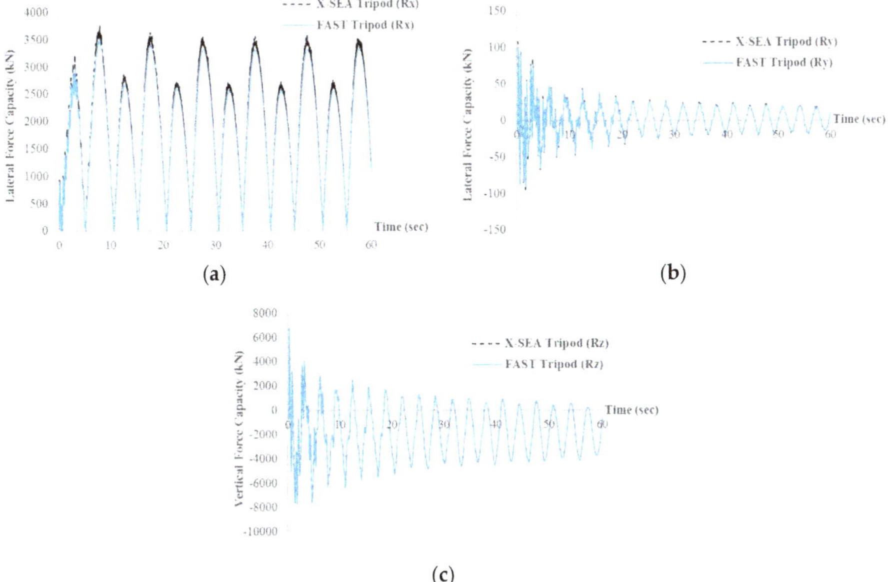

Figure 8. A comparison of the dynamic lateral reaction forces in tripod support structure that resulted from X-SEA and FAST programs in the x-direction (**a**), y-direction (**b**), and z-direction (**c**).

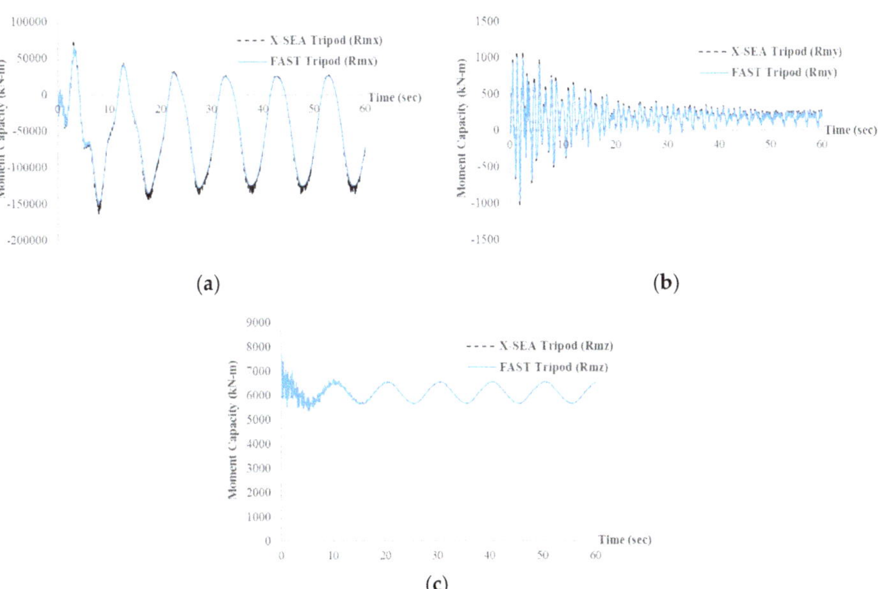

Figure 9. A comparison of the dynamic reaction moments of the tripod support structure that resulted from X-SEA and FAST programs in the x-direction (**a**), y-direction (**b**), and z-direction (**c**).

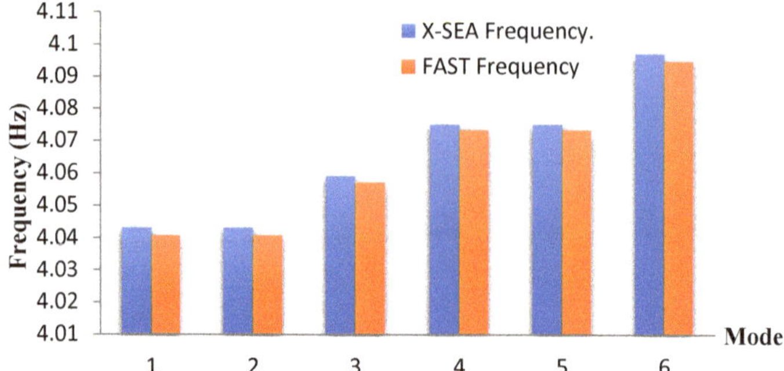

Figure 10. A comparison of natural frequency values of the tripod support structure resulting from the X-SEA and FAST programs.

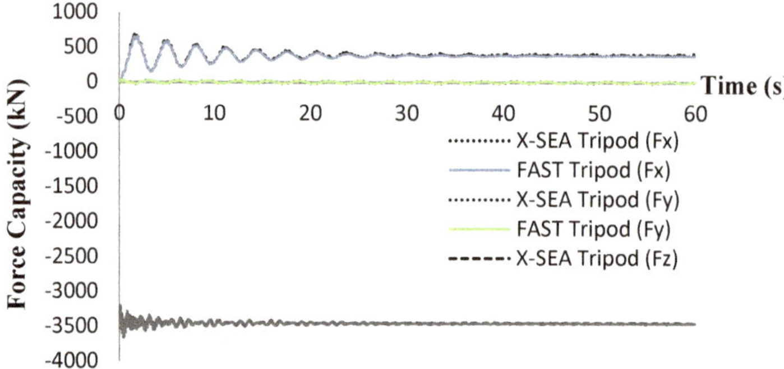

Figure 11. Comparison of the dynamic top-tower forces in x, y, and z directions of the tripod support structure that resulted from the X-SEA and FAST programs.

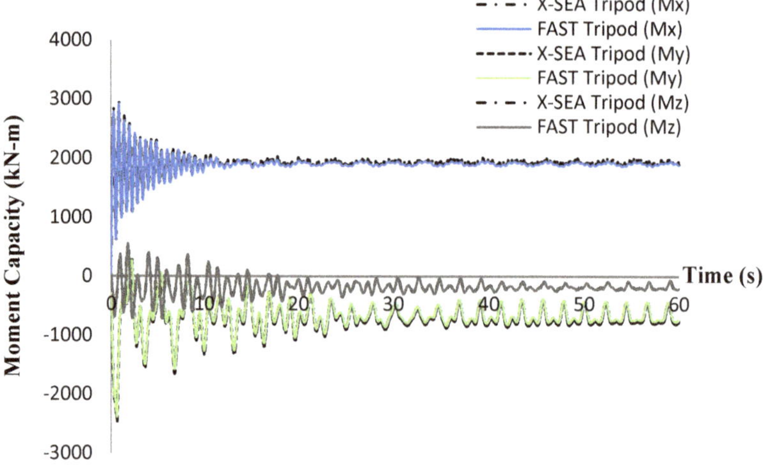

Figure 12. Comparison of the dynamic top-tower moments about the x, y, and z directions of the tripod support structure between the X-SEA and FAST programs.

In the first mode, the tripod support structure was oscillating at the local members connected to the pile head without considerable movement of the whole structure, as illustrated in Figure 13a. The second mode is the same as the first mode but the structure was oscillating in other directions and at different members, as illustrated in Figure 13b. The third to sixth modes are the in-plane and out-of-plane bending modes, which occurred at the local members, but they were oscillating in other directions as illustrated in Figure 13c–f, respectively.

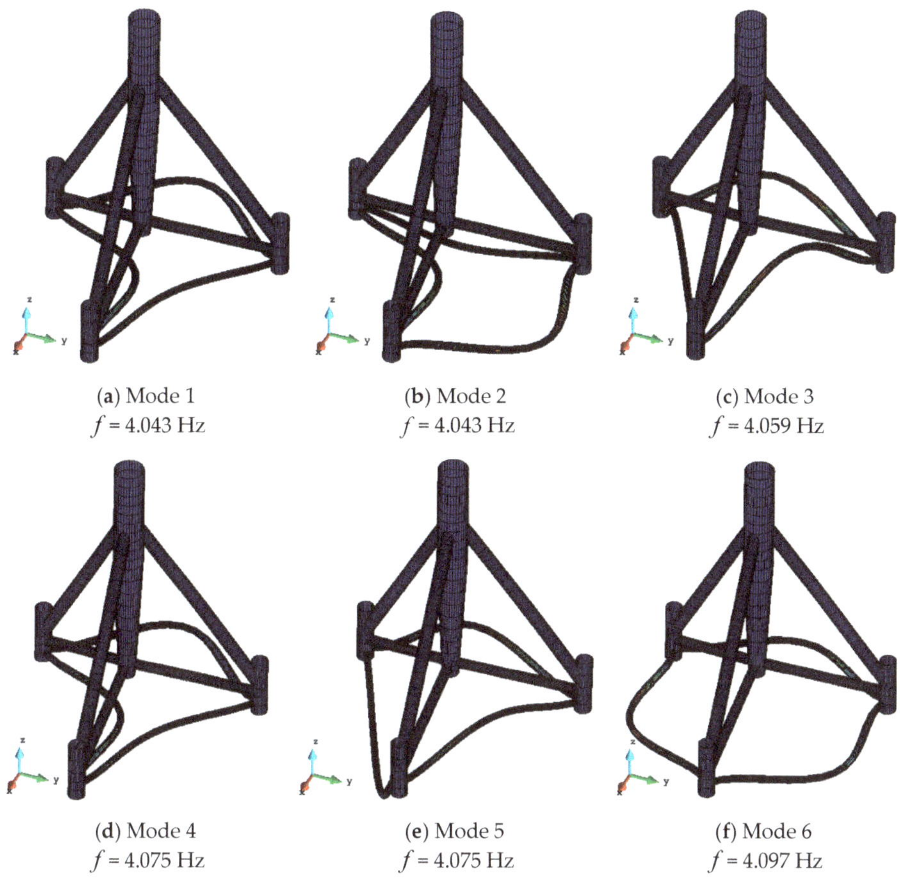

(**a**) Mode 1 $f = 4.043$ Hz	(**b**) Mode 2 $f = 4.043$ Hz	(**c**) Mode 3 $f = 4.059$ Hz
(**d**) Mode 4 $f = 4.075$ Hz	(**e**) Mode 5 $f = 4.075$ Hz	(**f**) Mode 6 $f = 4.097$ Hz

Figure 13. Natural frequencies and mode shapes of the tripod support structure.

3.2. Parametric Study of Tripod/Jacket-Supported Offshore Wind Turbines

In the coupled analysis method, the motions are interchanged between the FAST and X-SEA programs, and this considers the response of the tower corresponding to the turbine response. X-SEA determines the hydrodynamic forces and computes the response of the foundation and the support structure. The values are exchanged between these two programs by using the present study module. The important components in the equations of motion including displacement, velocity and acceleration are input to X-SEA. The six force components at the integral time step are then computed in X-SEA and returned to FAST.

In this example, a 5MW OWT supported by a tripod structure that was researched by NREL was used. The model is sitting on the seabed, is 55 m high, driven through 45 m of water and extends 10 m above the mean sea level, and takes account of environmental conditions including a wave height of 2.8 m and wave period of 6.07 s. In addition, the uncoupled analysis method was developed by taking six load components at the tower base produced by FAST. Those loads can be applied at the interface node using X-SEA to compute the responses and for comparing with the coupled analysis.

The responses in the x-direction and y-direction of the tripod support structure that resulted from the coupled analysis are considerably smaller than the results of the uncoupled analysis, as plotted in Figures 14 and 15. This can be attributed to the fact that in the coupled models, the interaction between the tower and support structures has been included in that results thanks to the additional coupling stiffness. This effectiveness of the coupled models in simulating more accurate responses has been observed in the dynamic analyses of jacket supported OWTs [21] and floating OWTs [29].

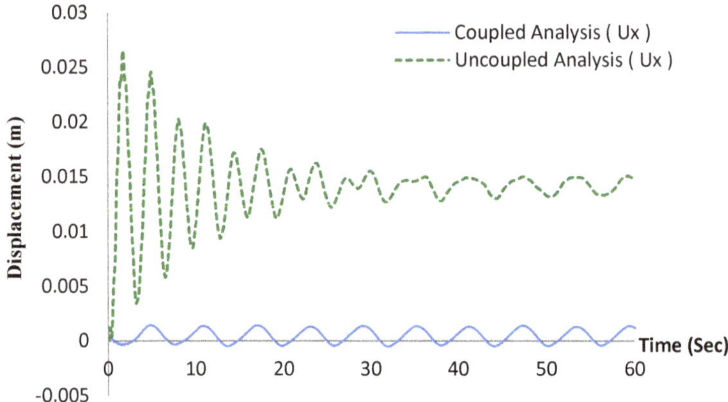

Figure 14. Comparison of deflections in the x- direction resulting from couple and uncouple analyses.

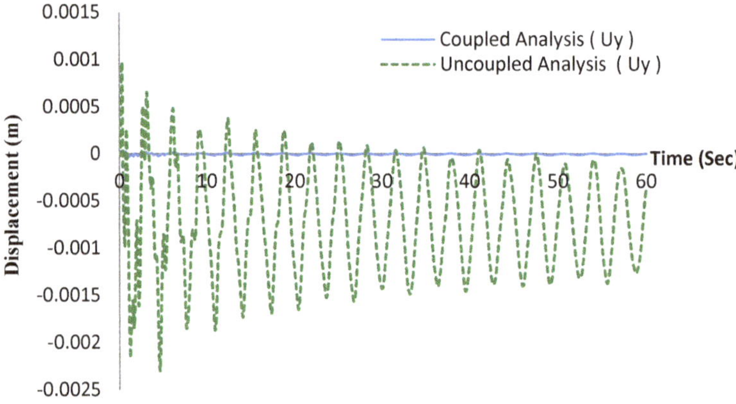

Figure 15. Comparison of displacements in the y-direction resulting from coupled and uncoupled analyses.

3.3. Coupled Analysis of OWT Support Structures Including the Pile-Soil-Structure Interaction

The previous section demonstrated the accuracy of the responses of fixed support structures for OWTs by using coupled models and analyses. In this section, the tripod support structure that was researched by NREL, with its geometry and properties described in Figure 7 and Table 1 is used for this parametric study. The environmental conditions are: 2.8 m wave height and 6.07 s wave period. Three models were investigated including the pile supported structure, pile superelement, and fixed support structure. The soil behavior is assumed to be nonlinear soil stiffness due to $P - Y$, $T - Z$, and $Q - Z$ curves in the specified offshore design standards. The pile of 1.5 m diameter and 0.05 m thickness penetrated the soil and embedded 55 m deep. The five soil layers with their soil data properties are listed in Figure 16. Another advanced technique introduced in this paper is the condensation of the stiffness matrix to reduce the degree of freedom size and computational time in the finite element analysis.

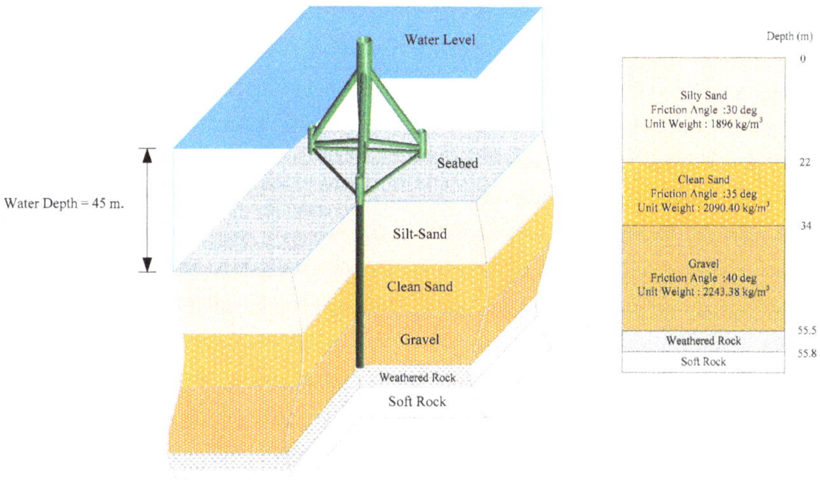

Figure 16. The tripod support structure for pile-soil-structure interaction (PSSI) modeling.

One of the primary purposes of free vibration and dynamic analysis is to avoid the cases where a non-stationary load can resonate with the structural system when the structure is excited by a loading frequency close to a natural frequency of the system. Thus, a design chart provides the first to sixth mode shapes and the corresponding natural frequencies in Figures 17 and 18. A scaling factor of 25 was used to amplify the mode shapes and only the odd modes are displayed in Figures 17 and 18 since the even modes have the same frequencies that vibrate in the orthogonal planes.

In the first mode shape, the piled support structure is oscillating globally in the x-direction as a bending mode of the whole structure with a natural frequency of 3.521 Hz, as illustrated in Figure 18a. The second mode has a similar shape to the first mode and a natural frequency of 3.539 Hz. It oscillates in large movements in the x-direction as shown in Figure 18b while the third mode shape with a natural frequency of 3.848 Hz is oscillating in the opposite x-direction, as illustrated in Figure 18c. The fourth mode shape has natural frequencies of 3.857 Hz and the movement of a local member. which is connected to the pile head. This mode is oscillating in the negative x-direction as shown in Figure 18d. In the fifth and sixth mode shapes, the main tubular stays still and both of the local members, which are connected to the pile head, oscillate with frequencies of 4.005 Hz and 4.047 Hz, respectively, as illustrated in Figure 18e,f.

By comparing the natural frequencies that resulted from the fixed support structure model, it is seen that the natural frequencies resulting from the piled support structure model are increased 1.148, 1.142, 1.055, 1.057, 1.017, and 1.012 times as illustrated in Figure 17. These significant differences in natural frequency relate to the self-weight or mass and stiffness of the supporting pile and soils. The piled-support models were found to represent the tripod structure more realistically and to produce more accurate natural frequencies.

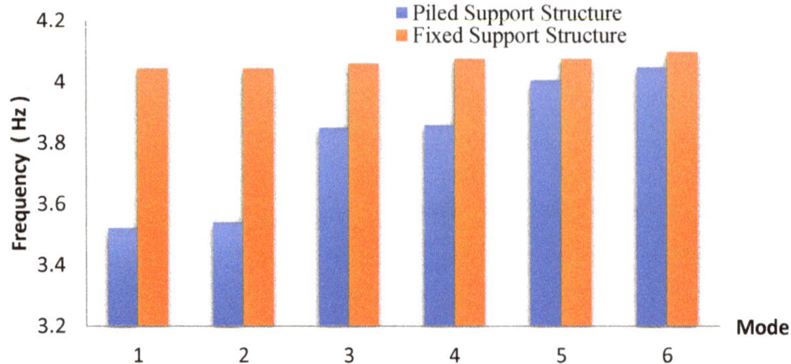

Figure 17. The comparison of natural frequencies that resulted from the piled support structure model and the fixed support structure model.

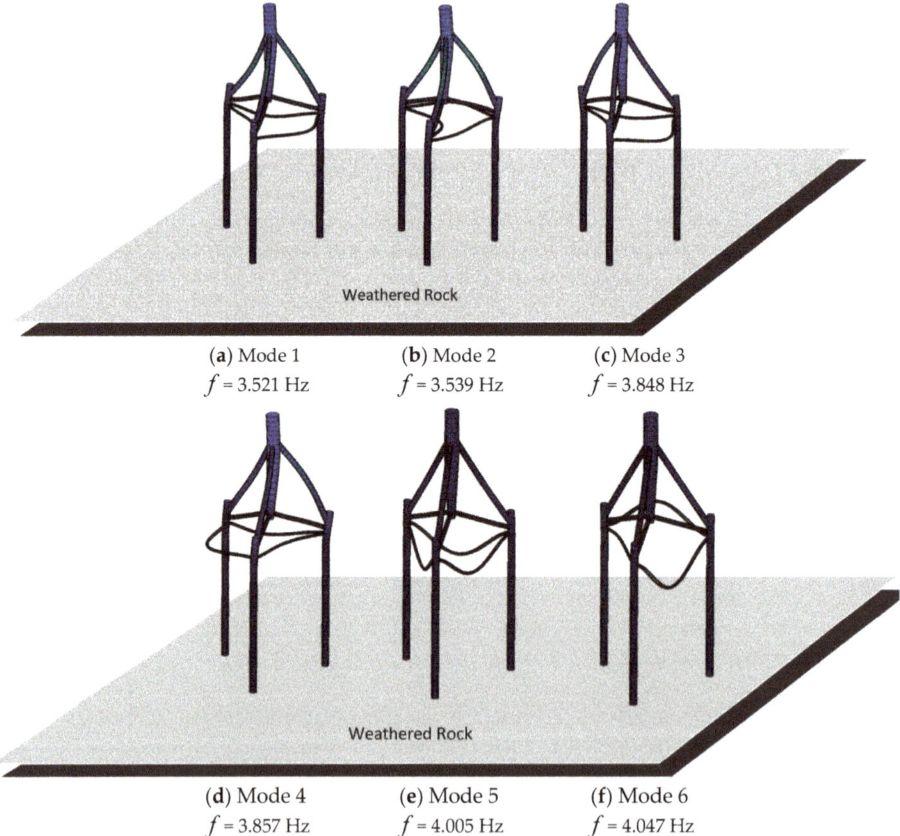

(**a**) Mode 1
f = 3.521 Hz

(**b**) Mode 2
f = 3.539 Hz

(**c**) Mode 3
f = 3.848 Hz

(**d**) Mode 4
f = 3.857 Hz

(**e**) Mode 5
f = 4.005 Hz

(**f**) Mode 6
f = 4.047 Hz

Figure 18. Natural frequencies and mode shapes resulting from the piled-supported structure model.

The displacements in the x-direction that resulted from the fixed support structure model and the piled support structure model are plotted in Figure 19. It can be noted that the piled support structure and pile superelement model produced responses 1.399 times more flexible that those of the fixed support structure model. It is important to note in Figure 20 that the displacements in y-direction from both the piled support and superelement models have a similar trend in the x-direction but are

1.731 times larger than that of the fixed support model. It is shown that, besides the physical factors such as environmental loads, turbine operation, mechanical and material properties of structures and soils, the simulated responses of the OWT and support structure are significantly influenced by the chosen model. The inclusion of pile-soil-structure interaction and 3D nonlinear soil stiffness make the prediction of the OWT-support structure responses more accurate and closer to the real behaviour.

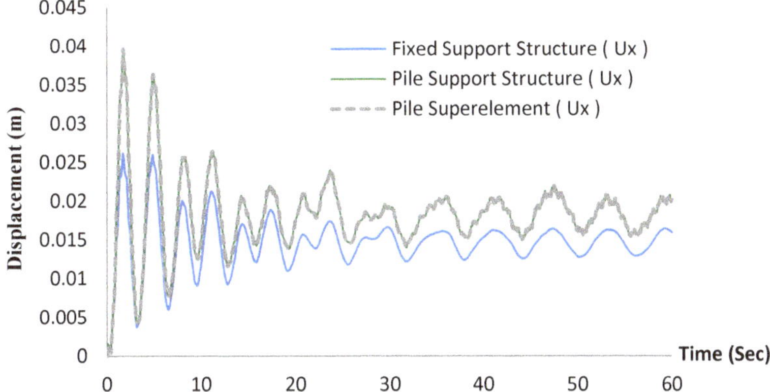

Figure 19. A comparison of the displacements in x-direction of the fixed support structure model, piled support structure model and piled superelement model.

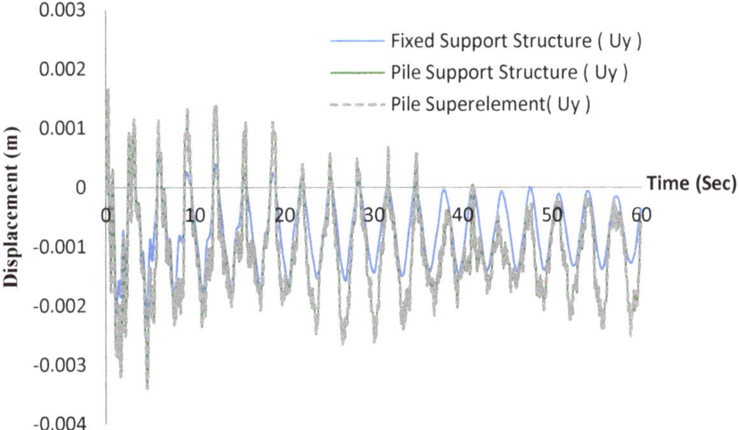

Figure 20. A comparison of the displacements in y-direction of the fixed support structure model, piled support structure model, and piled superelement model.

4. Concluding Remarks

Theoretical aspects and simulations of the coupled analysis of an offshore wind turbine (OWT) and its tripod support structure using a number of advanced techniques have been presented. The pile-soil-structure interaction was used to accurately represent the 3D nonlinear behavior of soil. The superelement of piles and condensation of the stiffness matrix was used for both computational efficiency and maintaining the numerical accuracy. By replacing the sub-structural module in FAST with the component of offshore substructures in X-SEA, the reaction forces and wind turbine loads and responses were calculated in each time step. The following conclusions are drawn from the numerical examples:

- For the tripod structure itself, the first to sixth modes are oscillating at a local member connected to the pile head. The sixth mode is the globally bending mode. The reaction forces and moments, natural frequencies and top-tower forces that resulted from X-SEA were in good agreement with those that resulted from FAST.

- The physical interaction between the tower and support structure has been included in the coupled models as additional coupling stiffness, which results in considerably smaller responses compared to the uncoupled models. This demonstrates that the coupled model should be used in the analysis and design of offshore wind turbine structures.

- For the piled-support tripod structures, the first to third modes are global whereas the fourth to sixth mode shapes are at a local member connected to the pile head. Their natural frequencies are considerably larger than those of the fixed support model as the physical interactions and infinite stiffness of the supporting pile-soil have been accounted for. This demonstrates the higher accuracy and validity of the piled-support structure model.

- The piled-support model produces larger responses than the fixed support model but is identical to the pile superelement model. Besides the physical factors, the simulation of OWT structures is significantly influenced by the chosen model, because it needs condensed stiffness information for the foundation base.

For future studies, physical models or measurements from real structures are recommended. The inclusion of 3D nonlinear soil stiffness into pile-soil-structure interaction would make the simulation more accurate and realistic. Other approaches to modeling soil-structure interaction, especially the inclusion of soil damping where energy dissipates during vibration could be considered.

Author Contributions: Formal analysis, P.P.; investigation, P.P. and V.N.D.; Validation, P.P.; Conceptualization, V.N.D. and K.-D.K.; Methodology, V.N.D. and K.-D.K.; Writing—original draft preparation, P.P. and V.N.D.; Writing—review and editing, V.N.D. and K.-D.K.; Funding acquisition, K.-D.K.

Funding: This work is supported by Konkuk University, Seoul, Korea.

Conflicts of Interest: The authors declare no conflict of interest.

References

1. Esteban, M.D.; López-Gutiérrez, J.-S.; Negro, V. Gravity-based foundations in the offshore sector. *J. Mar. Sci. Eng.* **2019**, *7*, 64. [CrossRef]
2. Oh, K.-Y.; Nam, W.-C.; Ryu, M.-S.; Kim, J.-Y.; Epureanud, B.I. A review of foundations of offshore wind energy convertors: Current status and future perspectives. *Renew. Sustain. Energy Rev.* **2018**, *88*, 16–36. [CrossRef]
3. Ryu, M.S.; Kim, J.-Y.; Lee, J.-S. Comparison of two meteorological tower foundations for offshore wind turbines. In Proceedings of the Twenty-Sixth International Offshore and Polar Engineering Conference, Rhodes, Greece, 26 June–2 July 2016.
4. Kim, S.-R.; Hung, L.C.; Oh, M. Group effect on bearing capacities of tripod bucket foundations in undrained clay. *Ocean Eng.* **2014**, *79*, 1–9. [CrossRef]
5. Schaumann, P.; Böker, C. Can jackets and tripods compete with monopiles? In Proceedings of the Copenhagen Offshore Wind, Copenhagen, Denmark, 6–28 October 2005.
6. Zaaijer, M.B. *Comparison of Monopile, Tripod, Suction Bucket and Gravity Base Design for a 6 MW Turbine*; Delft University of Technology: Delft, The Netherlands, 2003.
7. Ryu, M.S.; Kim, J.-Y.; Kang, K.-S. The first met-mast for offshore wind farm in Korea and its remote sensing system. In Proceedings of the Twenty-First International Offshore and Polar Engineering Conference, Maui, HI, USA, 19–24 June 2011.
8. Chen, D.; Huang, K.; Bretel, V.; Hou, L. Comparison of structural properties between monopile and tripod offshore wind-turbine support structures. *Adv. Mech. Eng.* **2013**, *5*, 1–9. [CrossRef]
9. Hung, L.C.; Kim, S.-R. Evaluation of combined horizontal-moment bearing capacities of tripod bucket foundations in undrained clay. *Ocean Eng.* **2014**, *85*, 100–109. [CrossRef]

10. Tran, N.X.; Hung, L.C.; Kim, S.-R. Evaluation of horizontal and moment bearing capacities of tripod bucket foundations in sand. *Ocean Eng.* **2017**, *140*, 209–221. [CrossRef]

11. Yang, H.; Zhu, Y.; Lu, Q.; Zhang, J. Dynamic reliability based design optimization of the tripod sub-structure of offshore wind turbines. *Renew. Energy* **2015**, *78*, 16–25. [CrossRef]

12. Margariti, G.; Papadopoulos, A.; Barmpas, D.; Gantes, C.J.; Gkologiannis, C.P. Design of monopile and tripod foundation of fixed offshore wind turbines via advanced numerical analysis. In Proceedings of the 8th GRACM International Congress on Computational Mechanics, Volos, Greece, 2–15 July 2015.

13. Hübler, C.; Hafele, J.; Gebhardt, C.G.; Rolfes, R. Experimentally supported consideration of operating point dependent soil properties in coupled dynamics of offshore wind turbines. *Mar. Struct.* **2018**, *57*, 18–37. [CrossRef]

14. Häfele, J.; Hübler, C.; Gebhardt, C.G.; Rolfes, R. An improved two-step soil-structure interaction modeling method for dynamical analyses of offshore wind turbines. *Appl. Ocean Res.* **2016**, *55*, 141–150. [CrossRef]

15. Dinh, V.N.; McKeogh, E. Offshore Wind Energy: Technology Opportunities and Challenges. In *Lecture Notes in Civil Engineering, Proceedings of the Vietnam Symposium on Advances in Offshore Engineering, Hanoi, Vietnam, 25 September 2018*; Springer: Singapore, 2018. [CrossRef]

16. Ong, M.C.; Bachynski, E.E.; Økland, O.D.; Passano, E. Dynamic responses of a jacket-type offshore wind turbine using decoupled and coupled models. In Proceedings of the ASME 2014 33rd International Conference on Ocean, Offshore and Arctic Engineering (OMAE), San Francisco, CA, USA, 8–13 June 2014.

17. Voormeeren, S.N.; van der Valk, P.L.C.; Nortier, B.P.; Molenaar, D.-P.; Rixen, D.J. Accurate and efficient modeling of complex offshore wind turbine support structures using augmented superelements. *Wind Energy* **2014**, *17*, 1035–1054. [CrossRef]

18. Kim, K.D.; Plodpradit, P.; Manovachirasan, A.; Sinsabvarodom, C.; Kim, B.J. Analysis of offshore structures for wind turbines and oil and gas using X-SEA software. In Proceedings of the 11th World Congress on Computational Mechanics (WCCM XI), 5th European Conference on Computational Mechanics (ECCM V), 6th European Conference on Computational Fluid Dynamics (ECFD VI), Barcelona, Spain, 20–25 July 2014.

19. Kim, K.-D.; Vachirapanyaku, S.; Plodpradit, P.; Dinh, V.-N.; Park, J.-H. Development of offshore structural analysis software X-SEA coupled with FAST. In Proceedings of the 38th International Conference on Ocean, Offshore & Artic Engineering, ASME 2019 OMAE 2019-96778, Glasgow, Scotland, UK, 9–14 June 2019.

20. Jonkman, J. FAST v8. National Renewable Energy Laboratory (NREL). USA, 2018. Available online: https://nwtc.nrel.gov/FAST8 (accessed on 19 September 2017).

21. Plodpradit, P.; Dinh, V.-N.; Kim, K.-D. Coupled analysis of offshore wind turbine jacket structures with pile-soil-structure interaction using FAST v8 and X-SEA. *Appl. Sci.* **2019**, *9*, 1633. [CrossRef]

22. Dinh, V.N.; Basu, B.; Brinkgreve, R.B.J. Wavelet-based evolutionary response of multi-span structures including wave-passage and site-response effects. *J. Eng. Mech.* **2014**, *140*, 8. [CrossRef]

23. Failla, G.; Santangelo, F.; Foti, G.; Scali, F.; Arena, F. Response-spectrum uncoupled analyses for seismic assessment of offshore wind turbines. *J. Mar. Sci. Eng.* **2018**, *6*, 85. [CrossRef]

24. Basu, B.; Staino, A.; Dinh, V.N. Vibration of wind turbines under seismic excitations. In Proceedings of the 5th Asian-Pacific Symposium on Structural Reliability and its Applications, Singapore, 23–25 May 2012. [CrossRef]

25. Fitzgerald, B.; Basu, B. A monitoring system for wind turbines subjected to combined seismic and turbulent aerodynamic loads. *Struct. Monit. Maint.* **2017**, *4*, 175–194. [CrossRef]

26. Kim, K.-D. *X-SEA User Manual*; Konkuk University: Seoul, Korea, 2016.

27. Kim, K.D.; Suthasupradit, S.; Kim, Y.H.; Lomboy, G.R.; Dinh, V.N. New development of XFINAS software for nonlinear dynamic and seismic analysis of structure. In Proceedings of the Third Asian-Pacific Congress on Computational Mechanics (APCOM'07) & the Eleventh International Conference on the Enhancement and Promotion of Computational Methods in Engineering and Science (EPMESC XI), Kyoto, Japan, 3–6 December 2007.

28. Kim, B.J.; Plodpradit, P.; Kim, K.D.; Kim, H.G. Three-dimensional analysis of prestressed concrete offshore wind turbine structure under environmental and 5-MW turbine loads. *J. Mar. Sci. Appl.* **2018**, *17*, 625–637. [CrossRef]

29. Dinh, V.N.; Basu, B.; Nielsen, S.R.K. Impact of spar-nacelle-blade coupling on the edgewise response of floating offshore wind turbines. *Coupled Syst. Mech.* **2013**, *2*. [CrossRef]
30. Damiani, R.; Jonkman, J.; Hayman, G. *SubDyn User's Guide and Theory Manual*; National Renewable Energy Laboratory (NREL): Golden, CO, USA, March 2015.

Article

Semi-Active Structural Control of Offshore Wind Turbines Considering Damage Development

Arash Hemmati and Erkan Oterkus *

Department of Naval Architecture, Ocean & Marine Engineering, University of Strathclyde, Glasgow G4 0LZ, UK; arash.hemmati-topkanloo@strath.ac.uk
* Correspondence: erkan.oterkus@strath.ac.uk; Tel.: +44-141-548-3876

Received: 29 June 2018; Accepted: 31 August 2018; Published: 5 September 2018

Abstract: High flexibility of new offshore wind turbines (OWT) makes them vulnerable since they are subjected to large environmental loadings, wind turbine excitations and seismic loadings. A control system capable of mitigating undesired vibrations with the potential of modifying its structural properties depending on time-variant loadings and damage development can effectively enhance serviceability and fatigue lifetime of turbine systems. In the present paper, a model for offshore wind turbine systems equipped with a semi-active time-variant tuned mass damper is developed considering nonlinear soil–pile interaction phenomenon and time-variant damage conditions. The adaptive concept of this tuned mass damper assumes slow change in its structural properties. Stochastic wind and wave loadings in conjunction with ground motions are applied to the system. Damages to soil and tower caused by earthquake strokes are considered and the semi-active control device is retuned to the instantaneous frequency of the system using short-time Fourier transformation (STFT). The performance of semi-active time-variant vibration control is compared with its passive counterpart in operational and parked conditions. The dynamic responses for a single seismic record and a set of seismic records are presented. The results show that a semi-active mass damper with a mass ratio of 1% performs significantly better than a passive tuned mass damper with a mass ratio of 4%.

Keywords: offshore wind; structural control; semi-active; tuned mass damper; earthquake

---

1. Introduction

The wind industry has attracted attention and is growing rapidly due to the environmental concerns over conventional energy resources. The offshore wind industry, in particular, is becoming more attractive because of its advantages over onshore wind. Offshore wind turbines (OWT) are subject to undesirable vibrations caused by environmental loadings, seismic excitations, and rotor frequency excitations, and these excessive vibrations need to be minimized to increase serviceability and lifetime of the system. A number of structural control devices using vibration control mechanisms have been developed to mitigate the aforementioned excessive vibrations. Tuned mass dampers (TMDs) and tuned liquid dampers (TLDs) are two main vibration control devices widely used.

Various vibration control mechanisms originally from civil engineering field have been proposed to control the level of vibrations in the wind industry. Three main vibration control methods such as passive, semi-active, and active exist. The applications and descriptions of these methods utilized in buildings and wind turbine structures were reviewed by Symans and Constantinou [1] and Chen and Georgakis [2]. The passive control system improves damping and stiffness of the main structure without the need of employing external forces [3,4]. The vibrations are not tracked via sensors in this method as curations of this system are constant. This method is widely used due to easy implementation and maintenance. Active vibration control system is a more sophisticated method in which not only mechanical properties are adjusted in the time domain but also external forces are employed. Thus, active control method requires the presence of active forces from external sources, resulting in high cost and complexity of the system. Semi-active vibration control system is the modified version of the passive control system with the capability of adjusting the properties of the system in the time domain with respect to certain properties of vibration forces such as frequency content and amplitudes. Vibration amplitudes and frequencies are tracked down using sensors and signal processing techniques in order to adjust the structural properties. Therefore, semi-active system optimizes vibration control capacity without employing external forces. In other words, semi-active system enjoys the best of both active and passive systems; therefore, it can be a more reliable and economically viable option for offshore wind turbines which are subjected to changes in their natural frequencies.

There is a considerable amount of literature on passive vibration control devices for wind turbines. One of the early studies in this field was done by Enevoldsen and Mørk [5] in which effects of passive tuned mass dampers on a 500 kW wind turbine were studied and a cost-effective design was achieved owing to the implementation of structural control devices. Later on, Murtagh et al. [6] investigated the use of tuned mass dampers (TMD) for mitigating along-wind vibrations of wind turbines. They concluded that the dynamic responses could be reduced providing that the device is tuned to the fundamental frequency. Colwell and Basu [7] examined effects of tuned liquid column dampers (TLCDs) on offshore wind turbine systems to suppress the excessive vibrations and found that TLCD can minimize vibrations up to 55% of peak responses of OWTs compared to the uncontrolled system. Stewart and Lackner [8] examined the impact of passive tuned mass dampers (PTMD) considering wind–wave misalignment on offshore wind turbine loads for monopile foundations. The results demonstrated that TMDs are effective in damage reduction of towers, especially in side–side directions. Stewart and Lackner [9] in another study investigated the effectiveness of TMD systems for four different types of platforms including monopile, barge, spar buoy, and tension-leg and they observed tower fatigue damage reductions of up to 20% for various TMD configurations. In addition, passive multiple tuned mass dampers (MTMD) were proposed to improve the effectiveness of the vibration control system [10]. Dinh and Basu [10] investigated the use of passive MTMDs for structural control of nacelle and tower of spar floating wind turbines and concluded that MTMDs are more effective in displacement reduction. Whilst passive control methods can reduce the dynamic vibrations to some extent provided they are tuned properly, they can be easily off-tuned as soon as the dominating frequency of the system changes, resulting in ineffectiveness of the system and even increased vibrations compared to the uncontrolled system. Highly dynamic nature of offshore wind turbines subjected to a number of dynamic loadings and interacting with nonlinear soil conditions results in fluctuations in fundamental frequency. Abrupt pulsed nature excitations such as earthquake motions can lead to degradation of soil stiffness and, consequently, reduction in natural frequency. Furthermore, the cyclic vibration of surrounding soil can cause natural frequency reduction. In addition, natural frequency increase might occur when there is stiffening phenomena in certain soil types. Therefore, this raises many questions regarding the use of passive control devices in offshore wind turbines for the whole lifetime in which there are variations in dominating frequency and the application of more sophisticated vibration control devices needs to be examined.

Active vibration control has also been studied for the application of wind turbines by a number of researchers [11–17]. Most studies of active control systems for wind turbines were focused on the vibration control of the blades. For example, Stanio and Basu [14] proposed an active vibration control system based on active tendons for wind turbines. The results of their numerical simulations showed that the proposed control approach is robust in improving the blade responses under vibrations due to the change of rotational speed of the blades. In addition, Fitzgerald and Basu [17] proposed a cable connected to an active tuned mass damper for the reduction of in-plane blade vibration and they found that the proposed control system mitigates the vibrations of large and flexible blades more effectively. In addition, Kim et al. [18] introduced a robust modal control of lightly damped structures using an active dynamic vibration absorber. They used a single active electrical dynamic absorber and tested its effectiveness in control of multiple modes both experimentally and analytically.

The semi-active control mechanism is more suitable for the systems with high time-variant parameters such as offshore wind turbines. Semi-active vibration control devices for the application of buildings have been actively studied by a number of researchers [1,19–23] in the last few decades. However, their application in wind energy is a new field. One of the earliest studies on semi-active control mechanism for wind turbines was done by Kirkegaard et al. [24], in which they presented an experimental and numerical investigation of semi-active vibration control of offshore wind turbines equipped with a magnetorheological (MR) fluid damper. The authors claimed that using MR dampers for offshore wind turbines results in considerable reduction of the lateral displacement compared to the uncontrolled system. Later on, Karimi et al. [25] proposed a controllable valve in tuned liquid column dampers for the application of offshore wind turbines. In addition, the use of semi-active tuned mass dampers in control of flapwise vibrations of wind turbines was examined by Arrigan et al. [26]. The authors proposed a frequency-tracking algorithm for retuning the vibration control device and they observed significant vibration reductions owing to the semi-active mechanism. Furthermore, Weber [27] studied application of an adaptive tuned mass damper concept based on semi-active controller using MR dampers. Their results showed that the real-time controlled MR semi-active tuned mass damper is a robust device for reducing structural vibrations. Semi-active control mechanism for tuned liquid column dampers (TLCDs) was also studied by Sonmez et al. [28]. The authors used a control algorithm based on short-time Fourier transformation (STFT) and investigated the effectiveness of the proposed device under random excitations. More recently, Sun [29] explored semi-active tuned mass dampers for the NREL (National Renewable Energy Laboratory) 5 MW baseline wind turbine excited by environmental loadings in conjunction with seismic motions considering post-earthquake damage to soil and tower stiffnesses. The author demonstrated the superiority of semi-active vibration control over the passive one in multihazard conditions. Although Sun's [29] work is well founded, it is limited to only one earthquake record (1994 Northridge Newhall 90) and further study for a suite of earthquake records with different frequency contents and intensities is required. Another limitation of the aforementioned work is that soil–pile interaction was modeled using a simplified method (closed-form solution) in which the stiffness of embedded pile is considered with a constant rotation and lateral stiffness value in seabed level. More advanced soil–pile interaction model based on time-variant nonlinear stiffness considering soil damage phenomena can enhance the previous works. In addition, the effect of semi-active tuned mass dampers on other structural responses such as base shear and base moment should be investigated.

To fill this gap, this study investigates semi-active tuned mass dampers for offshore wind turbines under multihazard conditions considering time-variant nonlinear soil–pile interaction properties and time-variant damage. A detailed model of the modern NREL 5 MW wind turbine equipped with semi-active tuned mass dampers (STMD) is developed. Stochastically wave and wind loadings in conjunction with seismic loadings are applied to the system and dynamic responses such as displacement, base shear, and base moments are investigated. Compared to the previous models, the developed model has the capacity to consider soil–pile interactions more realistically. Furthermore, a suite of seismic records is used with the aim to consider a wider range of seismic characteristics in the simulations. This paper is organized into five sections. In Section 2, the numerical model of the system including tuned mass dampers, the wind turbine, soil–pile interaction, and the control algorithm is presented. Section 3 defines the loading sources including wind, wave, and earthquake. The numerical results and discussions are presented in Section 4 and conclusions are made in Section 5.

2. Model Description

2.1. Tuned Mass Damper Systems

A tuned mass damper (TMD) is a structural control device that consists of a mass, a damper, and a mass attached to a primary structure to control excessive vibration of the primary structure by dissipating energy. The key feature of a TMD is that its frequency is tuned to a particular structural frequency to mitigate the vibrations when that frequency is excited. The theory of multiple degrees of freedom (MDOF) systems using tuned mass dampers are illustrated and presented in the following section.

The governing equations for the MDOF system in Figure 1 are given as:

$$m_1 \ddot{u}_1 + c_1 \dot{u}_1 + k_1 u - k_2(u_2 - u_1) - c_2(\dot{u}_2 - \dot{u}_1) = p_1 - m_1 \ddot{u}_g$$

$$m_2 \ddot{u}_2 + c_2(\dot{u}_2 - \dot{u}_1) + k_2(u_2 - u_1) - k_3(u_3 - u_2) - c_3(\dot{u}_3 - \dot{u}_2) = p_2 - m_2 \ddot{u}_g$$

$$\vdots$$

$$m_d \ddot{u}_d + c_d \dot{u}_d + k_d u_d = -m_d(\ddot{u}_n + \ddot{u}_g) \tag{1}$$

where m_i, c_i, k_i, u_i, and p_i are mass, damping, stiffness, deflection, and point load for different degrees of freedom of main structure ($i = 1, 2, \ldots, n$), and m_d, c_d, k_d, and u_d are mass, damping, stiffness, and deflection for the TMD attached to the primary structure. u_g is the absolute ground motion due to seismic loadings. The optimal tuned frequency of the TMD, ω_d is defined as:

$$\omega_d = \gamma \omega \tag{2}$$

in which ω is the natural frequency, and γ is the optimal tuning ratio which is determined by the following formula:

$$\gamma = \frac{1}{1 + \mu} \tag{3}$$

where μ is the mass ratio as given by the following formula:

$$\mu = \frac{m_d}{\sum\limits_{i=1}^{n} m_i} \tag{4}$$

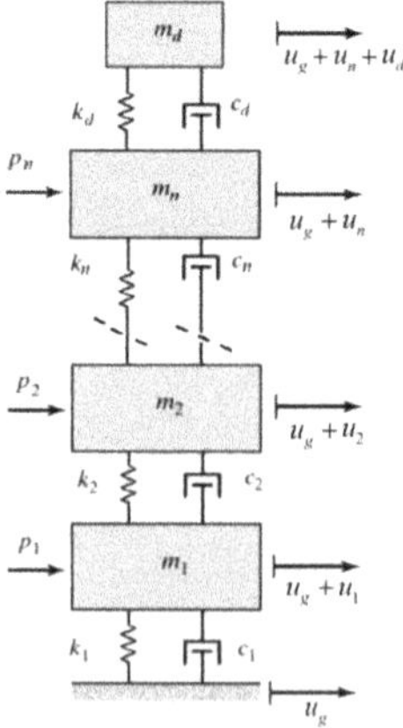

Figure 1. Multi degrees of freedom system equipped with a tuned mass damper (TMD).

The schematic configuration of a TMD inside the nacelle is shown in Figure 2.

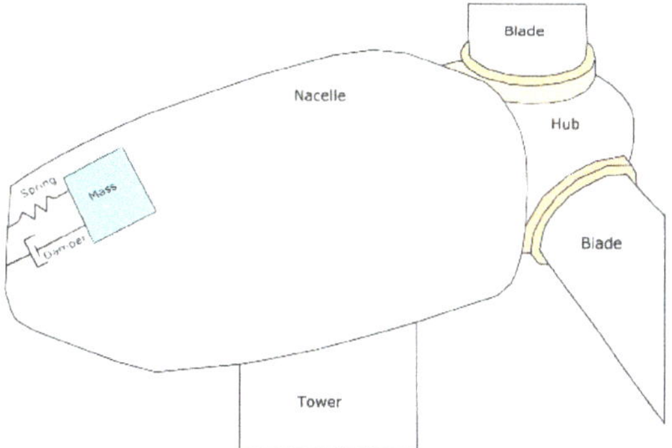

Figure 2. Schematic figure of TMD in the nacelle.

2.2. Semi-Active Vibration Control Algorithm

There are three main parameters that define a tuned mass damper: mass, stiffness, and damping. Mass of vibration control device cannot be changed in time domain due to practical reasons and only stiffness and damping of the device are altered in time domain depending on instantaneously structural properties of the system and instant dynamic responses. There have been studies on algorithms for time-variant properties of semi-active tuned mass dampers by [21,23,26,28]. In most of the previous studies, the stiffness of semi-active TMD is tuned according to instantaneously identified frequency using short-time Fourier transform and the damping parameters are modulated based on the TMD deflection in each time step.

2.2.1. Varying Stiffness

Stiffness of the semi-active tuned mass damper can be modified based on the identified dominant frequency using short-time Fourier transformation (STFT) function as suggested in the previous studies such as [27,29,30]. Unlike the standard Fourier transform, short-time Fourier transformation adds a time dimension to the base function parameters. A signal $x(\tau)$ is multiplied by a moving window function as $h(\tau - t)$:

$$\hat{x}(\tau) = x(\tau)h(\tau - t) \tag{5}$$

in which $\hat{x}(\tau)$ is a weighted signal, τ is the moving time and t is the fixed time.

The spectrum $S(t, \omega)$ at the fixed time can be defined by applying Fourier transform to $\hat{x}(\tau)$:

$$S(t, \omega) = \frac{1}{2\pi} \int e^{-j\omega t} \hat{x} = \frac{1}{2\pi} \int e^{-j\omega t} x(\tau)h(\tau - t) \tag{6}$$

and the power spectral density $P(t, \omega)$ of time t is calculated as

$$P(t, \omega) = |S(t, \omega)|^2 = S(t, \omega) . \overline{S(t, \omega)} \tag{7}$$

Then, the dominant frequency at time t can be identified using following equations:

$$\omega_{inst} = \{\omega | P(t_i, \omega) = \max\{P(t_i, \omega)\}\} \tag{8}$$

$$\omega_{id} = \frac{\sum_{k=\max\{1, i-m+1\}}^{i} \omega_{inst}(t_k)\max\{P(t_k, \omega)\}}{\sum_{k=\max\{1, i-m+1\}}^{i} \max\{P(t_k, \omega)\}} \tag{9}$$

where ω_{id} is the dominant frequency at time t_i determined through finding the average of instantaneous frequencies over m time steps ($m = 3$), and ω_{inst} is the instantaneous frequency. In this study, a moving window of 500 time steps ($n = 500$) with a Hamming window is used. The length of the hamming window is taken 1024, L, resulting in the vector of P_i with the size of $N \times 1$, where $N = (0.5 * L) + 1$. The dominant frequency at each time step is calculated and then stiffness of the tuned mass is retuned using the dominant frequency as

$$k_d^t = k_d^{t=0} \left(\frac{\omega_{id}}{\omega_n} \right)^2 \tag{10}$$

in which k_d^t is the time-variant stiffness of tuned mass damper that can be realized through a variable stiffness device, $k_d^{t=0}$ is the initial stiffness of tuned mass damper at the time of zero, and ω_n is the predamage fundamental frequency of the system in which the TMD was tuned to before the development of any damages.

2.2.2. Varying Damping

The damping of tuned mass dampers can be altered according to the dynamic responses in order to increase the effectiveness of the device. In the previous studies by Abe and Igusa [31], the authors investigated the time-variant damping for tuned mass dampers and concluded that TMD can improve its performance if the damping of TMD is time-dependent in such a way that its damping value changes to zero for the duration in which the relative displacement of TMD is increasing. This results in an increase in the efficiency of the system for controlling excessive vibrations. This time-dependent damping algorithm has been used in other works [23,29,32]. In this method, the relative displacement of TMD is tracked in each time step and if it is larger than that of the previous time step, the damping of TMD is set to zero, $c_d^t = 0$; otherwise the damping value is set to $2c_{opt}$, $c_d^t = 2c_{opt}$. c_{opt} is the optimal value of TMD's damping which can be determined from an estimation method suggested by Sadek et al. [33].

2.3. NREL 5 MW Wind Turbine

The 5 MW NREL wind turbine is considered as it is widely used as the turbine for benchmark studies [34]. This turbine is supported by the baseline monopile foundation developed in the second phase of Offshore Code Comparison (OC3) project conducted by NREL [35]. The geometric configuration of the turbine is shown in Figure 3. The tower and monopile are modeled by three-dimensional Timoshenko beam theory.

The particulars of the offshore wind turbine are listed in Table 1. Table 2 provides the material properties of the steel used in the tower and monopile of the offshore wind turbine. To consider additional weight of welds, bolts, and paint, the density of the steel in the tower is assumed 8% higher than that of the regular steel based on the value given in [34].

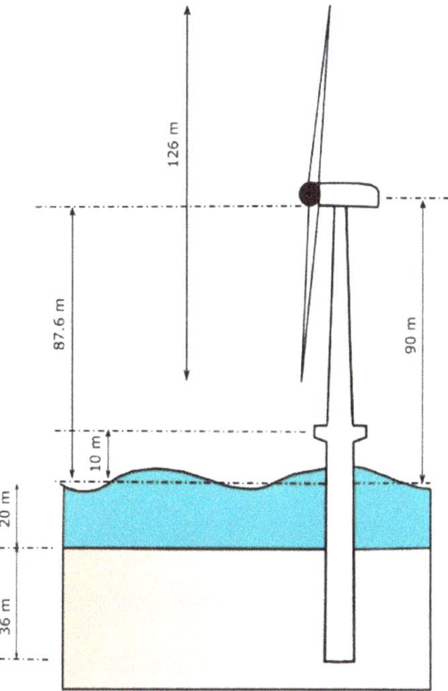

Figure 3. Schematic configuration of the offshore wind turbine.

Table 1. Properties of NREL 5 MW baseline turbine.

Turbine	Rated Power, Rotor Orientation	5 MW, Upwind, 3 Blades
	Control System	Variable Speed, Collective Pitch
Blade	Rotor Diameter, Hub Height	126 m, 90 m
	Cut-In, Rated, Cut-Out Wind Speed	3 m/s, 11.4 m/s, 25 m/s
	Cut-In, Rated Rotor Speed	6.9 rpm, 12.1 rpm
	Hub mass, Blade mass	56,780 kg, 17,740 kg
Nacelle	Nacelle Dimensions	18 m × 6 m × 6 m
	Nacelle Mass	240,000 kg
Tower	Base diameter, base thickness	6.0 m, 27 mm
	Top diameter, top thickness	3.87 m, 19 mm
	Tower mass	347,460 kg

Table 2. Material properties.

Component	Density (kg/m^3)	Young's Modulus (GPa)	Poisson's Ratio
Tower	8500	210	0.3
Monopile	7850	210	0.3

2.4. Soil–Pile Interaction

The nonlinear lateral soil resistance–deflection relationship for sand layers can be defined by a hyperbolic tangent function [36]:

$$P = A p_u \tanh\left[\frac{kH}{A p_u} y\right] \tag{11}$$

where P is the soil reaction at a given depth, A is a calibration factor which equals to 0.9 for cyclic loading, y is the lateral deflection of soil layers, and k is the initial stiffness coefficient which is determined from a function of the angle of internal friction, ϕ' [36]. H is depth and p_u is the ultimate lateral bearing capacity determined by the following equation:

$$p_u = \min \begin{cases} p_{us} = (C_1 H + C_2 D)\gamma H \\ p_{ud} = C_3 D \gamma H \end{cases} \tag{12}$$

where p_{us} is shallow ultimate resistance, p_{ud} is deep ultimate resistance, D is the pile diameter, γ is the effective soil weight, and C_1, C_2, and C_3 are coefficients determined from API standard [36]. Soil layer properties are shown in Figure 4a. The nonlinear resistance–deflection curves constructed based on the aforementioned method for different soil depths and layers (top, middle, and bottom of each layer) are illustrated in Figure 4b.

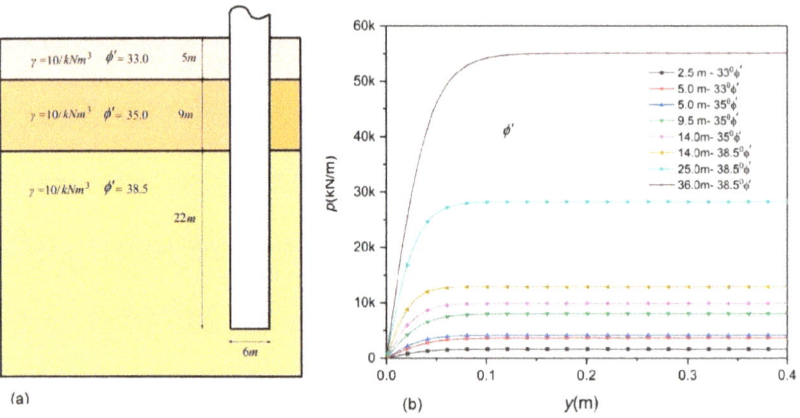

Figure 4. (**a**) Soil layer properties, (**b**) nonlinear lateral resistance–deflection curves.

3. Loading

Stochastically generated wind and wave loadings in conjunction with seismic motions are applied to the structure. Each of these loadings is introduced as follows.

3.1. Wind

Wind loading is the main external force as it generates large overturning moments due to its long moment arm, especially for offshore wind turbines with high hub heights. The wind speed acting on the system can be represented by a constant mean wind load $\bar{v}$ and a turbulent wind component $\vartheta(t)$, $v(t) = \bar{v} + \vartheta(t)$. In this work, the mean wind velocity $\bar{v}(z)$ as a function of height is determined by means of a logarithmic law as [37]:

$$\bar{v}(z) = V_{ref} \frac{\ln(z/z_0)}{\ln(H_{ref}/z_0)} \tag{13}$$

where V_{ref} is the mean wind velocity at the reference height $H_{ref} = 90$ m, z is the vertical coordinate, and z_0 is a surface roughness length parameter.

Turbulence is defined as random deviation imposed on the mean wind speed which is caused when the kinematic energy of wind is transformed to thermal energy. Turbulence of wind is expressed in terms of turbulence intensity, which is the ratio of the standard deviation of wind speed to the mean wind speed. Kaimal spectrum $S_v(f)$ [38] is adopted in this study to calculate the turbulent wind velocity as follows:

$$S_v(f) = \frac{4I^2 L_k}{(1 + 6fL_k/\bar{v})^{5/3}} \tag{14}$$

where I is the wind turbulence intensity, f is the frequency (Hz), and L_k is an integral length scale parameter. In this study, the stochastic wind profile is generated using Turbsim code [39] over a rectangular grid with 961 points (31×31) based on Equations (13) and (14). Then, the aerodynamic loadings on the blades are computed in FAST (Version 8.15, NREL, Golden, CO, USA) [40] code using the aforementioned wind profile based on blade element momentum (BEM) theory. Finally, the generated wind loading time history is used in the developed code to consider aerodynamic loadings.

3.2. Sea Wave Load

Wave loading acting on cylindrical structural members of fixed platforms can be obtained using Morison's equation [37] as the sum of inertia and drag forces. The transverse sea wave force acting on a strip of a length dz is given by the sum of inertia and drag force terms as the following equation [41]:

$$dF = \frac{\rho_w}{2} C_d D v|v|dz + \frac{\pi D^2}{2} C_m \rho_w \dot{v} dz \tag{15}$$

where C_d and C_m are the drag and inertia coefficients, respectively ($C_d = 1.2$ and $C_m = 2$ in the current study), D is the diameter of the member, $\dot{v}$ and v are horizontal acceleration and velocity of fluid particles induced by wave excitations, and ρ_w is water density ($1025 \, \text{kg/m}^3$).

The spectrum developed through the Joint North Sea Wave Observation Project (JONSWAP) is used to generate wave time histories [42] as follows:

$$S_{\eta\eta}(f) = \frac{\alpha g^2}{f^5} \exp\left[-\frac{5}{4}\left(\frac{f_m}{f}\right)^4\right] \gamma^{\exp\left[-\frac{(f-f_m)^2}{2\sigma^2 f_m^2}\right]} \tag{16}$$

in which $S_{\eta\eta}(f)$ is JONSWAP spectrum, η is the function of water surface elevation, γ is the peak enhancement factor (3.3 for the north sea), g is the acceleration of gravity, and f is the wave frequency (Hz). The constants in this equation can be defined as

$$\alpha = 0.076\left(U_{10}^2/Fg\right)^{0.22} \tag{17}$$

$$f_m = 11(v_{10}F/g^2)^{-1/3}/\pi \tag{18}$$

and

$$\sigma = \begin{cases} 0.07, & f \le f_m \\ 0.09, & f > f_m \end{cases} \tag{19}$$

where U_{10} is the mean wind velocity at 10 m from the sea surface, and F is the fetch length in which the wind blows without any change of direction.

Then, total wave force acting on the structural members can be calculated as

$$F_f(t) = \int_0^d dF\phi_f(z)dz \tag{20}$$

where dF is the wave loading on the member mentioned in Equation (15), and ϕ_f is the shape function of the offshore structure subjected to wave loading, d is the depth of the water surface, and z is the vertical direction.

3.3. Seismic Excitation

To simulate seismic excitations on the offshore wind turbines, time series of acceleration of strong ground motions during past earthquake events are used. Two horizontal directions are selected to represent the behavior of earthquake events. In this study, sloshing of water surrounding the structure is ignored as it is believed to have insignificant effects. The seismic records are selected from the Pacific Earthquake Engineering Research centre (PEER) Database [43] and listed in Table 3. The magnitudes of the seismic ground motions selected in this study vary between 5.5 and 7.5.

Table 3. Seismic records.

ID	Earthquake	Magnitude	Year	Record Station	Soil Type
1	Kobe, Japan	6.9	1995	Kobe University	B
2	Northridge-01	6.69	1994	17645	D
3	Northridge-Landers	7.28	1992	17645 Saticoy St.	D
4	Northridge-Narrows-01	5.99	1987	17645 Saticoy St.	D
5	Tabas, Iran	7.35	1978	Tabas	C
6	Manjil, Iran	7.37	1990	Abbar	E
7	Manjil, Iran	7.37	1990	Abhar	D
8	Manjil, Iran	7.37	1990	Qazvin	C
9	Manjil, Iran	7.37	1990	Rudsar	D
10	Erzican, Turkey	6.69	1992	Erzincan	D
11	Loma Prieta	6.93	1989	Apeel 10-Skyline	D
12	Loma Prieta	6.93	1989	Apeel 2-Redwood City	E
13	Cape Mendocino	7.01	1992	Cape Mendocino	B
14	Cape Mendocino	7.01	1992	Eureka-Myrtle & West	C
15	Cape Mendocino	7.01	1992	Fortuna-Fortuna Blvd.	D
16	Cape Mendocino	7.01	1992	Petrolia	D
17	Cape Mendocino	7.01	1992	Shelter Cove Airport	D
18	Landers	7.28	1992	Amboy	C
19	Landers	7.28	1992	Baker Fire Station	D
20	Landers	7.28	1992	Bell Gardens-Jaboneria	D
21	Imperial Valley-06	6.53	1979	Aeropuerto Mexicali	C
22	Imperial Valley-06	6.53	1979	Agrarias	D
24	Imperial Valley-06	6.53	1979	Bonds Corner	D
24	Imperial Valley-06	6.53	1979	Brawley Airport	C
25	Imperial Valley-06	6.53	1979	Calexico Fire Station	D
26	Imperial Valley-06	6.53	1979	Calipatria Fire Station	D
27	Imperial Valley-06	6.53	1979	Cerro Prieto	D
28	Imperial Valley-06	6.53	1979	Chihuahua	D
29	Imperial Valley-06	6.53	1979	Coachella Canal #4	C
30	Imperial Valley-06	6.53	1979	Compuertas	C

4. Numerical Results and Discussions

4.1. Model Verification

In this section, the results of the natural frequency and dynamic analyses using the developed model in MATLAB (R2017a, MathWorks, Natick, MA, USA) are verified. To carry out natural frequency analysis, the stiffness of nonlinear soil–pile interaction is linearized by taking initial stiffness of the p–y curves [44]. The resulting first and second natural frequencies are listed in Table 4 and compared with the results of the model created by commercial finite element software ANSYS (16, Canonsburg, PA, USA) and the results from the literature [45]. There is good agreement between the results of natural frequency analyses.

Table 4. Natural frequency analysis results (Hz).

Mode	Code	ANSYS	Dong Hywan Kim et al. [45]
1nd Fore–aft	0.235	0.234	0.234
1nd Side-to-side	0.235	0.234	0.233
2nd Fore–aft	1.426	1.426	1.406
2nd Side-to-side	1.426	1.426	1.515

Then, a dynamic analysis for the offshore wind turbine subjected to a single earthquake motion (Northridge) is carried out and the results are compared with the results obtained from the dynamic analysis performed in ANSYS. Figure 5a shows the nonscaled time history of acceleration of Northridge earthquake starting from the instant of 50 s. Figure 5b illustrates the time history of nacelle displacement simulated with the code written in MATLAB and the corresponding results obtained from ANSYS and good matches are observed.

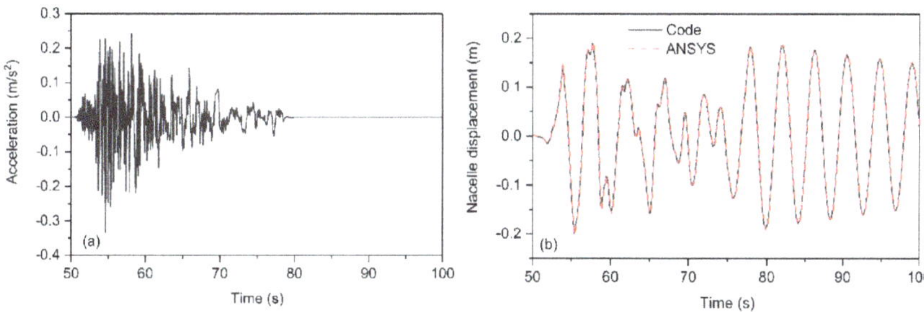

Figure 5. (**a**) Time history of acceleration of seismic excitation (Northridge), (**b**) time history of the nacelle displacement simulated with ANSYS and the developed code under seismic excitation (Northridge).

4.2. Damage Development

Dynamic performance of passive tuned mass dampers is threatened by changes in the natural frequency of the system. This change in the natural frequency can occur either gradually over the lifetime of the system due to soil degradation under long-term cyclic loading or rapidly over a short period of time due to seismic excitation. Figure 6a shows the effect of soil stiffness changes (damage or stiffening in soil) on the first and second natural frequency of the system. The Figure shows that a 50% reduction in soil stiffness leads to 2.2%, and 4.8% reduction in the first and second natural frequencies of the system, respectively. The Figure indicates that the second natural frequency changes more and degradation of soil stiffness has a larger effect on the frequency change rather than stiffening of soil. Similarly, the frequency change of the system due to tower damage is shown in Figure 6b. The Figure suggests that tower damage reduces the natural frequency to a greater extent. For example, 20% stiffness reduction of tower leads to 5.9% decrease in the first natural frequency of the system.

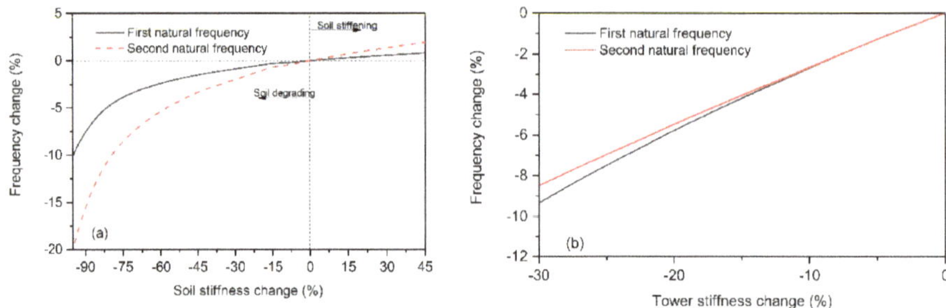

Figure 6. Frequency change due to (**a**) soil degrading/stiffening, (**b**) tower stiffness reduction.

The degradation stiffness model of monopile foundations under long-term cyclic loading was studied by Martin Achmus et al. [46]. Sun also considered soil and tower stiffness damage development for seismic loading using simplified linear stiffness reduction scenarios [29]. In this study, a rapid degradation stiffness model is assumed as the focus of the study is on the short-term damage development due to seismic excitation. Therefore, the damage development model similar to Sun [29] is assumed with the values in which a 5% reduction in natural frequency occurs. To model damage development, it is assumed that damage begins developing at the start of earthquake and soil stiffness and tower stiffness reduces linearly in 20 s as depicted in Figure 7. Tower stiffness and tower stiffness are assumed to reduce 30% and 15%, respectively. The reduction in the stiffness of the tower is assumed in the whole tower.

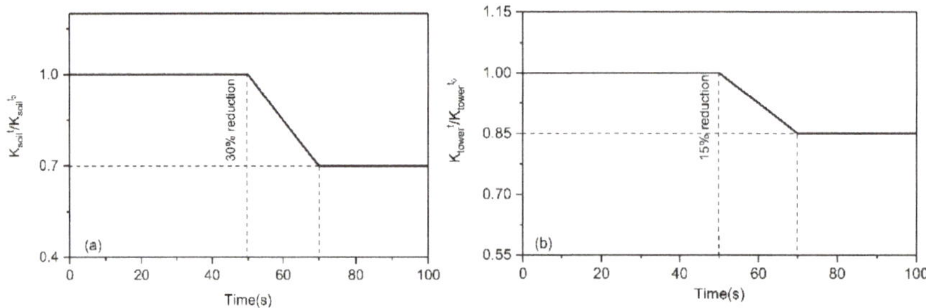

Figure 7. Damage development: (**a**) soil stiffness, (**b**) tower stiffness.

4.3. Response to a Single Seismic Record

To give a preliminary insight into the dynamic responses of offshore wind turbines equipped with semi-active and passive tuned mass dampers considering frequency change as a result of damage development, the responses to a single seismic record for different loading conditions are discussed in this section. Four loading conditions are adopted according to IEC (International Electrotechnical Commission) standards [47] and their properties are tabulated in Table 5. In the first loading condition scenario (LC1), the turbine is operating under steady wind loading at the rated wind speed. In the second loading condition (LC2), the parked turbine is subjected to a steady wind speed of 40 m/s. For both of these loading conditions, there is no wave loading which represents calm sea conditions. Loading conditions LC3 and LC4 are the same as LC1 and LC2 but with stochastic wind and wave loadings. For all these loading conditions, the seismic event in conjunction with damage development occurs at the instant of 50 s.

Table 5. Loading condition (LC) information.

	Wind		Wave		Seismic	
Loadcases	Wind Speed at the Hub Height (m/s)	Turbulence Intensity (%)	Wave Period (s)	Significant Wave Height (m)	Starting Instant	Damping
LC1	11.4 (Operational)	0	-	-	50 s	1%
LC2	40.0 (Parked)	0	-	-	50 s	5%
LC3	11.4 (Operational)	14.5	9.5	5.0	50 s	1%
LC4	40.0 (Parked)	11.7	11.5	7.0	50 s	5%

The identified dominant frequency according to short-time Fourier transport function is calculated in each time step and the stiffness of semi-active tuned mass damper is retuned according to Equation (10). The identified dominant frequency and retuned frequencies are depicted in Figure 8. The reason that they are not equal is that the optimal tuning ratio, γ, is multiplied by the identified frequency in order to obtain the optimal retuned frequency of TMD. The figure shows that there is 5.2% reduction in the natural frequency of the system and consequently retuned frequency of tuned mass damper. This retuned frequency can be realized with 10% decrease in the stiffness of tuned mass damper.

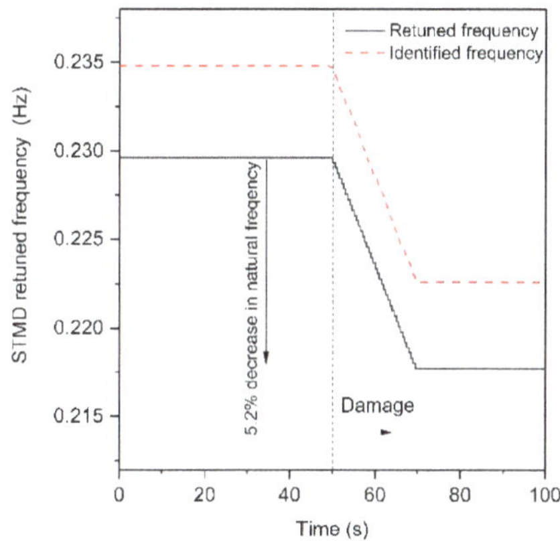

Figure 8. Retuned frequency of semi-active tuned mass damper (STMD).

The dynamic responses of the offshore wind turbine subjected to a single seismic record (Kobe) are discussed here. In the following section, baseline denotes uncontrolled system. For the controlled systems, PTMD and STMD denote passive TMD and semi-active TMD, respectively. The parameters of the tuned mass dampers used in this section are tabulated in Table 6. Figure 9 compares the nacelle displacement responses of the turbine under steady wind loadings. At first glance, it is clear that for LC1 and LC2, STMD is superior to PTMD. In Figure 9a, the peak of nacelle displacement decreased from 0.96 m to 0.91 m for operational loading LC1 and the dynamic response of PTMD is nearly as much as the baseline system especially after the end of earthquake and damage development. This shows that the PTMD becomes off-tuned and unable to control the vibration. However, STMD can retune to the new frequency and mitigate the dynamic responses. The displacement reductions are more pronounced for the parked condition (LC2) in which the peak of nacelle displacement for STMD is 0.19 m compared to 0.26 m of the passive tuned mass damper, nearly 16% more reductions compared to the baseline system.

Table 6. TMD parameters.

Mass (kg)	k_d (N/m)	c_d (N/(m/s))	ω_d (Hz)
20,000	41,657	10,000	0.229

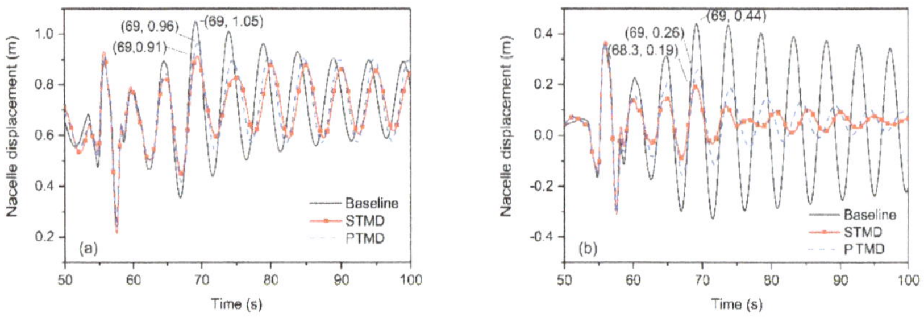

Figure 9. Time history of nacelle displacement under steady wind loading and seismic excitation considering damage development. (**a**) LC1, (**b**) LC2. PTMD: passive tuned mass damper.

To scrutinize the energy spectrum of dynamic responses, the power spectral density (PSD) of the nacelle displacements for LC1 and LC2 is obtained and presented in Figure 10. Fast Fourier transformation based on Hamming window is used to capture a smooth PSD curve. Figure 10a indicates that there are two distinct peaks corresponding to the energy of wind loading and turbine frequency (1P) whose frequencies are around zero and 0.2 Hz, respectively. It is clear that the PSD for energy from wind loading (frequencies close to zero) shows negligible changes for PTMD and STMD as these devices are tuned to the first natural frequency of the system. However, a 37% reduction in the peak of the power spectrum for the STMD system can be observed for the turbine frequency (1P). Similarly, power spectral density of the nacelle displacement for LC2 (parked condition) is shown in Figure 10b. Compared to the operational condition (Figure 10a), energy spectrum corresponding to frequency of wind loading (close to zero) has much lower peak due to the fact that in parked condition the turbine absorbs a small portion of wind loading as a result of pitching mechanism in the blades and the energy is concentrated around the frequency range of first natural frequency of structure. This is expected because in the parked condition the vibration of structural modes dominates compared to the operational condition where the vibration due to external excitations dominates. The Figure also indicates that the peak of spectrum for the STMD system is reduced as much as 92% compared to the baseline system (uncontrolled system). However, this percentage reduction is lower for the PTMD with 76% reduction. These reductions for both PTMD and STMD are higher in the parked condition (Figure 10b) compared to the operational condition (Figure 10a) because in the parked conditions the aerodynamic damping is negligible and these structural control devices compensate for low total damping value of the system.

Figure 11 compares the nacelle displacement responses of the turbine under stochastic wave–wave loadings in conjunction with seismic ground motion and damage development (LC3 and LC4). Again it can be seen that the STMD system shows a better performance in mitigating vibrations and reducing peak displacements. For example, the peak value of displacement is decreased from 1.5 m of the baseline system to 1.27 m of the STMD system for the operational condition (LC3), with 15% reduction. This reduction percentage for the PTMD system is lower, as much as only 6%. Therefore, the STMD's effectiveness in reducing the peak values is more than twice that of the PTMD. For the parked condition (LC4), higher vibration reductions are observed. For example, the peak nacelle displacement reduced from 1.91 m to 1.01 m as a result of the implementation of the semi-active tuned mass damper. This means that the semi-active achieves 47% reduction in the peak nacelle displacement.

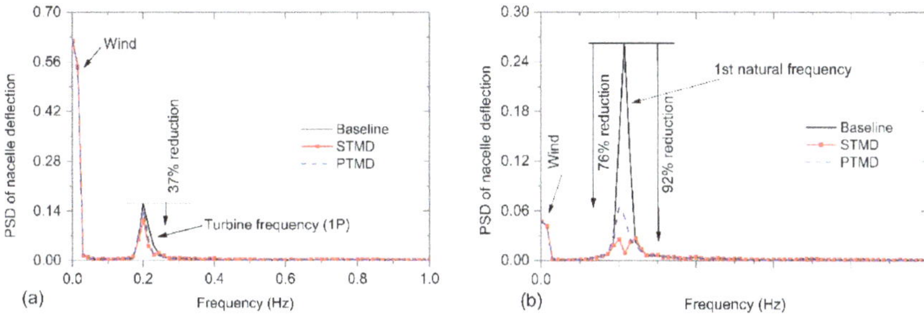

Figure 10. Power spectral density (PSD) of nacelle displacement under steady wind loading and seismic excitation considering damage development. (**a**) LC1, (**b**) LC2.

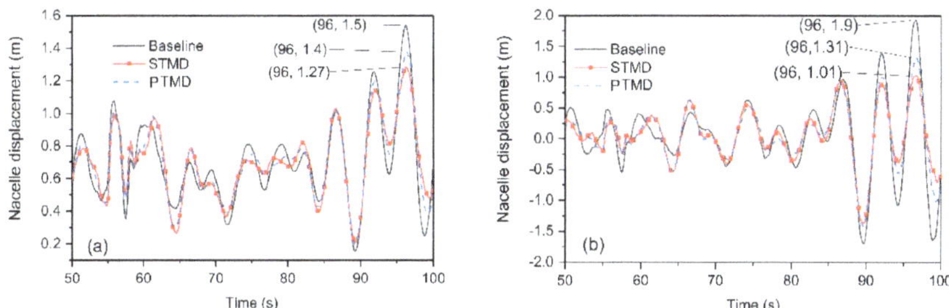

Figure 11. Time history of nacelle displacement under stochastic wind–wave loadings and seismic excitation considering damage development. (**a**) LC3, (**b**) LC4.

Looking into power spectral density of the nacelle displacement for LC3 and LC4 in Figure 12, some energy is concentrated around the frequency of 0.1 Hz that corresponds to the energy of wave loadings. Similar to Figure 10, more PSD reduction is observed for parked condition (LC4). However, the difference between the reduction in PSD for PTMD and STMD is less than 10%.

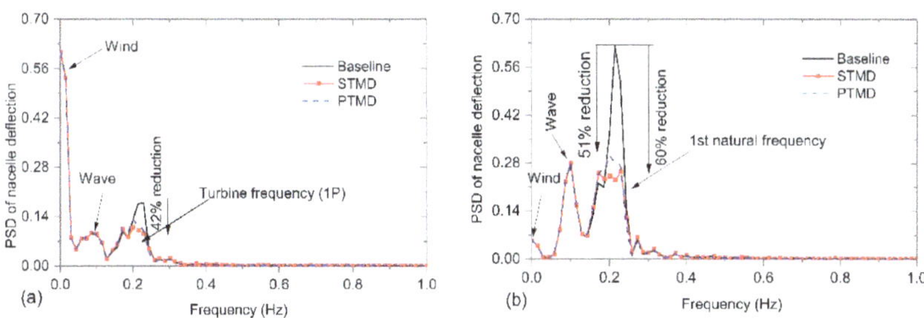

Figure 12. PSD of nacelle displacement under stochastic wind–wave loadings and seismic excitations considering damage development. (**a**) LC3, (**b**) LC4.

Figures 13 and 14 show a representative 50 s window time history of the base shear force for steady (LC1 and LC2) and stochastic loadings (LC3 and LC4), respectively. For steady wind loadings (LC1 and LC2), larger base shear is obtained during ground motion and both passive and semi-active TMDs have slight effects on the dynamic responses during the ground motion and damage development. However, the displacements after the damage development are reduced owing to the vibration control devices. For both load cases, STMD is superior to the PTMD. On the other hand, for stochastic loading (Figure 14), changes in the base shear due to tuned mass dampers are insignificant.

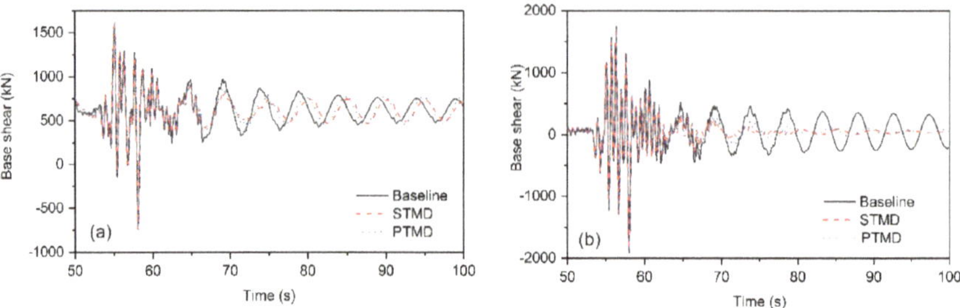

Figure 13. Time history of fore–aft base shear force under only steady wind loading and seismic excitation considering damage development. (**a**) LC1, (**b**) LC2.

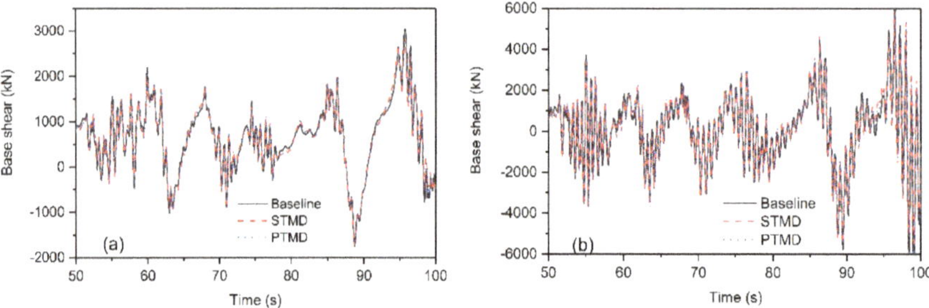

Figure 14. Time history of fore–aft base shear force under stochastic wind–wave loadings and seismic excitations considering damage development. (**a**) LC3, (**b**) LC4.

Figures 15 and 16 compare the base overturning moment time histories for the controlled and uncontrolled systems. Figure 15 shows that the vibration control devices mitigate the base moment values after the development of damage for steady loading. For load case (LC1), the peak values of the base moment, which occur at 69 s, reduce 10% and 15% when PTMD and STMD are used, respectively. This reduction is higher for the parked condition (LC2), where PTMD and STMD reduce the base moment values up to 43% and 57%, respectively. For stochastic loading (Figure 16), the effect of the vibration control devices is less significant for this single seismic motion record.

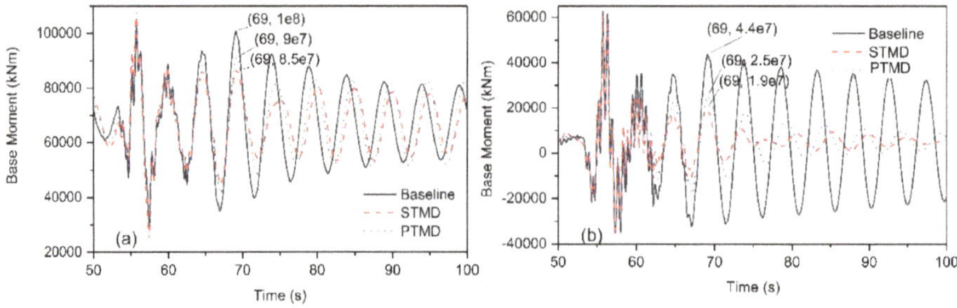

Figure 15. Time history of the fore–aft base moment under only steady wind loading and seismic excitation considering damage development. (**a**) LC1, (**b**) LC2.

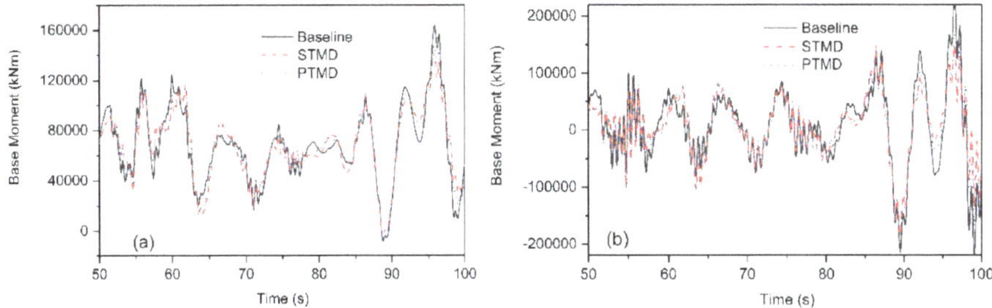

Figure 16. Time history of the fore–aft base moment under stochastic wind–wave loadings and seismic considering damage development. (**a**) LC3, (**b**) LC4.

It should be noted that the semi-active tuned mass damper used in this study has both varying stiffness and damping and the combined effect of them is shown in the results. Individual effects of them were also investigated to determine the contribution of each in the response reduction and the results showed that the contribution of each effect individually varies depending on the load cases. However, for all load cases, the contribution of varying damping is larger than the contribution of varying stiffness. For example, the contribution of the varying damping in the response reduction (nacelle deflection) is 57%, whereas the corresponding contribution of varying stiffness is 43%.

4.4. Response to a Seismic Record Set

In this section, more analyses based on a set of ground motion records with different soil and intensity properties as listed in Table 3 are performed and the influences of the structural control devices on the dynamic responses are systematically investigated. Standard deviation and peak values of each time history are taken for the systems equipped with optimal PTMD and STMD with mass ratios ranging between 1% and 4% and compared with the baseline system (uncontrolled system) as the percentage of reduction (improvement). The positive values mean a reduction in the responses which can be defined as the effectiveness of the structural control device. On the other hand, negative values denote increases in the response which means that the vibration control device worsens the vibration performance. The standard deviations and peak values of fore–aft displacements of the nacelle are tracked as they are representative of serviceability and fatigue lifetime of the system, respectively. Then, these values are compared with those of the uncontrolled system (offshore wind turbine without any structural control devices) as the percentage of reduction.

$$\text{Peak Response Reduction} = \frac{Peak_{uncontrolled} - Peak_{controlled}}{Peak_{uncontrolled}} \tag{21}$$

$$\text{Std Response Reduction} = \frac{Std_{uncontrolled} - Std_{controlled}}{Std_{uncontrolled}} \tag{22}$$

in which Peak and Std denote peak and standard deviation of deflections, respectively. Controlled denotes the offshore wind turbine equipped with structural control devices, and Uncontrolled denotes the baseline offshore wind turbine without any vibration control devices. Figure 17a,b illustrates the standard deviation reduction of the nacelle displacement for STMD and PTMD, respectively, for loading condition LC3. For STMD, it is clear that dynamic responses reduce as the mass ratio increases for most ground motions. However, a different trend for PTMD (Figure 17b) is observed in which negative performances are seen for most ground motion records and even increasing the mass ratio of TMD does not improve the performance. This behavior is expected since the passive tuned mass damper is unable to mitigate the vibrations as it becomes off-tune by changing the frequency of the system due to damage development.

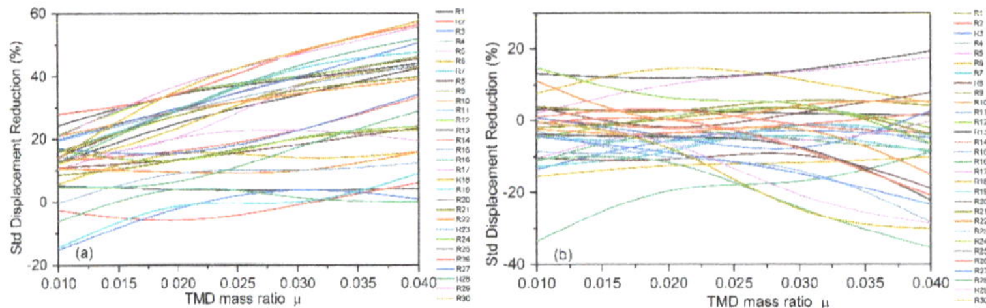

Figure 17. Standard deviation reduction of fore–aft displacement for LC3 under a set of seismic records. (**a**) STMD, (**b**) PTMD.

Figure 17 shows standard deviation reduction for all seismic records for only nacelle displacement under LC3 loading conditions. For the sake of brevity, the dynamic response reductions for all the seismic records are averaged for each loading condition and presented in Figures 18 and 19. The average of reduction percentage in the standard deviation of dynamic responses for all ground motion records for load cases LC3 and LC4 (stochastic wind and wave loading in conjunction with seismic excitation and damage development) is obtained and plotted in Figure 18. Dashed lines correspond to semi-active tuned mass dampers and solid lines are for passive tuned mass dampers. For the operational condition (Figure 18a), the standard deviation of nacelle displacements reduces by 20% for STMD with 1% mass ratio and this reduction percentage increases to 39% by increasing the mass ratio to 4%. On the other hand, the PTMD with 1% mass ratio leads to only 10% standard deviation reduction, half of its STMD counterpart. It is interesting that the performance of PTMD becomes worse when the mass ratio increases up to 4%, resulting in 10% increase in the standard deviation of deflection. This suggests that increasing the mass ratio of PTMD cannot improve its dynamic performance and even it worsens the dynamic performance due to the controller becoming off-tune as well as the reduction in the natural frequency of system as a result of the additional mass of tuned mass damper. From the results shown in Figure 18, it is concluded that a semi-active mass damper with a mass ratio of 1% shows much better performance than a passive tuned mass damper with a mass ratio of 4% for the case when there is a change in natural frequency of the system. This means that STMD with a very low mass ratio is more effective than a PTMD with a large mass ratio. Similar trends can be observed for base shear force and base moment responses; however, it should be noted that base shear force and base moment experience lower dynamic response reduction with the vibration control devices. For example, the standard deviation of the base shear force shows a maximum of 7% reduction for STMD with a mass ratio of 3%. Therefore, it could be concluded that the considered structural control devices have more influence on nacelle displacement and base overturning moments rather than base shear force.

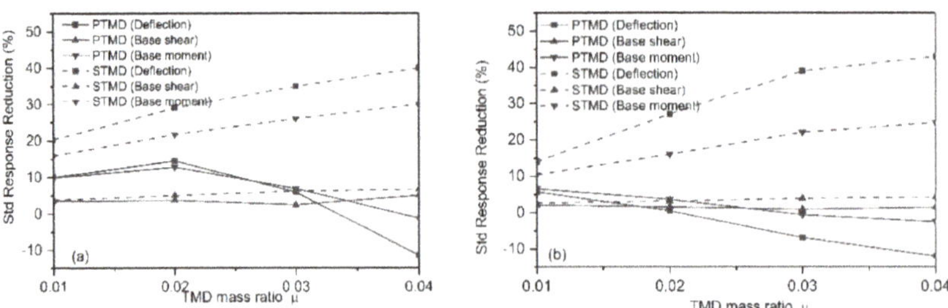

Figure 18. Standard deviation reduction of fore–aft displacement for stochastic wind–wave loadings and seismic excitation. (**a**) LC3, (**b**) LC4.

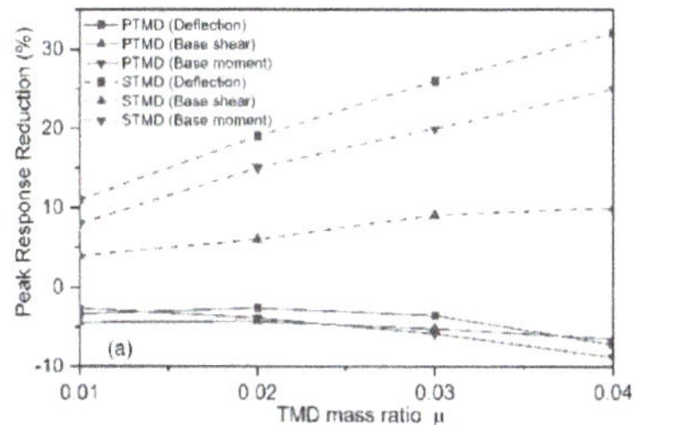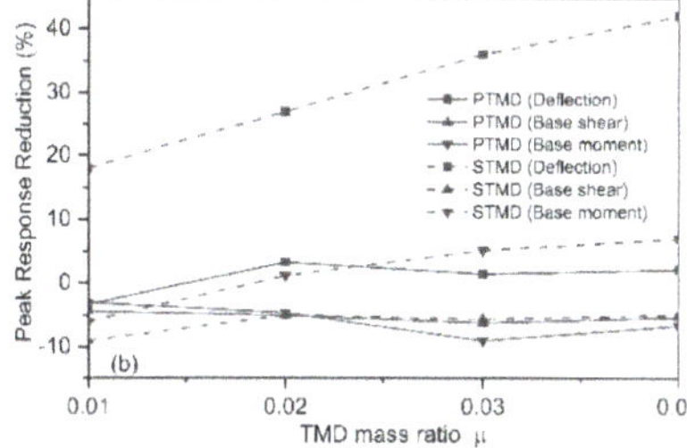

Figure 19. Peak response reduction of fore–aft displacement for stochastic wind–wave loadings and seismic excitation. (**a**) LC3, (**b**) LC4.

Similarly, the average reduction percentage in peak values of dynamic responses for all ground motion records is plotted in Figure 19. Similar trends to the results of standard deviations are observed with some differences. For example, for the operational loading LC3, the reduction in the peak value of the nacelle displacement is 11% for the STMD with a mass ratio of 1% and it increases to 32% with a fourfold increase in the mass ratio. It is worthy to note that the PTMD's effectiveness in mitigating the peak values of dynamic responses is very low for all the mass ratios. This means that PTMD systems have negative impacts on the peak values, resulting in a deterioration in serviceability of the system. Similar trend for LC4 can be seen in Figure 19b; however, the peak response reductions are higher for STMDs compared to the operational loading.

Since changes in the natural frequency of the system are inevitable due to various reasons and a number of measurement campaigns in operational wind farms have observed a difference between design natural frequency and real natural frequency, semi-active tuned mass dampers are a better option for massive tuned mass dampers. Therefore, implementation of this kind of vibration control device can mitigate undesired vibrations and reduce the dynamic response, especially the displacements and base overturning moments to a great extent. Consequently, it has the potential to reduce fatigue damages and increase the lifetime of the system, resulting in an improvement in lifecycle of wind turbines and reduction in the cost of energy production. In terms of practicality of implementation, MR dampers can be used as time-variant damping devices and variable stiffness devices can be used in order to change the stiffness of the device.

5. Conclusions

In this paper, a numerical model for an offshore wind turbine controlled by semi-active tuned mass dampers (STMD) subjected to wind, wave, and earthquake excitations considering time-variant damage development are presented. Nonlinear soil–pile interaction is considered. Short-time Fourier transform (STFT) is utilized to identify the changes in the natural frequency of the system and to retune the semi-active tuned mass damper. Time-variant stiffness and damping of STMD are modified in each time step according to a vibration control algorithm based on short-time Fourier transform. Numerical analyses are carried out for operational and nonoperational conditions to investigate the performance of STMD compared to the PTMD under multihazards. Dynamic responses for a single earthquake record as well as a set of earthquake records are presented. The results show that STMDs perform significantly better than PTMDs, especially when there is a change in natural frequency of the system. A semi-active mass damper with a mass ratio of 1% shows much better performance than a passive tuned mass damper with a mass ratio of 4%. A semi-active tuned mass damper with a mass ratio of 2% can reduce the standard deviation of the displacement and base overturning moment up to 20% and 16%, respectively. However, its passive counterpart increases the dynamic responses. This significant difference between the performances of the passive and semi-active devices is for the case when the natural frequency is shifted by up to only 5% and is even higher for the case with higher frequency changes. The results highlight the significance of implementation of a semi-active tuned mass damper for offshore wind turbines which are subjected to varying natural frequency due to gradual or sudden damage development. To implement the aforementioned structural control devices in the design of offshore wind turbines, more comprehensive studies with a focus on the experimental investigations and practicalities of these devices are needed. Furthermore, the maintenance requirements of these devices should be investigated in the scope of total maintenance regime of the offshore wind turbine.

Author Contributions: A.H. and E.O. conceived and designed the simulations. A.H. developed the simulation tool. A.H. and E.O. analyzed the data and wrote the paper.

Acknowledgments: This research was made possible by a Ph.D. scholarship from the University of Strathclyde scholarship.

Conflicts of Interest: The authors declare no conflict of interest.

References

1. Symans, M.D.; Constantinou, M.C. Semi-active control systems for seismic protection of structures: A state-of-the-art review. *Eng. Struct.* **1999**, *21*, 469–487. [CrossRef]
2. Chen, J.-L.; Georgakis, C.T. Spherical tuned liquid damper for vibration control in wind turbines. *J. Vib. Control.* **2013**, *21*, 1875–1885. [CrossRef]
3. Housner, G.W.; Bergman, L.A.; Caughey, T.K.; Chassiakos, A.G.; Claus, R.O.; Masri, S.F.; Skelton, R.E.; Soong, T.T.; Spencer, B.F.; Yao, J.T.P. Structural Control: Past, Present, and Future. *J. Eng. Mech.* **1997**, *123*, 897–971. [CrossRef]
4. Soong, T.T.; Spencer, B.F. Supplemental energy dissipation: state-of-the-art and state-of-the-practice. *Eng. Struct.* **2002**, *24*, 243–259. [CrossRef]
5. Enevoldsen, I.; Mørk, K.J. Effects of a Vibration Mass Damper in a Wind Turbine Tower*. *Mech. Struct. Mach.* **1996**, *24*, 155–187. [CrossRef]
6. Murtagh, P.J.; Ghosh, A.; Basu, B.; Broderick, B.M. Passive control of wind turbine vibrations including blade/tower interaction and rotationally sampled turbulence. *Wind Energy* **2008**, *11*, 305–317. [CrossRef]
7. Colwell, S.; Basu, B. Tuned liquid column dampers in offshore wind turbines for structural control. *Eng. Struct.* **2009**, *31*, 358–368. [CrossRef]
8. Stewart, G.M.; Lackner, M.A. The impact of passive tuned mass dampers and wind–wave misalignment on offshore wind turbine loads. *Eng. Struct.* **2014**, *73*, 54–61. [CrossRef]
9. Stewart, G.; Lackner, M. Offshore wind turbine load reduction employing optimal passive tuned mass damping systems. *IEEE Transact. Control. Syst. Technol.* **2013**, *21*, 1090–1104. [CrossRef]
10. Dinh, V.N.; Basu, B. Passive control of floating offshore wind turbine nacelle and spar vibrations by multiple tuned mass dampers. *Struct. Control. Health Monit.* **2014**, *22*, 152–176. [CrossRef]
11. Lackner, M.; Rotea, M.; Saheba, R. Active structural control of offshore wind turbines. In Proceedings of the 48th AIAA Aerospace Sciences Meeting Including the New Horizons Forum and Aerospace Exposition, Orlando, FL, USA, 4–7 January 2010.
12. Fitzgerald, B.; Basu, B.; Nielsen, S.R.K. Active tuned mass dampers for control of in-plane vibrations of wind turbine blades. *Struct. Control. Health Monit.* **2013**, *20*, 1377–1396. [CrossRef]
13. Fitzgerald, B.; Basu, B. Active tuned mass damper control of wind turbine nacelle/tower vibrations with damaged foundations. *Key Eng. Mater.* **2013**, *569–570*, 660–667. [CrossRef]
14. Staino, A.; Basu, B. Dynamics and control of vibrations in wind turbines with variable rotor speed. *Eng. Struct.* **2013**, *56*, 58–67. [CrossRef]
15. Krenk, S.; Svendsen, M.N.; Høgsberg, J. Resonant vibration control of three-bladed wind turbine rotors. *AIAA J.* **2012**, *50*, 148–161. [CrossRef]
16. Maldonado, V.; Boucher, M.; Ostman, R.; Amitay, M. Active vibration control of a wind turbine blade using synthetic jets. *Int. J. Flow Control.* **2009**, *1*, 227–238. [CrossRef]
17. Fitzgerald, B.; Basu, B. Cable connected active tuned mass dampers for control of in-plane vibrations of wind turbine blades. *J. Sound Vib.* **2014**, *333*, 5980–6004. [CrossRef]
18. Kim, S.-M.; Wang, S.; Brennan, M.J. Optimal and robust modal control of a flexible structure using an active dynamic vibration absorber. *Smart Mater. Struct.* **2011**, *20*, 045003. [CrossRef]

19. Yalla, S.K.; Kareem, A.; Kantor, J.C. Semi-active tuned liquid column dampers for vibration control of structures. *Eng. Struct.* **2001**, *23*, 1469–1479. [CrossRef]

20. Hrovat, D.; Barak, P.; Rabins, M. Semi-active versus passive or active tuned mass dampers for structural control. *J. Eng. Mech.* **1983**, *109*, 691–705. [CrossRef]

21. Sun, C.; Nagarajaiah, S. Study on semi-active tuned mass damper with variable damping and stiffness under seismic excitations. *Struct. Control. Health Monit.* **2013**, *21*, 890–906. [CrossRef]

22. Karnopp, D.; Crosby, M.J.; Harwood, R. Vibration control using semi-active force generators. *J. Eng. Ind.* **1974**, *96*, 619–626. [CrossRef]

23. Nagarajaiah, S.; Sonmez, E. Structures with semiactive variable stiffness single/multiple tuned mass dampers. *J. Struct. Eng.* **2007**, *133*, 67–77. [CrossRef]

24. Kirkegaard, P.H.; Nielsen, S.R.K.; Poulsen, B.L.; Andersen, J.; Pedersen, L.H.; Pedersen, B.J. Semiactive vibration control of a wind turbine tower using an MR damper. In Proceedings of the Fifth European Conference on Structural Dynamics, Munich, Germany, 2–5 September 2002.

25. Karimi, H.R.; Zapateiro, M.; Luo, N. Semiactive vibration control of offshore wind turbine towers with tuned liquid column dampers using H output feedback control. In Proceedings of the 2010 IEEE International Conference on Control Applications, Yokohama, Japan, 8–10 September 2010.

26. Arrigan, J.; Pakrashi, V.; Basu, B.; Nagarajaiah, S. Control of flapwise vibrations in wind turbine blades using semi-active tuned mass dampers. *Struct. Control. Health Monit.* **2010**, *18*, 840–851. [CrossRef]

27. Weber, F. Dynamic characteristics of controlled MR-STMDs of Wolgograd Bridge. *Smart Mater. Struct.* **2013**, *22*, 095008. [CrossRef]

28. Sonmez, E.; Nagarajaiah, S.; Sun, C.; Basu, B. A study on semi-active Tuned Liquid Column Dampers (sTLCDs) for structural response reduction under random excitations. *J. Sound Vib.* **2016**, *362*, 1–15. [CrossRef]

29. Sun, C. Semi-active control of monopile offshore wind turbines under multi-hazards. *Mech. Syst. Signal Proc.* **2018**, *99*, 285–305. [CrossRef]

30. Song, B.; Yi, Y.; Wu, J.C. Study on Seismic Dynamic Response of Offshore Wind Turbine Tower with Monopile Foundation Based on M Method. *Adv. Mater. Res.* **2013**, *663*, 686–691. [CrossRef]

31. Abé, M.; Igusa, T. Semi-Active Dynamic Vibration Absorbers For Controlling Transient Response. *J. Sound Vib.* **1996**, *198*, 547–569. [CrossRef]

32. Sun, C. Mitigation of offshore wind turbine responses under wind and wave loading: Considering soil effects and damage. *Struct. Control Health Monit.* **2018**, *25*, 1–22. [CrossRef]

33. Sadek, F.; Mohraz, B.; Taylor, A.W.; Chung, R.M. A method of estimating the parameters of tuned mass dampers for seismic applications. *Earthq. Eng. Struct. Dyn.* **1997**, *26*, 617–636. [CrossRef]

34. Jonkman, J.; Butterfield, S.; Musial, W.; Scott, G. *Definition of a 5-MW Reference Wind Turbine for Offshore System Development*; Office of Energy Efficiency and Renewable Energy: Washington, DC, USA, 2009; Technical Report No. NREL/TP-500-38060.

35. Jonkman, J.; Musial, W. Offshore code comparison collaboration (OC3) for IEA task 23 offshore wind technology and deployment. *Off. Sci. Technol. Inf. Technol. Rep.* **2010**, *303*, 275–3000.

36. American Petroleum Institute (API). *RP 2A-WSD: Recommended Practice for Planning, Designing and Constructing Fixed Offshore Platforms Working Stress Design*; API Publishing Services: Short Hills, NJ, USA, 2000.

37. DNV. *DNV-OS-J101 Design of Offshore Wind Turbine Structures*; DNV: Oslo, Norway, 2004.

38. Kaimal, J.; Wyngaard, J.; Izumi, Y.; Cote, O. Spectral characteristics of surface-layer turbulence. *Q. J. R. Meteorol. Soc.* **1972**, *98*, 563–589. [CrossRef]

39. Jonkman, B.J. TurbSim user's guide: version 1.50. *Nrel Rep. Natl Renew. Energy Lab.* **2009**, *7*, 58.

40. Jonkman, J.M.; Buhl, M.L. *SciTech Connect.: FAST User's Guide-Updated August 2005*; Technical Report No. NREL/TP-500-38230; SciTech: Perth, Australia, 2005.

41. DNV. *DNV-RP-C205: Environmental conditions and environmental loads*; DNV: Oslo, Norway, 2010; pp. 9–123.

42. Hasselmann, K. Measurements of wind-wave growth and swell decay during the Joint North Sea Wave Project (JONSWAP). *Dtsch.hydrogr.z* **1973**, *12*, 1–95.

43. Chiou, B.; Darragh, R.; Gregor, N.; Silva, W. NGA Project Strong-Motion Database. *Earthq. Spectr.* **2008**, *24*, 23–44. [CrossRef]

44. Doherty, P.; Gavin, K. Laterally loaded monopile design for offshore wind farms. *Proc. Inst. Civ. Eng. Energy* **2012**, *165*, 7–17. [CrossRef]

45. Kim, D.H.; Lee, S.G.; Lee, I.K. Seismic fragility analysis of 5 MW offshore wind turbine. *Renew. Energy* **2014**, *65*, 250–256. [CrossRef]

46. Achmus, M.; Kuo, Y.-S.; Abdel-Rahman, K. Behavior of monopile foundations under cyclic lateral load. *Comput. Geotech.* **2009**, *36*, 725–735. [CrossRef]

47. International Electrotechnical Commission. *IEC 61400-1: Wind Turbines Part 1: Design Requirements*, 3rd ed.; IEC: Geneva, Switzerland, 2005.

Journal of
Marine Science and Engineering

Article

Towing Operation Methods of Offshore Integrated Meteorological Mast for Offshore Wind Farms

Hongyan Ding [1,2,3], **Ruiqi Hu** [3], **Conghuan Le** [1,3,*] **and Puyang Zhang** [1,3]

1   State Key Laboratory of Hydraulic Engineering Simulation and Safety, Tianjin University, Tianjin 300072, China; dhy_td@163.com (H.D.); zpy_td@163.com (P.Z.)
2   Key Laboratory of Coast Civil Structure Safety, Ministry of Education, Tianjin University, Tianjin 300072, China
3   School of Civil Engineering, Tianjin University, Tianjin 300072, China; ruiqi_hu@163.com
*   Correspondence: leconghuan@163.com

Received: 22 January 2019; Accepted: 3 April 2019; Published: 11 April 2019

Abstract: An offshore integrated meteorological mast (OIMM) is introduced which has great application potential for the development of offshore wind turbine power. This innovative OIMM features in two aspects: the integrated construction and the integrated transportation. Its integrated techniques enable this OIMM to be prefabricated onshore and transported by a relatively small tugboat to the installation site. It is efficient in construction, rapid in transportation and saving in cost. The towing process is an important section for the integrated transportation, which makes the towing operation necessary to investigate. With the numerical simulation software MOSES, the hydrodynamic behavior of the towing operation is investigated. Two special wet towing methods (surface towing and submerged towing) are adopted and analyzed in terms of the towing resistance, towing speed, fairlead position and the motion response. The results show that for both towing methods, to obtain a higher speed by increasing the towing force is uneconomic since the towing resistance increases a much higher percentage than the towing speed dose. Surface towing has a smaller resistance but larger motion response compared to submerged towing. The submerged towing shows a clear descending heave motion. The heave and pitch motions are smaller with the lower fairlead position and fluctuate less with deeper submerged depth.

Keywords: offshore integrated meteorological mast (OIMM); towing method; towing resistance; fairlead position; motion response

---

1. Introduction

The development of offshore wind turbine power characteristics is based on the anemometer technology including both meteorological mast structures and measurement devices [1]. While the primary problem for an offshore meteorological mast is associated with the structure design. Offshore meteorological masts are usually required to be installed on a fixed or a floating foundation. It can be expensive to install such a foundation or platform and maintain the instrument in an offshore location [1]. Tianjin University and Dao Da Heavy Industry Company (DDHI) developed an offshore meteorological mast with an innovative foundation as an integrated structure (OIMM) that reduces costs by shortening the installation process and narrowing down the required amount of marine operation equipment [2,3]. This offshore integrated meteorological mast (OIMM) is shown in Figure 1 with the transporting operation. The OIMM excels in two aspects: the integrated construction and the integrated installation. It can be manufactured onshore, hoisted into the sea and wet towed to an installation site. Since the structure has a big triangular floating tank, thus it can be self-floating and wet towed. As can be seen in Figure 1, the OIMM has three major components: the upper meteorological mast, the floating tank, and the anti-slide skirt plates (bottom).

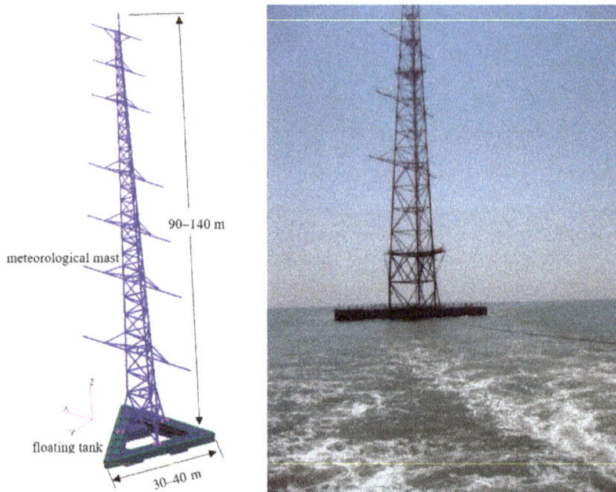

Figure 1. Offshore integrated meteorological mast and towing operation.

The design concept of this OIMM is based on many former research [4–6] about the one-step-installation technology applied in the lager-scale composite bucket foundation (CBF). However, the construction and installation process for the OIMM are more simplified and different, because it has a relatively small dimensional size of infrastructure and a much higher superstructure. Generally, the entire OIMM weights about 300 t, and it has a height varying from 90–140 m, which can be adjusted based on specific requirements. The huge floating tank usually has multiple subdivisions inside to function as individual ballast compartments and to facilitate the stability adjustment during the wet-towing. It can also function as a gravity type foundation during the service life. The upper structure is a meteorological mast that carries many wind measurement devices. The bottom anti-slide skirt plates penetrate the sea bed working as suction caissons, and providing horizontal righting moment by the lateral resistance.

Many publications are focused on the offshore structure installation technology to understand the behavior of the structure and estimate the feasibility of this installation process. The basic dynamics of a subsea template, assembled system, were studied by calculating the dynamic forces and displacements by using time integration in SIMO (simulation of marine operations, SESAM package) [7,8]. They estimated the tension of a lift wire by SIMO and a simplified method simulating this dynamic system which agreed well with the experimental investigation. The feasibility of dry transport a TLP (tension leg platform) by a heavy lift transport vessel was studied, and the motion behavior of TLP structure and its transport vessel were evaluated [9]. Eriksson and Kullander [10] conducted a wet towing of a substructure with special equipment and vessels in Sweden. Some researchers focus on the towing stability feature and important influence factors in terms of the different sea conditions for a floating platform and offshore wind turbines [11–14]. A dynamic analysis of the wet tow for a floating wind turbine based on a multi-body model in SESAM has been performed by Ding et al., and the towline force and the stability features are studied [15]. While a comparison of the cost estimation between the dry towing and the wet towing for a semisubmersible platform, conducted by Kim et al. [16], suggests that the wet towing is more cost-saving for this structure. More installation solutions, especially for deepwater structures, backed by engineering tools and numerical simulation methods are discussed by Wang et al. [17] with main characteristics and critical challenges.

Little research has been done on the offshore meteorological mast, especially for the transport operation. Based on the above research, this paper adopts two special wet towing methods to the innovative OIMM and compares the corresponding dynamic performance by using the numerical simulation software MOSES (offshore platform design and installation software). The purpose is to obtain the feasibility of the two towing methods with respect to its reliability and operability.

Figure 2 demonstrates the detailed information of the applied towing methods: the surface towing and the submerged towing (or subsurface towing). As for the submerged towing method, the mooring point can either be attached to the floating tank (fairlead 1) or to the mast (fairlead 2). While for the surface towing method, the mooring point is attached to the floating tank (fairlead 1). The OIMM is designed to be suitable for a water depth around 50 m. Therefore, when performing a submerged towing simulation, the submerged depth of the OIMM is defined varying within 0–40 m. For the surface towing method, it is simple with only one submerged depth (draft). The towing resistance based on these two methods is obtained and compared between the MOSES simulation and the manual computation method. The towing speed and hydrodynamic motion responses relating to the mooring positions are also considered and compared. The results can provide recommendations and references for the estimation of the feasibility of the two methods and on the optimization of the towing operation for OIMM.

The paper is organized as follows. Section 2 introduces the relevant methods for the computation of the towing resistance and the hydrodynamic theory. A comparison of towing resistance between the numerical method and manual method is shown in Section 3, as well as the detailed explanations about the towing speed and the mooring positions. Section 4 illustrates the results based on the discussion in Section 3, and conclusions are drawn in Section 5.

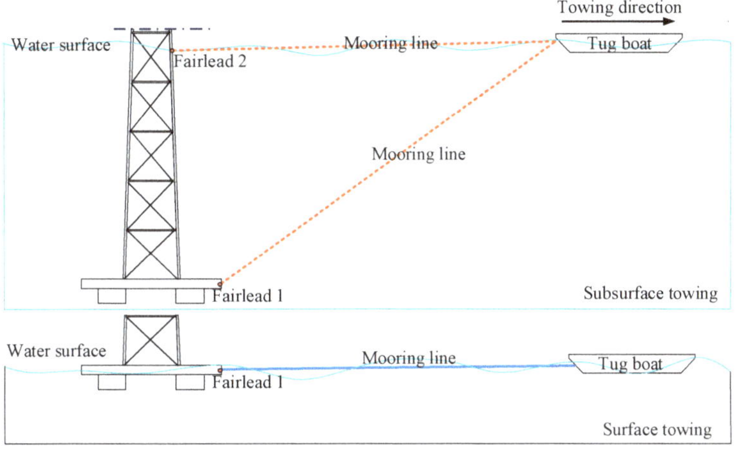

Figure 2. Surface towing and submerged towing methods.

2. Theory Background

The towing resistance is a basic parameter, on which the towing system depends. The estimation methods have been investigated by many researchers [18,19] in terms of the speed reduction and numerical calculation. The total resistance for different sea states can be described by four different types of resistance: surface friction, energy loss due to wave generating, energy loss due to current, wind resistance. For low speed towing, the surface friction counts for a majority. According to DNV-RP-H103 [20], the tug will have the maximum towing force at zero velocity and in the absence of wind, waves and current, and the available towing force decreases with forward velocity. In this paper, the towing resistance is obtained from two methods: the numerical computation by MOSES, and the manual computation by the Guidelines for Towage at Sea (CCS code) [21].

2.1. Towing Resistance By MOSES

In MOSES, the towing resistance is primarily influenced by the wind force and the viscous drag force. The viscous drag force is illustrated as a portion of the system which is the velocity squared term in Morison's equation, the viscous roll damping, or the viscous drag on a piece. This can be either an excitation due to wave or current, or a damping in still water [22]. Generally, MOSES employs three hydrodynamic theories: Morison's equation, strip theory, and 3D-diffraction theory. The Morison's Equation is used for beam element and the viscous drag will be computed. For panel model, the 3D-diffraction theory is applied.

The drag force on a cylinder in an unsteady viscous flow is determined by Morison's Equation, which consists of a combination of an inertia term and a drag term:

$$dF = \rho\pi\frac{D^2}{4}C_M\ddot{V} + \frac{\rho}{2}C_DD|V|V \tag{1}$$

where dF is the force per unit length of a cylinder [N/m], D is the diameter of cylinder [m], ρ is the density of fluid [kg/m^3], C_M is the mass coefficient, C_D is the drag coefficient, $\ddot{V}$ is the acceleration [m/s^2].

Drag force on submerged element consists mainly of friction resistance and pressure resistance. Friction resistance can be expressed as:

$$F = \frac{1}{2}C_F\rho V^2 S \tag{2}$$

where F is the friction resistance on object [N] and V is the towing velocity [m/s]. C_F is the dimensionless friction coefficient depended on the Reynolds number. S is the wet surface area of the object [m^2].

The pressure distribution on panel element can be found by velocity potential function and Bernoulli's equation with boundary conditions [23].

$$V = \nabla\Phi \equiv i\frac{\partial\Phi}{\partial x} + j\frac{\partial\Phi}{\partial y} + k\frac{\partial\Phi}{\partial z} \tag{3}$$

$$p + \rho gz + \rho\frac{\partial\Phi}{\partial t} + \frac{\rho}{2}V{\cdot}V = C \tag{4}$$

where p is the pressure in Bernoulli's equation, C is an arbitrary function of time.

2.2. Towing Resistance By Code

As for the manual computation, the Guidelines for Towage at Sea (CCS code) provides an estimation of ocean towage resistance considering the towing speed, by the following empirical equation:

$$R_T = 1.15[R_f + R_B + (R_{ft} + R_{Bt})] \text{ kN} \tag{5}$$

$$R_f = 1.67A_1V^{1.83} \times 10^{-3} \text{ kN} \tag{6}$$

$$R_B = 0.147\delta A_2 V^{1.74+0.15V} \text{ kN} \tag{7}$$

where: R_f and R_B are the friction and the residual resistance of towed vessel, in kN; R_{ft} and R_{Bt} are the friction and the residual resistance of towing vessel, in kN; A_1 is the wetted surface area under waterline of vessel or surface structure, in m^2; V is the towage velocity, in m/s; δ is the block coefficient; A_2 is the submerged transverse section area amidships, in m^2. R_f is the residual resistance of towing vessel, in kN.

For drilling units or other surface structures with huge wind area, the towage resistance is also to be calculated as follows, taken whichever is greater:

$$\sum R = 0.7(R_f + R_B) + R_a \text{ kN} \tag{8}$$

$$R_a = 0.5\rho V^2\sum C_x A_i \times 10^{-3} \text{ kN} \tag{9}$$

where: R_f and R_B are the same as (1) above, and the R_a is the air resistance; ρ is the air density, to be taken as 1.22 kg/m^3; V is the wind velocity, to be taken as 20.6 m/s; A_i is the wind area, to be taken as upwind, in m^2; C_x is the configuration coefficient of wind area.

Equations (5)–(9) are applied to compute the total towing resistance of the floating tank and the upper meteorological mast. The resistance of the towing vessel (tugboat) is not included in the total resistance so that it can be compared with numerical results for which the towing resistance is given separately.

The sea states in MOSES are defined according to the items in CCS guidelines. The wind velocity is 20.6 m/s, and the current velocity is 0.5 m/s. The towing speed is 3 m/s with a significant wave height of 2 m.

The results of towing resistance based on this code will be analyzed in Section 3.

2.3. The Hydrodynamic Theory

The motion of the OIMM is featured by the free surface boundary condition, which is nonlinear and makes the description of the body motion complex. Based on the potential theory, the dynamic responses of marine structures are widely studied. The problem of hydrodynamics can be formulated in terms of potential theory based on the following assumptions:

The sea water is assumed incompressible and inviscid. The fluid motion is irrotational. Viscous effects are neglected and the fluid is assumed to be incompressible. The depth h is finite and constant and the free surface is infinite in all directions. Assumed the motion of the body and the fluid to be small for a linearization of the boundary condition and free surface condition.

A velocity potential ϕ can be used to describe the fluid velocity vector $V(x, y, z) = (u, v, w)$ at time t at the point $X = (x, y, z)$ in a cartesian coordinate system fixed in space [23].

It is convenient to decompose the total velocity potential in the alternative forms [24]:

$$\phi = \phi_0 e^{-iwt} + \phi_7 e^{-iwt} + \sum_{i=1}^{6} \phi_i \dot{\eta}_i \tag{10}$$

where $\phi_0 e^{-iwt}$ is the velocity potential of the incident wave system; $\phi_7 e^{-iwt}$ is the scattered field due to the presence of the body; the diffraction problem can be solved by considering the ϕ_0 and the ϕ_7 together; ϕ_i, $i = 1$–6, is the contribution to the velocity potential from the ith mode of motion. Therefore the ϕ_i ($i = 1$–6) is called the radiation potential.

The towing dynamic problem in this paper is analyzed separately by whether or not considering the forward speed. To achieve that, we change the forward speed by applying different towing force. In two ways, the system has no forward speed. One is to tow the structure with a limited force and keep the OIMM to its initial position without moving forward. Another way is that the towing tugboat is fixed in the global system and the OIMM connected with the tugboat by a mooring line, and then the OIMM can oscillate under the environment loadings.

3. Comparisons and Analyses

3.1. Towing Resistance in Different Towing Methods

In Section 2, the numerical method and the manual computation method have been introduced to obtain the towing resistance. The results are shown in this section and analyzed in Section 4. The towing resistance is compared based on the surface towing and the submerged towing. As for the submerged towing, the towing resistance of the OIMM is different if the initial submerged depth changes. Therefore, four submerged depths are considered and compared by the manual method and the numerical methods. For the surface towing, only one submerged depth (draft) is considered because the vertical height of the floating tank is small (shown in Figure 1). The OIMM can be self-floating with a draft of 0.85 m which has been verified to be able to maintain adequate floating stability. Thus, this draft is used for the surface towing.

The towing resistance for submerged towing computed by the manual method and the numerical method has been tabled in four groups corresponding to four submerged conditions (10 m, 20 m, 30 m, 40 m). For both numerical and manual methods, the resistance caused by the mooring line is not considered. The wind force listed in Table 1 is based on the CCS code and the wind force on the floating tank is not considered. The detailed computation of the towing resistance is listed in Table 2.

Table 1. Wind force of OIMM (meteorological mast).

Elevation/m	90–80	80–70	70–60	60–50	50–40	40–30	30–20	20–10	0–10
Area/m^2	5.19	8.55	10.71	10.82	13.66	14.39	17.12	20.56	33.09
Wind force/kN	1.923	3.165	3.799	4.843	3.649	4.245	4.875	5.855	8.565

Note: The elevation is positive in vertical upward direction.

Table 2. Towing resistance by CCS code.

Items	Wave & Current/t		Wind Force/t	Total/t
	Floating Tank	**Tower**		
Surface towing	11.66	0.00	4.09	15.75
Submerged 10 m	20.28	5.58	3.89	29.75
Submerged 20 m	20.28	9.04	3.58	32.90
Submerged 30 m	20.28	11.92	3.20	35.40
Submerged 40 m	20.28	14.22	2.72	37.22

Note: The submerged depth and blockage impacts are not considered here.

In MOSES, a time domain simulation was conducted and the required towing force was adjusted according to the final velocity of the OIMM. The towing resistance is compared under the same towing speed. The wind force and the towing resistant by MOSES are shown in Table 3.

Table 3. Towing resistance by MOSES.

Items	Wind Force/t	Towing Resistance/t
Surface towing	5.34	17.94
Submerged 10 m	3.01	32.56
Submerged 20 m	2.32	39.35
Submerged 30 m	1.73	48.42
Submerged 40 m	1.51	51.21

The wind forces from Tables 2 and 3 are generally agreed. More detailed discussion is presented in Section 4.1.

3.2. Towing with Forward Speed

In this section, the relation among the towing resistance, towing methods, and the resulted speed are analyzed. The analyses results will provide the towing operation with suggestions on feasible environment conditions and the choice of possible towing equipment.

In Section 2.2, it has been illustrated that the total potential can be investigated in terms of the radiation potential and the diffraction potential separately. The radiation problem does not consider an incident wave. Therefore, the towing operation can be investigated by whether including the wave and current loads. This can be achieved directly by the numerical method.

In addition, the towing operation performed in the calm sea condition with a forward speed will be compared in terms of the surface towing and the submerged towing. For both towing operations, optional towing velocity is around 3–5 knots [7]. Therefore, four different velocities (0 m/s, 1 m/s, 2 m/s,

3 m/s) are chosen to compare the towing resistance and towing stability in complex loading states. As has been mentioned, the zero forward speed can be achieved by exerting a limited towing force.

Moreover, another towing operation is performed as a comparison with zero forward speed achieved by fixing the tugboat in the global system. The simulation is performed in both surface towing and submerged towing. The wave, current and wind loads are taken in to account. The towing force is not needed when the tugboat is fixed. The results of the comparison are shown in Section 4.2.

3.3. Mooring Positions

This section focuses on the feasibility and superiority of the mooring positions (fairlead positions) based on two towing methods. Generally, the surface towing method is applied to the self-floating objects and the large structures will adopt the submerged towing [25]. Considering the high-rise feature of the OIMM, both towing methods can be applied. For the surface towing, only one fairlead position, fairlead 1 is taken. As for the submerged towing, both fairlead 1 and fairlead 2 can be applied. The dynamic performance of this towing system relates to the mooring positions, because the orientation and the position of the applied towing force are associated with the fairlead positions.

To eliminate the additional effect by the towline and tugboat, the towing system is simplified by substitute the towline with a constant horizontal force, which will provide the driving force to OIMM. In this simplified system, the only interaction is the environment loading with OIMM.

In addition, for each mooring position, the dynamic features of the submerged towing are further compared in terms of the submerged depth. This comparative analysis will be a more valuable reference when the OIMM is higher and the sea is deeper. With different submerged depths, not only the towing resistance but also the motion responses will change. The impact caused by fairlead positions on the dynamic motion responses of the OIMM are compared in heave and pitch motions. The results are discussed in Section 4.3 which may be benefit the optimization of the towing operation and maneuverability of the towed structures.

4. Results and Discussions

4.1. Towing Resistance

In this section, the towing resistance obtained by two methods (MOSES and CCS code) is discussed. Two towing methods are distinguished by the submerged depth. Specifically, the surface tow corresponds to the shallowest submerged depth and the submerged tow includes four different depths (10 m, 20 m, 30 m, 40 m). The comparison is shown in Figure 3.

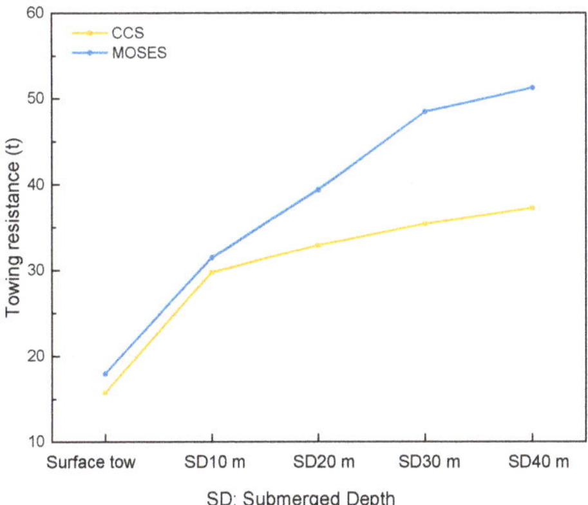

Figure 3. Towing resistances by two method (MOSES and CCS code) under surface tow and submerged tow.

Figure 3 depicts the variations of the towing resistance between numerical method and manual computation (CCS). Basically, the surface towing method has the lowest towing resistance suggesting that the surface friction force caused by current and wave accounts for a large part of the towing resistance. In Section 3.1, from Tables 2 and 3, the wind forces by MOSES and CCS code are basically agreed. Thus, the total resistance is dominated by wave and current loadings. However, the growth of towing resistance by numerical method is much higher than that by CCS code, especially for the submerged towing.

The difference is quite small for the surface tow and shallow submerged tow. The maximum different is around 37% corresponding to the submerged depth of 40 m. This increased difference between MOSES and CCS methods can be accounted for by the submerged area of the slender structure (meteorological mast) which increase as the depth goes deeper. That is, the Morison's Equation was used in MOSES to compute the wave and current interaction on beam element (meteorological mast), while for CCS code, the entire submerged part adopted the same estimation method. The figure implies that the CCS code comparatively underestimates the towing resistance, especially for submerged conditions with slender part.

Other reasons accounting for the overestimation may because that the sea environment performed in MOSES is irregular wave condition and the current induced force is more sensitive to the panel model for diffraction. Moreover, the resistance growth by CCS code mainly results from the increase of the wet surface of the meteorological mast which is quite limited and the friction and the residual resistance of the floating tank does not change under the water. While by the numerical method, a slender structure will make a difference because of the Morison's Equation.

For this high-rise structure, the wind force can be significant compared with the wave and current loads, which should be considered. To analyze the individual effects of environment loadings, the surface towing was performed. It means the water plane and wet surface were the same, and then the towing resistance caused by the wave and current is the same. Besides the wind load is at the maximum since the entire mast is above the surface. The towing force is balanced with the resistance and results in a stable forward speed. In Figure 4, wind, wave and current effects are separately investigated in terms of the induced forward speed and they are compared with the total effect. The total effect means all environment loads are considered. The towing force applied to the system varies from 5–25 t.

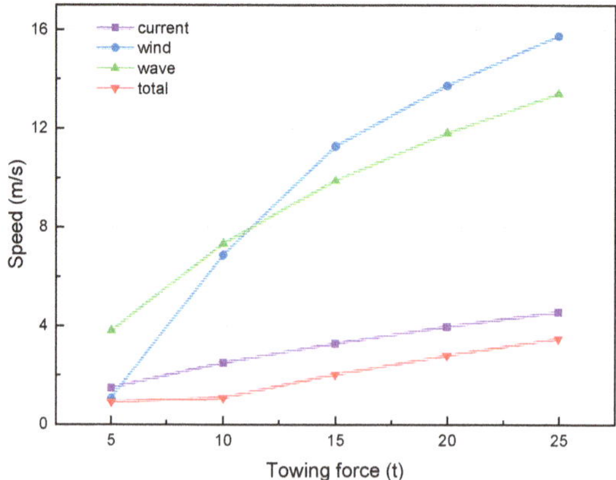

Figure 4. Comparison between the towing speed under different sea loads varying with the towing force by MOSES.

As shown in Figure 4, generally the current induced speed line is closest to the total effect line. Wave occupies much less impact when the forward speed is higher. The wind load has a dominant effect only when the towing force at the lowest level and it declines significantly afterward. In general,

Figure 4 suggests that for OIMM the wave and current loads are dominant for normal towing speeds (above 1 m/s), which supports the increased difference shown in Figure 3. Since the submerged part of the mast is not considered in manual results and the current and wave loads on this part are considerable.

4.2. Forward Speed

In this section, the towing resistance under different forward speeds are compared. The towing resistance is defined as the viscous drag force in MOSES. The resistance variations are compared in a time domain for both surface towing and subsurface towing. The comparison is considered from two perspectives: zero-speed and none-zero speed. This has been illustrated in Section 3.2. Two loading environments are defined for the none-zero speed towing. One does not consider the wave and current loadings (the still water condition). The other is normal including all environment loadings. Notice that the towing force applied to the still water condition was the same as that applied to the speed condition of 3 m/s.

Figure 5 shows the time domain results of the viscous drag force for the surface towing. Viscous drag curves fluctuate significantly as towing speed increases because of the free surface effects on the floating tank. The dynamic motion under higher speed conditions is larger because the wet surface area varies significantly and accounts for this fluctuation. The fixed condition shows a similar variation as the 0 m/s speed condition. The viscous drag forces for the fixed and the 0 m/s speed conditions are the lowest, which is reasonable because the resistance is in proportion to the second order of the velocity term. That is, the lower speed corresponds to lower sea loads. The still water condition has a close and smaller viscous resistance as the 3 m/s speed condition since the same towing force was applied for them and the latter has wave and current induced resistance. Without the wave and current, the viscous drag curve is smooth. In addition, the viscous resistance increases by around 20% for every speed increase of 1 m/s.

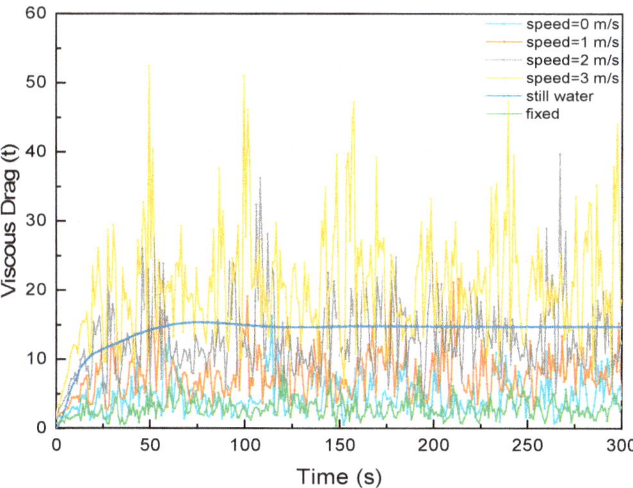

Figure 5. Comparison between the time history curves of the viscous drag force under different towing speeds for surface towing.

Figure 6 illustrates the time domain results of the viscous drag force of the submerged towing. The submerged depth is 40 m. The resistance curve fluctuates much less than that of the surface towing, which is because the water plane is much smaller. The resistance force increases significantly as each additional growth of the towing speed. More specifically, the resistance force nearly doubled for each additional towing speed. The fixed condition presents a similar variation as that of the

0 m/s speed condition. As for the still water condition, basically it has a close and smaller viscous resistance similar to that of the 3 m/s speed condition since the same towing force was applied for them. Likewise, the time history curve is smooth for the still water condition. Generally, the towing process for submerged towing is more stable, which can be beneficial from a practical perspective.

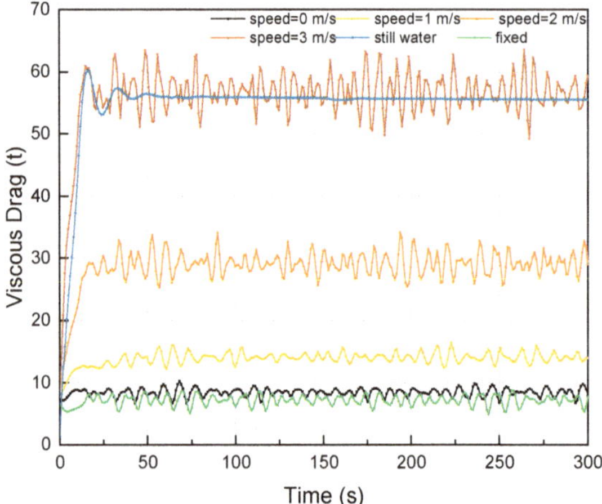

Figure 6. Comparison between the time history curves of the viscous drag force under different towing speeds for submerged towing (submerged depth: 40 m).

In Figure 7, the applied towing force and the balanced viscous drag resistance for both surface towing and submerged towing are compared. The submerged depth is 40 m for subsurface towing and six groups are chosen based on the towing speed. For fixed body condition, no towing force is applied and only the balanced resistance is shown. Compared with the surface towing, to perform a submerged towing requires larger towing force so as to achieve a similar towing speed, especially for high speed towing. The required towing force and balanced viscous drag force are proportional to the second order of the towing speed. For the surface towing, the water plane area is similar and the distinction between the viscous drag force and towing force is quite clear for the still water condition. The resistance is smaller compared with the condition of speed 3 m/s, which is reasonable. Because the wave and current effects are eliminated.

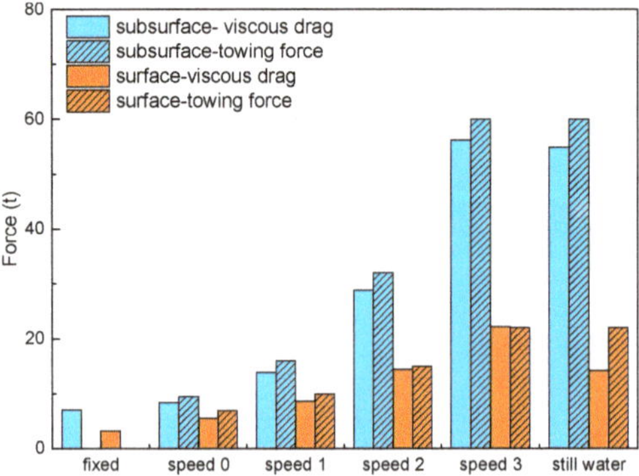

Figure 7. Comparison between the surface towing and the submerged towing under different towing speeds in terms of the viscous drag force and the towing force.

Generally, based on the above figures, it suggests that to obtain a higher speed by increasing the towing force is low efficient in terms of the transportation cost since the towing resistance will increase by a larger margin.

4.3. Mooring Position

In this section, the impacts of the mooring positions are investigated from the viscous resistance and the motion response perspectives. The submerged towing was performed with an initial depth of 40 m. As has been shown in Figure 2, there are two mooring positions named as fairlead 1 attached to the floating tank; and fairlead 2 attached to the tower near to the surface, respectively. The two fairlead positions excel at different aspects. For instance, the fairlead 2 is near to the free surface and can be conveniently connected with a towline. While the fairlead 1 attached the floating tank in a height near to the center of gravity has a smaller force arm, which is benefit for a relatively small motion response amplitude. Thus, it is necessary to analyze the dynamic behavior under these two conditions. Figure 8 depicts the motion responses and the sea environment with two fairlead positions.

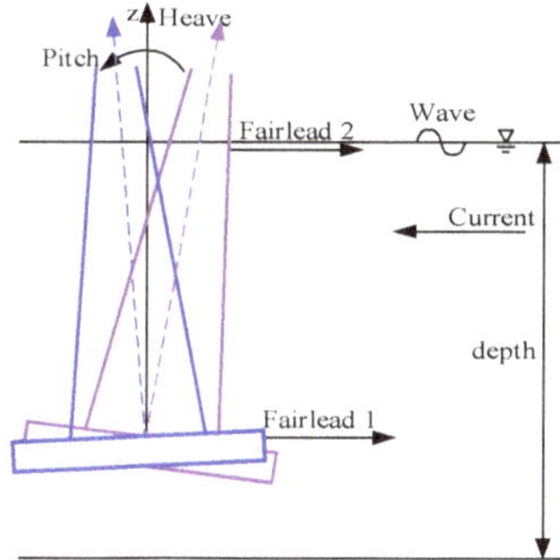

Figure 8. Schematic view of the pitch and heave motion directions, the loading environments and the fairlead positions.

In Figure 9, it is shown that the surface towing method has a much lower resistance but significant fluctuation since it has a larger water plane area. While for the submerged towing method, the fairlead 1 has a smaller resistance level compared with the fairlead 2. The difference of viscous drag between fairlead 1 and 2 is related to various reasons, for example, the initial submerged depth and the towing speed. Therefore, to verify this regulation, more simulations with different submerged depths are compared with the same forward speed, as shown in Figures 10 and 11.

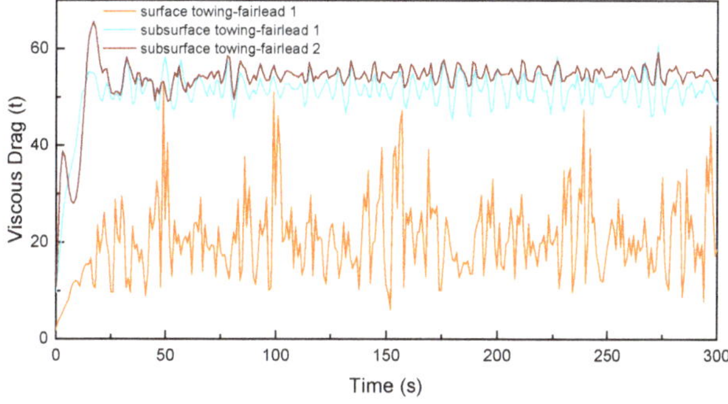

Figure 9. Comparison between the time history curves of the viscous drag force for surface towing and subsurface towing in terms of the fairlead position (towing speed: 3 m/s).

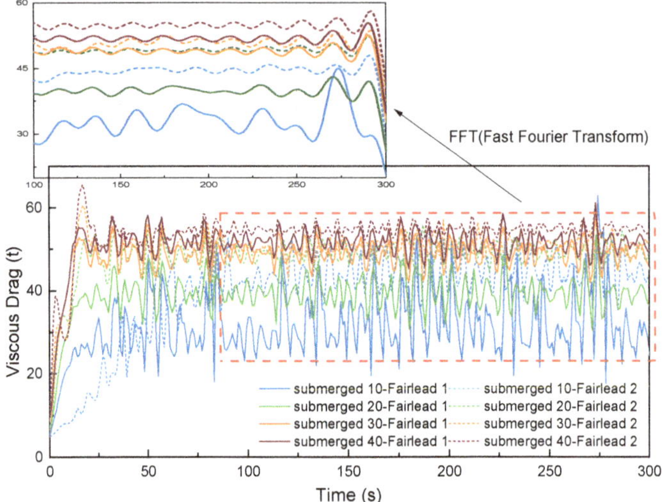

Figure 10. Comparison between the time history curves of the viscous drag force under different submerged depths in terms of the fairlead position (towing speed: 3 m/s).

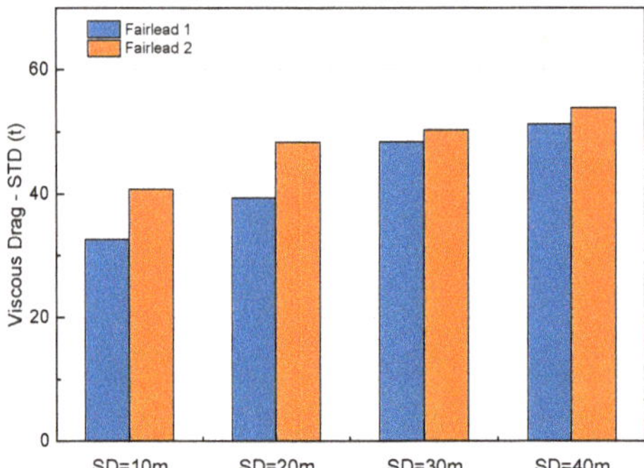

Figure 11. Statistical comparison of viscous drag force.

In Figure 10, the time history curves for viscous drag force with four submerged depths are compared in terms of the fairlead positions. These curves are smoothed by FFT (fast fourier transform). To eliminate the impacts due to the towing speed, the same towing speed (3 m/s) is chosen for all submerged conditions.

This shows that the difference of viscous drag force caused by fairlead position is more significant when the submerged depth is shallow, and the difference becomes moderate for the increase of submerged depth. Besides, these curves under shallow-submerged condition obviously have greater fluctuations and lower viscous resistance than the deep ones. This can be observed clearly from Figure 11 with the standard deviation (STD) of the viscous resistance. This is related to the wet surface and the water plane area. For shallow submerged depth, the submerged part is smaller, thus the viscous drag is smaller; while the water plane is larger, corresponding to a heavier fluctuation. Since the water plane area is much smaller when it was towed in a deeper depth, the difference of the viscous drag due to the fairlead positions becomes limited. In conclusion, for practical operations, when the same level of the towing speed is required, adopting the mooring position of fairlead 1 is reasonable, since the towing resistance is lower.

The pitch and heave motion responses for the submerged towing are compared in Figures 12 and 13. The wind force is not considered here to simplify the loading environment and the same towing speed is maintained. The motion responses are clearly influenced by two fairlead positions.

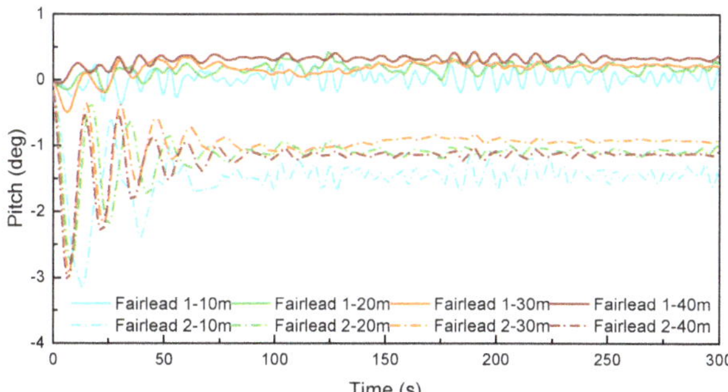

Figure 12. Pitch motion versus fairlead positions under different submerged depths.

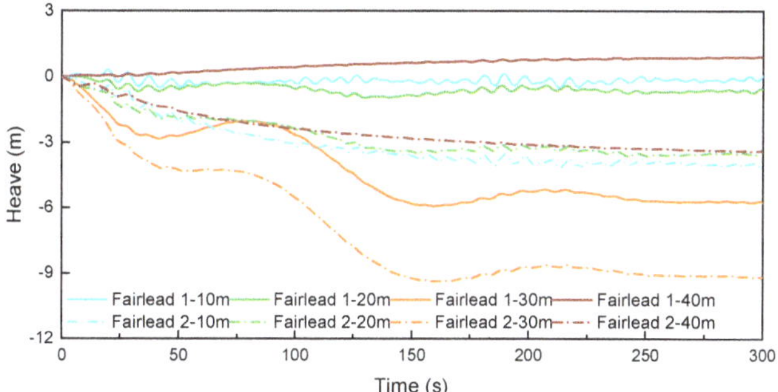

Figure 13. Heave motion versus fairlead positions under different submerged depths.

In Figure 12, pitch motions are depicted. As can be seen, the pitch angles are positive for fairlead 1 suggesting a backward rotation of the OIMM. For deeper submerged depth, pitch angles are larger, but its fluctuations are smaller. As for fairlead 2, it shows a quite opposite feature. The negative pitch angle means the OIMM leans forward. The oscillations are more stable, and the amplitude is smaller at a deeper submergence depth. In general, the pitch motion for the fairlead 2 is larger than that of the fairlead 1.

As for Figure 13, the heave motion of the submerged towing with fairlead position 1 shows a slight descending motion except for the deepest one which ascends slowly by around 1 m. The submerged depth of 30 m presents the most significant heave motion and ends up with a relative stable vertical position. As for fairlead 2, the heave motions are negative for all depths and the amplitude is larger. The deepest submerged condition shows the lowest heave motion. Likewise, the submerged depth of 30 m with fairlead 2 shows the most significant heave motion. In general, with a mooring position of the fairlead 2, the heave motion is larger than that of the fairlead 1.

The pitch and heave motion responses prove that fairlead 1 is superior to fairlead 2 because of the smaller amplitude. As shown in Figure 13, the descending heave motion can be explained that the OIMM leans forward and the wave and current force may result in a relative downward force on the OIMM. Especially when the mooring position is near to the surface and the towing force may aggravate the lean angle. Besides, the current velocity is different on the top and the bottom surfaces of the floating tank accounting for the pressure difference, which may further result in this descending motion. The ascending motion for fairlead 1 can be explained for the same reason. The submerged depth is near to the sea bed where the current velocity decreases rapidly. Thus, the current velocity decreases near to the bottom of the floating tank and the induced pressure on bottom surface of is larger than that of the upper surface of the floating tank. Therefore, the OIMM could be lifted slightly.

5. Conclusions

This study deals with the dynamic characteristics of an innovative integrated offshore structure OIMM in terms of the towing resistance and motion responses. Two special towing methods are adopted for this high-rise structure and the hydrodynamic software MOSES is applied to simulate the towing process in time domain. This paper estimates the feasibility of the towing operations and provides suggestions for potential problems when conducting an offshore wet towing for the OIMM.

First, the towing resistance is obtained by MOSES and CCS code. The numerical result is verified with manual computation. The manual result of towing resistance is agreed with the result by MOSES for surface towing and shallow submerged towing. For deep submerged towing, the difference between two methods is caused by the applied theory in MOSES and CCS. The potential theory and Morison equation are applied in numerical method, while empirical equations are used by CCS code. The current impact is the most important component since the induced speed is closest to the overall induced speed.

Secondly, the surface and submerged towing methods are compared in terms of different forward speeds and the results suggest that to obtain a higher speed by increasing the towing force is unreasonable with respect to the transportation cost, since the towing resistance increases much higher. Larger towing force is required for submerged towing to achieve the same speed level as surface towing, particularly when the towing speed becomes higher.

Thirdly, based on the fairlead positions, time domain simulations are conducted to estimate the surface and the submerged towing methods with respect to the resistance and dynamic responses. It suggests that higher mooring position is likely to result in larger viscous resistance. Deeper submerged depth is preferred when the lower motion amplitude is required. Fairlead 1 excels in two aspects: lower resistance and lower motion responses. The submerged towing should be performed with sufficient distance away from the seabed because the OIMM is likely to descend by a large distance.

Finally, it can be concluded that the numerical method by MOSES can be applied for the preliminary estimation of the towing operations for OIMM considering complex sea environment. More accurate and detailed estimation will require to consider the properties of the mooring line and the impacts of the tugboat in the system. These conclusions can provide recommendations and references for the estimation of the feasibility of the towing methods and on the optimization of the towing operation for OIMM and resembling structures.

Author Contributions: Conceptualization, H.D., P.Z. and C.L.; Methodology, R.H. and C.L.; Software, C.L. and R.H.; validation, H.D., P.Z. and C.L.; Formal Analysis, R.H. and C.L.; Writing-Original Draft Preparation, R.H. and C.L.; Writing-Review & Editing, C.L., R.H. and P.Z.; supervision, H.D. and P.Z; project administration, H.D. and P.Z; funding acquisition, H.D., P.Z. and C.L.

Funding: This research was funded by the National Natural Science Foundation of China (Grant Nos. 51679163 & 51779171), Innovation Method Fund of China (Grant No. 2016IM030100), Tianjin Municipal Natural Science Foundation (Grant No. 18JCYBJC22800) and State Key Laboratory of Coastal and Offshore Engineering (Grant No. LP1709).

Conflicts of Interest: The authors declare no conflict of interest.

References

1. Assessment Wind Measurement Program North Sea. Available online: https://offshorewind.rvo.nl/file/view/31040402/assessment-wind-measurement-program-north-sea-dnv-gl (accessed on 6 March 2015).
2. Ding, H.; Han, Y.; Zhang, P. Numerical simulation on floating behavior of buoyancy tank foundation of anemometer tower. *Trans. Tianjin Univ.* **2014**, *20*, 243–249. [CrossRef]
3. Zhang, P.; Hu, R.; Ding, H.; Le, C. Preliminary design and hydrodynamic analysis on the offshore integrated anemometer mast during wet-towing. Structures, Safety, and Reliability. In Proceedings of the ASME 2018 37th International Conference on Ocean, Offshore and Arctic Engineering, American Society of Mechanical Engineers, Madrid, Spain, 17–22 June 2018.

4. Ding, H.; Lian, J.; Li, A.; Zhang, P. One-step-installation of offshore wind turbine on large-scale bucket-top-bearing bucket foundation. *Trans. Tianjin Univ.* **2013**, *19*, 188–194. [CrossRef]

5. Zhang, P.; Han, Y.; Ding, H.; Zhang, S. Field experiments on wet tows of an integrated transportation and installation vessel with two bucket foundations for offshore wind turbines. *Ocean Eng.* **2015**, *108*, 769–777. [CrossRef]

6. Zhang, P.; Ding, H.; Le, C.; Zhang, S.; Huang, X. Preliminary analysis on integrated transportation technique for offshore wind turbines. In Proceedings of the ASME 2013 32nd International Conference on Ocean, Offshore and Arctic Engineering, American Society of Mechanical Engineers, Nantes, France, 9–14 June 2013.

7. Jacobsen, T. Subsurface Towing of a Subsea Module. Master's Thesis, Norwegian University of Science and Technology, Trondheim, Norway, 2010.

8. Jacobsen, T.; Leira, B.J. Numerical and experimental studies of submerged towing of a subsea template. *Ocean Eng.* **2012**, *42*, 147–154. [CrossRef]

9. Zhang, D.; Sun, W.; Fan, Z. Application of a newly built semi-submersible vessel for transportation of a tension leg platform. *J. Mar. Sci. Appl.* **2012**, *11*, 341–350. [CrossRef]

10. Eriksson, H.; Kullander, T. *Survey of Swedish Suppliers to a Floating Wind Turbine Unit Equipped with Pumps for Oxygenation of the Deepwater*; BOX-WIN Technical Report no. 7, C101; University of Gothenburg: Goteborg, Sweden, 2013; ISSN 1400-383X.

11. Jurewicz, J.M. Design and Construction of an Offshore Floating Nuclear Power Plant. Master's Thesis, Massachusetts Institute of Technology, Cambridge, MA, USA, 2015.

12. Kang, W.H.; Zhang, C.; Yu, J.X. Stochastic extreme motion analysis of jack-up responses during wet towing. *Ocean Eng.* **2016**, *111*, 56–66. [CrossRef]

13. Li, C.; Liu, Y. Fully-nonlinear simulation of the hydrodynamics of a floating body in surface waves by a high-order boundary element method. In Proceedings of the ASME 2015 34th International Conference on Ocean, Offshore and Arctic Engineering, American Society of Mechanical Engineers, St. John's, NL, Canada, 31 May–5 June 2015.

14. Zhao, Z.; Tang, Y.; Wu, Z.; Li, Y.; Wang, Z. Towing Analysis of Multi-Cylinder Platform for Offshore Marginal Oil Field Development. In Proceedings of the ASME 2016 35th International Conference on Ocean, Offshore and Arctic Engineering, American Society of Mechanical Engineers, Busan, Korea, 30 June 2016.

15. Ding, H.; Han, Y.; Le, C.; Zhang, P. Dynamic analysis of a floating wind turbine in wet tows based on multi-body dynamics. *J. Renew. Sustain. Energy* **2017**, *9*, 033301. [CrossRef]

16. Kim, B.; Kim, T.W. Scheduling and cost estimation simulation for transport and installation of floating hybrid generator platform. *Renew. Energy* **2017**, *111*, 131–146. [CrossRef]

17. Wang, Y.; Duan, M.; Liu, H.; Tian, R.; Peng, C. Advances in deepwater structure installation technologies. *Underw. Technol.* **2017**, *34*, 83–91. [CrossRef]

18. Zelazny, K. A method of calculation of ship resistance on calm water useful at preliminary stages of ship design. *Zesz. Nauk./Akad. Morska Szczec.* **2014**, *38*, 125–130.

19. Putra, I.N.; Susanto, A.D.; Suharyo, O.S. Comparative analysis results of towing tank and numerical calculations with harvald guldammer method. *Int. J. Appl. Eng. Res.* **2017**, *12*, 10637–10645.

20. Veritas, D.N. Modelling and Analysis of Marine Operations. In *Offshore Standard*; DET NORSKE VERITAS (DNV): Akershus, Norway, 2014.

21. *Guidelines for Towage at Sea*; China Classification Society: Beijing, China, 2012.

22. MOSES. *Reference Manual for Moses*; Ultramarine Inc.: Houston, TX, USA, 2009.

23. Faltinsen, O.M. *Sea Loads on Ships and Offshore Structures*; Cambridge University Press: New York, NY, USA, 1993.

24. Newman, J.N. Wave effects on multiple bodies. *Hydrodyn. Ship Ocean Eng.* **2001**, *3*, 3–26.

25. Berg, P.W.S. A Discussion of Technical Challenges and Operational Limits for Towing Operations. Master's Thesis, Norwegian University of Science and Technology, Trondheim, Norway, 2017.

Journal of
*Marine Science
and Engineering*

Article

Failure Analysis of Topside Facilities on Oil/Gas Platforms in the Bohai Sea

Songsong Yu [1], Dayong Zhang [2,*] and Qianjin Yue [2]

[1] Faculty of Vehicle Engineering & Mechanics, Dalian University of Technology, Dalian 116024, China; 931679960@mail.dlut.edu.cn

[2] School of Ocean Science & Technology, Dalian University of Technology, Panjin 124221, China; yueqj@dlut.edu.cn

* Correspondence: zhangdayong_2001@163.com

Received: 22 February 2019; Accepted: 25 March 2019; Published: 27 March 2019

Abstract: The jacket platform in the Bohai Sea oilfield is an important engineering development, offering design alternatives in this economically important region. However, ice-induced vibration in cold areas threatens the safety and operation of these platforms. On two occasions, intense ice-induced vibrations triggered the rupture of a well's blow down pipeline, loosening the flanges on the Bohai platform, and leading to the ejection of high-pressure natural gas. Subsequent mechanical analysis of the failed parts helped define the mechanism of failure and identified the failure criteria, based on a prototype structure monitoring system. The analysis revealed that the deck's inertial force, which resulted in ice-induced steady-state vibration, was the major cause of the accident. Three fixed cones were thus installed on the platform the following winter, effectively reducing the vibrations.

Keywords: jacket platform; ice-induced vibration; failure mechanism; ice-resistant cone

1. Introduction

The shallow waters of the Bohai Sea are home to an abundance of oil reserves. More than 100 offshore platforms have been built on this gulf, making it the second-largest offshore oil production base in China. However, the Bohai Sea is a seasonal icing sea region and the ice is frequently driven by tides. Natural conditions, such as water depth, temperature, and presence of waves cause severe ice conditions in winter. Owing to the specific characteristics of marginal oilfields, the Bohai Sea oil and gas platform is an economically important jacket structure, to which the threat from sea ice is far greater than environmental hazards (wind, waves, currents, tides, earthquakes, etc.). Additionally, sea ice in cold regions poses higher risks to offshore oil platforms [1–3].

This paper discusses two serious accidents that took place on a Bohai platform, caused by intense ice-induced vibration. During the events, the unmanned wellhead platform, which was designed as a three-leg vertical jacket structure (Figure 1), was exposed to steady-state vibration that lasted for more than 10 minutes, resulting in the rupture of a well's blow down pipeline (Figure 2a). High-pressure natural gas was subsequently ejected, leading to automatic shutdown of the platform. An inspection of the site revealed a loose flange, which was the cause of the gas leak (Figure 2b). It was also made known that the currents occurred during slack tide, when ice speeds were slow. The thickness of the ice was found to be 8 cm and 11 cm, respectively, when the accidents occurred.

Figure 1. The natural gas platform where the accident took place.

(a)

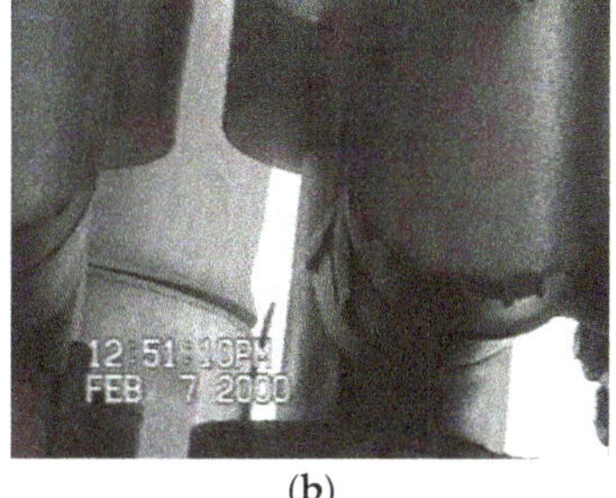

(b)

Figure 2. Sea-ice damage to platforms in Bohai: (**a**) Rupture of a blow down pipeline. (**b**) The loose flange.

This paper addresses the risk of oil and gas exploration in the Bohai Sea during winter. Based on data obtained by field monitoring systems, case studies were conducted on gas leakage accidents to illustrate the ice-induced failure of the oil and gas platform, and clarify the criteria of failure. It provides a theoretical basis to study the safety and security of oil and gas development in cold regions of Bohai Sea.

2. Failure Mechanism Analysis of the Upper Substructure of the Platform

2.1. Failure Analysis of Pipeline Fracture

Natural gas pipelines are an important component of offshore platforms. During a period of intense ice formation, the blow-down pipeline in the upper platform shook violently and was fractured. The ice-induced vibration resulted in the leakage of natural gas. In the vibration curve, obtained by field monitoring (Figure 3), it can be seen that there was a strong steady-state vibration of the platform when the accident occurred.

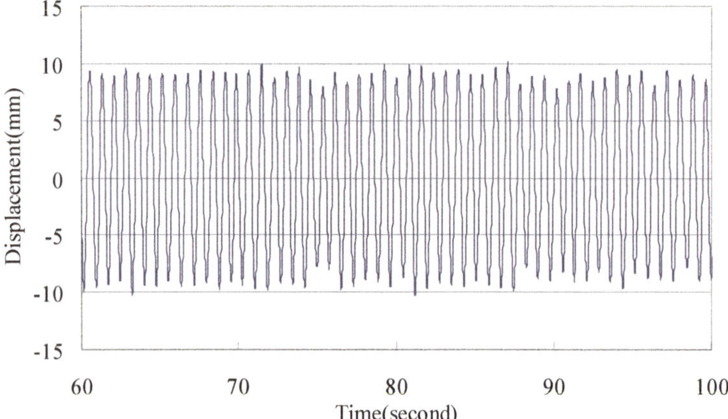

Figure 3. Displacement curve of the top deck when the pipeline fracture occurred.

Figure 4 shows a part of the blowdown pipeline after the accident. Mechanical analysis revealed that the blowdown pipe belongs to the cantilever structure and the fixed part is connected to the deck. The horizontal ice-induced shaking that occurred on the fixed part of the pipe was found to be equivalent to the bending vibration caused by a seismic load. This eventually caused a whiplash effect.

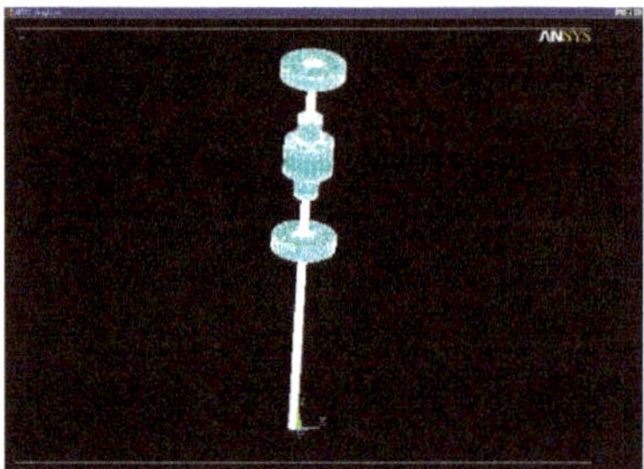

Figure 4. Partial structure of the blowdown pipe.

A fracture analysis of the affected section of the blowdown pipe (Figure 5) identified the following characteristics:

1. The fractured surface, which was a typical fatigue fracture, had an obvious fatigue source zone, a fatigue crack propagation zone, and a fatigue fracture zone.
2. The fracture was perpendicular to the axis, which meant that the fatigue fracture was caused by a bending load.
3. The fracture surface was uneven, with no obvious fatigue curve, and the distance between fatigue striations was large, illustrating low-cycle fatigue fracturing under high stress.

Study of the fixed part of the blowdown pipe indicates that the fatigue fracture load was a bending load. The following conclusions were made by analyzing the macro- and micro characteristics of the fracture: The fracture of the blowdown pipe had obvious fatigue features, and it exemplified a low-cycle fatigue fracture under a bending load. When the damage caused by alternating external loads reached a certain degree, it resulted in fatigue failure. The severe ice-induced vibration accelerated the fatigue failure of the blowdown pipe and eventually caused the ejection of high-pressure natural gas.

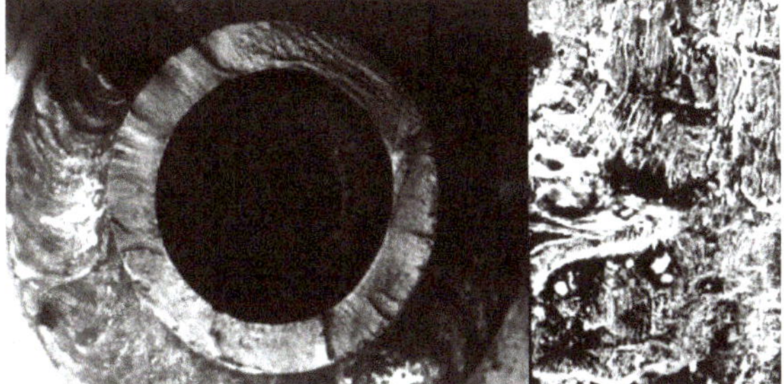

Figure 5. The fractured edge of the blowdown pipe.

2.2. Failure Analysis of Flange Looseness

Figure 6 shows the severe ice-induced vibration, which resulted in loosening of the flange of the platform's upper deck valve, and subsequent leakage of natural gas. An inspection of the platform after the accident revealed that the graphite spiral gasket flanges had 47% loose bolts, and the remaining had pretightening force of at least 50 N-m, after the system was fully depressurized. It was revealed that the reduction in flange pretightening force caused by bolts loose was the main reason for the flange failure.

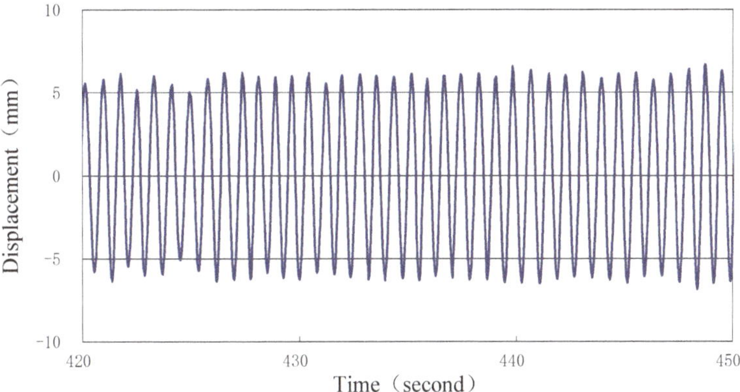

Figure 6. Displacement curve of the top deck when the accident occurred.

A flange is an important coupling part of industrial pipes, generally consisting of washers, flange bolts, and nuts. The flange's connection depends on the pretightening force between the bolt and the nut, transmitting pressure between the gasket and the flange; the washer undergoes elastic-plastic deformation to achieve a seal. It is liable to cause deteriorating connection and loose of the flange in the operating conditions where impact and vibration occur frequently. Several long-term studies have been conducted on the failure mechanism of vibration-induced bolt loosening. Goodier and Sweeney [4] conducted an experimental study on the loosening mechanism of bolted structures under an axial vibration load and found that self-loosening is caused by the relative movement of the thread and other contact surfaces. However, when a radial slip between the screw-thread occurs, the circumferential friction is zero. Because of the movement caused by normal contact stress, the screw–thread pair eventually slips in the circumferential direction. Gerhard Junker [5] studied the loosening mechanism of bolted structures under lateral loads and confirmed that transverse vibration was the main reason for it. Hess and Pai [6] from the University of South Florida used finite element analysis software to analyze contact state. They revealed that contact state can be divided into two types—local slip and complete slip—and cumulative local slips need much lower lateral loads. Jiang et al. [7,8] from the

University of Nevada studied the loosening of bolted structures under lateral vibration. In addition, research has shown that there is a stress limit for the threaded connection structure under lateral vibration. If this limit is adhered to, there will be no significant loosening on pressure from cyclic loads. If the bolt's pretightening force is increased, the loosening stress limit rises as well; however, this also increases the chances of fatigue fracturing of the bolt. Nassar and Housari [9] studied the bolt-loosening effect under various lateral vibration conditions. Studies have shown that the loosening torque and frictional force of the structure are affected by lateral vibration amplitude.

The inertial forces generated by the ice-induced vibration cause horizontal tremors in the topside facility, resulting in reduction of the pretightening force in the flange. When there is not much change in the static load and working temperature, the friction between the female and male screws keep the connection tight and satisfy the self-locking condition. The friction force between the spiral pairs instantaneously decreases or disappears when the platform experiences horizontal vibration. The threads on the bolt push into the thread surface of the nut due to the inclined surface, forcing the nut to rotate and release. Subsequently, when the pretightening force reaches a critical point, natural gas is free to escape. In order to further verify the looseness mechanism and process of the upper platform flange under ice-induced vibration, the study carried out a laboratory test of the system, based on results of the inspection performed after the accident.

It is difficult to simulate the pipeline flange situation in a laboratory, as the upper pipeline of the platform is complex and contains a long pipe. Considering that the vibrational load on the flange is a lateral one, the test intercepted a section of the flange and used the "four-point bending" model to simulate the situation. By using an actuator to control the amplitude and frequency, the bending stress at the flange and vibrational load on the bolt were revealed: both showed similar measurements to the actual situation, validating the rationality of the test.

Figure 7 shows the arrangement of the laboratory test: It comprised racks, a test pipeline flange, loading system actuator, and data acquisition system. The racks and flange form a beam structure that has concentrated forces at both fixed ends; the middle flange is stressed by bending. The amplitude and frequency of lateral alternating loads are controlled by actuators. Brackets and clamps are symmetrically applied to the pipeline on both sides of the flange, and the ends of the pipeline are fixed by a bolt and "v block" to the I-beam. This experiment simulated the accident's platform pipeline system under vibration using Class600 (11.0 MPa), 2-inch (DN50) nominal diameter, standard convexity welding neck flange, and 8 M16, high-strength bolts made of CrMoA. The metal–graphite, spiral-wound gasket consisted of a locating ring and an inner ring, and was made of 0Cr13.

Figure 7. Arrangement of the laboratory test, which simulated the characteristics of the actual accident.

Based on the results of long-term monitoring and the statistics of pipe flanges used on the platform, the most common vibration acceleration levels under ice-induced vibration were found to be 1.0 m/s and 2.0 m/s, corresponding to two different ice conditions, respectively. Through vibration amplitude and frequency control, the bending stress and horizontal vibration acceleration simulation of the flange under ice-induced vibration was realized.

Numerical analysis has shown that experimental amplitude should be 4 mm for the experimental flange to be under real bending stress. In addition, when vibration frequency is set as 2 Hz, the vibration acceleration of the experimental pipeline is 1.0 m/s; when vibration frequency increases to 6 Hz, the vibration acceleration of the experimental pipeline is 2.0 m/s. The bolts are preloaded at 15%, 30%, and 45% of yield strength (640 MPa). Table 1 shows four groups of experimental vibrational parameters.

Table 1. Four groups of vibrational parameters.

Initial Pretightening Force	Vibration Frequency	Amplitude	Acceleration	Time	Structure States	Final Pretightening Force
18 KN	2 Hz	4 mm	1.0 m/s^2	7 h	No	17.5 KN
18 KN	6 Hz	4 mm	2.0 m/s^2	8 h	Yes	190 N
36 KN	6 Hz	4 mm	2.0 m/s^2	7 h	No	35.4 KN
49 KN	6 Hz	4 mm	2.0 m/s^2	17 h	No	47.6 KN

The first set of pretightening force is kept at 15% of the yield strength (18 KN), wherein the lower pretightening force can ensure that the graphite gasket is not crushed. Experimental data has found that when the excitation vibration acceleration is small (2 Hz), the pretightening force of the bolt hardly changes. In contrast, when the excitation frequency and acceleration are large, the bolt may loosen. When it occurs suddenly, the pretightening force is directly reduced from the initial value to almost zero (Figure 8).

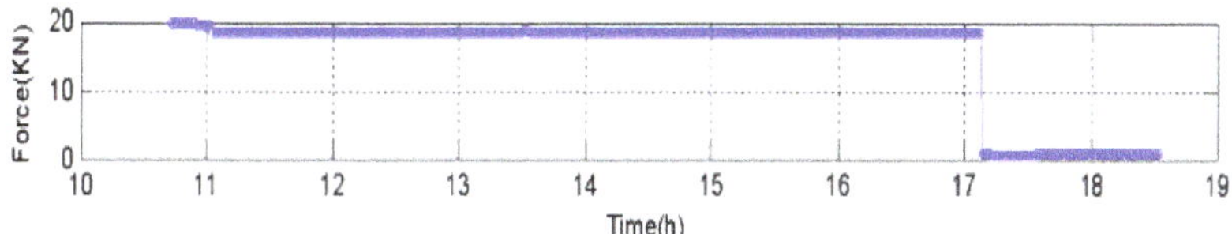

Figure 8. The pretightening force of the bolt when the initial pretightening force is 15% of the yield strength and excitation frequency is 6 Hz.

Figures 9 and 10 show the increase of the bolt's pretightening force to 30% and 45% of the yield strength, respectively. The analysis reveals that this force has a small decrease, even at a large vibration frequency. It is far from being loose, although the downward trend is obvious.

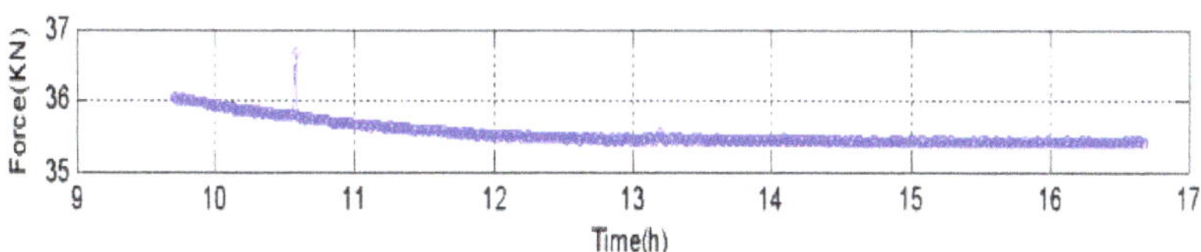

Figure 9. The pretightening force of the bolt when the initial force is 30% of the yield strength and the excitation frequency is 6 Hz.

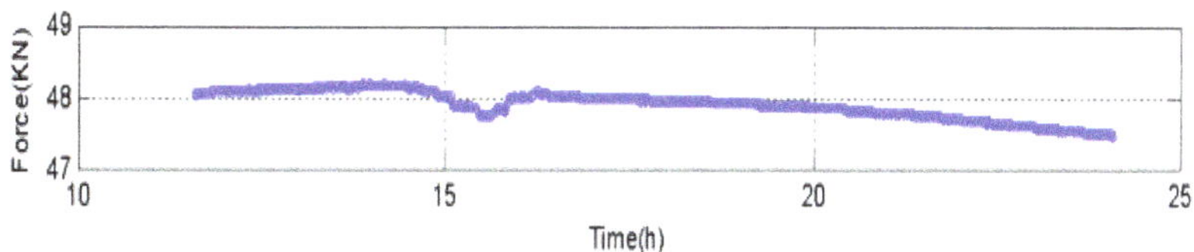

Figure 10. The pretightening force of the bolt when the initial force is 45% of the yield strength and the excitation frequency is 6 Hz.

Further tests on the flange in a vibrating environment show that a smaller initial pretightening force of the bolt can ensure a great sealing performance of the gaskets; however, the flange bolts will loosen under strong ice-induced vibration. Larger initial pretightening forces can delay or even prevent loosening, but excessive initial pretightening force may cause the gasket to yield prematurely, affecting its sealing performance.

3. Ice-Induced Steady-State Vibration Mechanism Analysis

Field monitoring has found that the periodic load generated by sea ice can evoke large acceleration responses in the platform. We performed spectral analysis of the above-mentioned offshore platform (Figure 11) using monitored response data. The results show that the vibration of the platform was focused on the first mode.

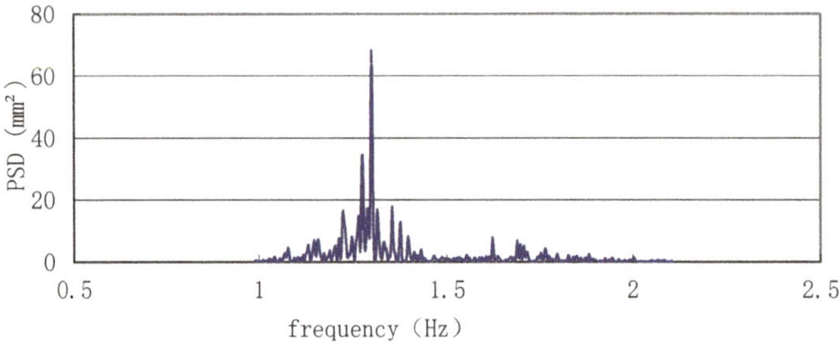

Figure 11. The studied offshore platform's main deck X-directional displacement spectrum.

The platform's vibration curve (Figures 3 and 6) shows that the tremor remained stable for a long time before the fracture of the blowdown pipe and gas leakage from the flange. The amplitude was consistently large and constant, about 30 mm, which means that a relatively strong steady-state vibration occurred in the early stage of the event. The frequency stabilized at 1.25 Hz, which is close to the platform's fundamental frequency, and the structure's vibration has obvious dynamic magnification.

Ice-induced vibrations of offshore structures pose a security challenge, and have been studied for decades in the regions of Cook Inlet, Beaufort Sea, and Baltic Sea. The steady-state vibration of the platform mentioned above is similar to the resonance in forced-vibration theory. The excitation frequency is close to the natural frequency of the structure, inducing dynamic magnification of the structural vibration. However, an analysis revealed that the change in frequency of the dynamic ice force depends upon the ice-breaking patterns and ice velocity. If such accidents are analyzed according to the forced-vibration theory, the frequency of dynamic ice forces will depend only on the ice. Thus, the steady-state alternating power is generated by sea ice, that is, it produces a breaking frequency. In fact, the mechanical property of natural sea ice has very large dispersion, and it is impossible to produce long-lasting stable sizes of crushing ice.

Blenkarn [10], Määttänen [11–13], and Engelbrektson [14,15] state that ice-induced steady-state vibration has typical self-excited characteristics. The steady-state vibrations of said platform occurred when the currents were in slack tide. Sea ice moves at a slow speed, taking into consideration the time currents take to change directions. Thus, the measured vibration curve analysis concludes that the structure and sea ice have similar speeds in a steady-state vibration period. The same direction and reverse movement of the structure and sea ice form a cyclic loading on the ice sheet and control its breaking period.

As a material, ice is very sensitive to the loading rate. Figure 12 shows that the uniaxial compressive strength of ice varies significantly with the strain rate, and the mechanical behavior of ice changes greatly when subjected to compressive loads. According to the different macroscopic mechanical behaviors of ice, the strain rate is divided into ductile region, ductile–brittle transitional region, and brittle region. Yue et al. [16] pointed out that different loading rates of sea ice, caused by the relative velocities of the ice and the structure, led to different vibrational modes of the jacket platform. The mechanism of ice-induced self-excited vibration lies in the ductile–brittle transition region of the ice sheet.

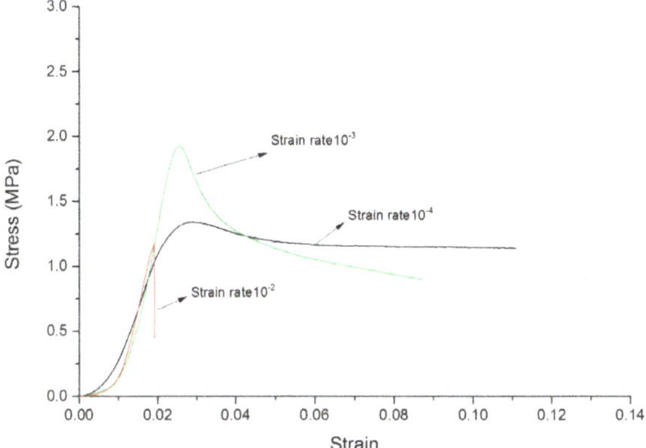

Figure 12. Sea ice load curves under different strain rates.

As shown in Figure 13, the cycle of self-excited vibration is divided into the loading phase, transition phase, and unloading phase. In the loading phase, the platform structure moves in the same direction as the sea ice at a similar speed. The loading speed of the ice sheet is slow, in the range of the ductile–brittle transition region. With loading of the ice force, new microcracks continue to be generated inside the ice, although they do not expand rapidly. As the loading rate continues to increase, the force is large enough to allow airfoil cracks to interpenetrate, and the ice sheet undergoes brittle failure. In the unloading stage, the direction of the structure's motion is opposite to that of the ice. The structure smashes the ice sheet, which was already damaged during the loading phase. It breaks down simultaneously, and the ice's force starts to unload. The structure swings back to the starting point of the loading phase, marking the end of a vibration cycle.

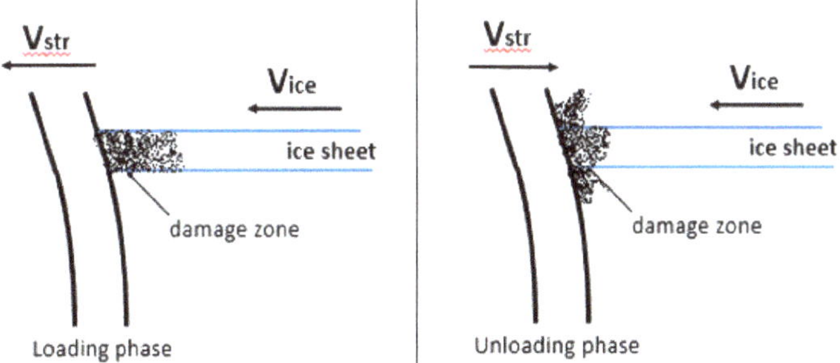

Figure 13. Ice-induced self-excited vibration physical mechanism.

The study revealed that since the loading rate of the ice sheet is in a ductile–brittle transition interval, ductile failure occurs on interaction between the ice sheet and the structure, and a large number of dislocation movements occur in the ice sheet [17]. With a relatively large compression deformation, contact between the ice sheet and the structure becomes more regular (Figure 14). The ice sheet ductile-to-brittle transition point marks the maximum uniaxial compressive strength corresponding to the loading rate (Figure 15) and the simultaneous breaking of sea ice, resulting from regular destruction, also aggravates the ice's force on the structure. That is the reason why the breaking frequency is stable and the amplitude is extremely large during the steady-state vibration of the platform.

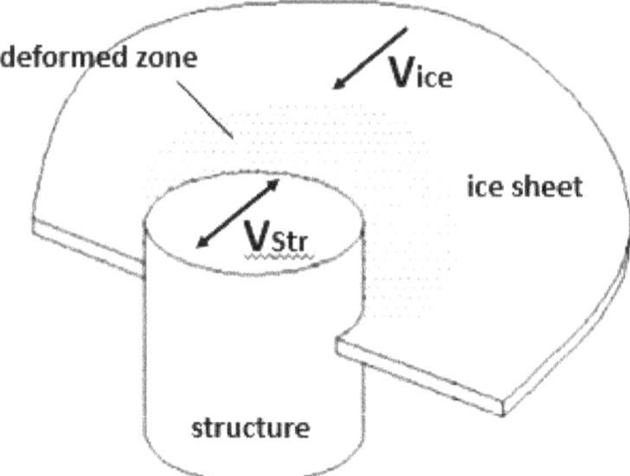

Figure 14. Outline of the ice sheet crushing the cylindrical structure.

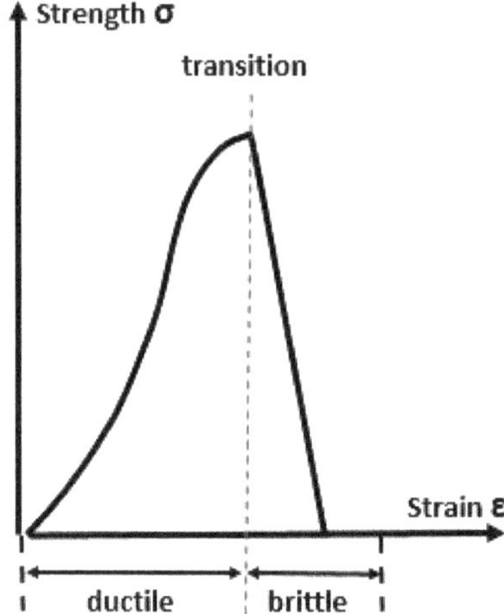

Figure 15. Loading curve of the ice sheet during strain rate increase.

The broken ice was able to easily bypass the platform's legs and avoid being stacked up due to the small diameter of the legs and the large amplitude and speed of the backswing, caused by the low rigidity of the platform. This results in synchronous crushing (loading and unloading processes) of the ice, and the ice's force also obtains a significant periodic variation in synchronization.

Ice-induced steady-state vibration on platforms is a major threat to safe oil and gas operations at Bohai Gulf. At a specific ice speed, ice extrusion produces a steady-state alternating excitation, which results in dynamic amplification of surface vibration. This further leads to fatigue and functional failure of the weak parts of the superstructure.

4. Evaluation of Ice-Resistant Modification

The conditions triggering the ice-induced self-excited vibrations are harsh. Although the frequency of this vibration is not high, it can generate a constant vibrational amplitude. Steady-state vibration is the main cause of ice-induced damage of marine structures, posing serious hazards.

The analysis discussed above resulted in an ice-resistant platform being created in the winter following the accident (Figure 16). By installing an ice-resistant cone, the ice sheet damage model changed from crushing destruction to bending destruction, thereby eliminating ice-induced self-excited vibration. In addition, continuous field-monitoring data was recorded, and the effect of the ice-resistant modification was evaluated.

Figure 16. The ice-resistant modification of the platform.

Ice-induced failures in offshore platforms are always accompanied by steady-state vibrations. In order to highlight the variation in response characteristics of the platform structure after the modification, the 10-minute maximum amplitude data of the two years (before and after the accident) were compared. The average amplitude is presented in Table 2.

Table 2. Average maximum amplitudes (1999–2000).

Vertical Structure		Conical Structure	
Sample Starting Time	**Average Maximum (mm)**	**Sample Starting Time**	**Average Maximum (mm)**
Jan 28—22:57	8.38	Jan 15—01:28	6.05
Jan 28—23:07	5.64	Jan 11—09:20	2.07
Jan 29—00:27	12.2	Jan 14—09:58	4.18
Jan 29—00:37	13.9	Jan 16—16:58	1.42
Jan 17—07:51	1.88	Jan 18—15:45	4.58
Jan 24—04:42	4.32	Jan 19—04:15	2.27
Jan 27—08:48	4.78	Jan 21—08:05	4.27
Mean value	7.3	Mean value	3.548

Liaodong Bay, which is one of the three bays in the Bohai Gulf, faced severe ice conditions in the winter following the ice-resistant modification. However, as a result of the installation of the cone, average peak values of the vibration significantly decreased, and a damping effect was obvious. With regard to the offshore platform's structural design, the modification caused a significant reduction in the average value of the peak vibration, thus vouching for its ice-resistant capability and fatigue life increase.

5. Conclusions

This paper addresses the ice-induced vibration risk of oil and gas exploration in the Bohai Sea through a case study of the pipeline and flange failure that occurred in Liaodong Bay. Based on the failure analysis of topside facilities on oil/gas platforms in the Bohai Sea, this paper recommends that the following criteria be considered for future design and risk assessment studies:

- Although the frequency of ice-induced, self-excited vibrations is not high, it is the main cause of ice-induced damage of marine structures, and must be paid more attention.
- Due to small leg diameter and low rigidity, the oil and gas platforms in Bohai are more likely to cause synchronous ice-crushing and ice-induced steady state vibration in a range of ice speeds.

- The inertial forces generated by the ice-induced steady-state vibration may lead to fatigue and functional failure of weak topside facilities.
- The addition of ice-resistant cones results in a significant decrease of the average peak values of the vibration on the platform.

This research provides the theoretical basis for structural design and safety of offshore structures in cold regions. The authors recommend that further research, such as fatigue analyses, completing the evaluation of ice-induced vibration failure, and risk warning and forecasting be conducted on related problems.

Author Contributions: Project administration: Q.Y.; investigation and original draft writing: S.Y.; review and editing: D.Z.

Funding: This study was supported by the National Natural Science Foundation of China (Grant No. 51679033), National Key Research and Development Plan (Grant Nos. 2017YFF0210700, 2016YFC0303400), and Fundamental Research Funds for Dalian University of Technology (DUT18LK49).

Acknowledgments: The authors are pleased to acknowledge the support of this work by This study was supported by the National Natural Science Foundation of China (Grant No. 51679033), National Key Research and Development Plan (Grant Nos. 2017YFF0210700, 2016YFC0303400), and Fundamental Research Funds for Dalian University of Technology (DUT18LK49).

Conflicts of Interest: The authors declare no conflict of interest.

References

1. Zhang, D.Y.; Yue, Q.J. Major challenges of offshore platforms design for shallow water oil and gas field in moderate ice conditions. *Ocean Eng.* **2011**, *38*, 1220–1224. [CrossRef]
2. Zhang, D.Y.; Xu, N.; Yue, Q.J.; Liu, D. Sea Ice Problems in Bohai Bay Oil and Gas Exploitation. *J. Coast. Res.* **2015**, *73*, 676–680. [CrossRef]
3. Wang, S.Y.; Yue, Q.J.; Wang, S.Y.; Yue, Q.J. Vibration Reduction of Bucket Foundation Platform with Fixed Ice-Breaking Cone in the Bohai Sea. *J. Cold Reg. Eng.* **2012**, *26*, 160–168. [CrossRef]
4. Goodier, J.N. Loosening by Vibration of Threaded Fastening. *Mech. Eng.* **1945**, *67*, 798–802.
5. Junker, G.H. Criteria for self loosening of fasteners under vibration. *Aircr. Eng. Aerosp. Technol.* **1972**, *44*, 14–16. [CrossRef]
6. Pai, N.G.; Hess, D.P. Three-dimensional finite element analysis of threaded fastener loosening due to dynamic shear load. *Eng. Fail. Anal.* **2002**, *9*, 383–402. [CrossRef]
7. Jiang, Y.Y.; Zhang, M.; Lee, C.H. A Study of Early Stage Self-Loosening of Bolted Joints. *J. Mech. Des.* **2003**, *125*, 518–526. [CrossRef]
8. Jiang, Y.Y.; Zhang, M.; Park, T.W.; Lee, C.H. An Experimental Study of Self-Loosening of Bolted Joints. *J. Mech. Des.* **2004**, *126*, 925–931. [CrossRef]
9. Nassar, S.A.; Housari, B.A. Study of the Effect of Hole Clearance and Thread Fit on the Self-Loosening of Threaded Fasteners. *J. Mech. Des.* **2006**, *129*, 1053–1062.
10. Blenkarn, K. Measurement and Analysis of Ice Forces on Cook Inlet structures. In Proceedings of the Conference on American Society of Civil Engineers, Houston, TX, USA, 22–24 April 1970; pp. 21–34.
11. Määttänen, M. Stability of self-excited ice-induced structural vibration. In Proceedings of the Fourth International Conference on Port and Ocean Engineering Under Arctic Conditions, Memorial University of Newfoundland, St. John's, NL, Canada, 26–30 September 1977; pp. 684–694.
12. Määttänen, M. The effect of structural properties on ice-induced self-excited vibrations. In Proceedings of the IAHR Symposium on Ice, Hamburg, Germany, 27–31 August 1984; Volume 2, pp. 11–20.
13. Määttänen, M. Ice-induced Vibrations in Structures—Self-Excitation. In Proceedings of the IAHR Symposium on Ice, Sapporo, Japan, 23–27 August 1988; Volume 2, pp. 658–665.
14. Engelbrektson, A. Dynamic ice loads on lighthouse structures. In Proceedings of the Fourth International Conference on Port and Ocean Engineering Under Arctic Conditions, Memorial University of Newfoundland, St. John's, NL, Canada, 26–30 September 1977; Volume 2, pp. 654–864.
15. Engelbrektson, A. An ice-structure interaction model based on observations in the gulf of bothnia. *POAC* **1989**, *89*, 504–517.

16. Yue, Q.J.; Guo, F.W.; Kärnä, T. Dynamic ice forces of slender vertical structures due to ice crushing. *Cold Reg. Sci. Technol.* **2009**, *56*, 77–83. [CrossRef]
17. Guo, F.W.; Yue, Q.J. Physical mechanism and process of ice induced steady state vibration. In Proceedings of the 20th International Conference on Port and Offshore Engineering under Arctic Conditions, Luleå, Sweden, 9–12 June 2009; Volume 34.

Journal of
*Marine Science
and Engineering*

Article

Cross-Flow Vortex-Induced Vibration (VIV) Responses and Hydrodynamic Forces of a Long Flexible and Low Mass Ratio Pipe

Xifeng Gao [1,2], Zengwei Xu [1], Wanhai Xu [1,2] and Ming He [1,*]

[1] State Key Laboratory of Hydraulic Engineering Simulation and Safety, Tianjin University, Tianjin 300350, China; gaoxifeng@tju.edu.cn (X.G.); xuzengwei@tju.edu.cn (Z.X.); xuwanhai@tju.edu.cn (W.X.)

[2] Collaborative Innovation Centre for Advanced Ship and Deep-Sea Exploration, Shanghai 200240, China

* Correspondence: minghe@tju.edu.cn; Tel.: +86-158-225-84004

Received: 10 May 2019; Accepted: 2 June 2019; Published: 5 June 2019

Abstract: Laboratory tests were carried out to investigate the cross-flow (CF) dynamic responses and hydrodynamic forces of a flexible pipe that subjected to vortex-induced vibration (VIV). The pipe had a critical mass ratio of 0.54 and an aspect ratio of 181.8. The uniform flow environment was realized by towing the pipe along a towing tank. The towing velocity ranged from 0.1–1.0 m/s with an interval of 0.05 m/s. Two axial pre-tension cases (200 N and 300 N) were enforced. The structural strains were measured at seven positions evenly distributed along the pipe. Then a modal analysis method was applied to reconstruct the displacement responses. It is revealed that the maximum CF displacement amplitude reached up to 2.18 pipe diameter and the strain response exhibited higher harmonic components. The CF dominant frequency gradually rises with the increase of reduced velocity and up to a three-order vibration mode can be observed. In addition, mean drag coefficient, lift force coefficient and added mass coefficient were also calculated to further investigate the fluid force feature of a low mass flexible pipe undergoing VIV.

Keywords: vortex-induced vibration; flexible pipe; critical mass ratio; cross-flow

---

1. Introduction

Vortex-induced vibration (VIV) is a typical fluid-structure interaction behavior, which has significant effects on slender structures, such as marine risers [1], tall buildings [2], cables of bridges [3], and receiver tubes of concentrated solar power plants [4]. In recent years, a great deal of effort has been made to reveal the mechanism of VIV and some milestone findings have been reported in several review works of Sarpkaya [5], Gabbai and Benaroya [6], Williamson and Govardhan [7], Wu, et al. [8], and Rashidi, et al. [9].

One key parameter, the mass ratio m^*, which characterizes the structural mass m_s relative to the mass of displaced fluid $\rho\pi D^2/4$ (where ρ is the fluid density and D is the cylinder diameter), has a significant influence on the VIV characteristic of a circular cylinder. It is well known that cylinders with low mass ratios have much broader lock-in ranges than those with high mass ratios. Furthermore, high mass ratio cylinders are less influenced by the variation of the added mass coefficient, because the added mass is a lower percentage of the total mass per unit length [7,10].

The vibration frequency and amplitude of an elastically mounted rigid circular cylinder with high mass ratio in air undergoing cross-flow (CF) VIV have been well characterized by Feng [11]. It was found that there were two types of amplitude response depending upon the mass-damping parameter Cn (where $Cn = 2m \times \zeta$, being ζ the structural damping ratio), namely the initial and lower branches. However, the VIV response at low mass ratio and low mass-damping are different. Comparisons of

VIV responses in water and air were made by Khalak and Williamson [12]. It was observed that a low mass ratio yielded a much higher peak amplitude. Moreover, the VIV response not only contains the initial and lower branches, but also includes a much higher "upper response branch" between them. Govardhan and Williamson [13,14] studied the CF VIV of an elastically-mounted rigid cylinder at low mass-damping conditions. Large-amplitude vibrations were observed once the mass ratio was less than a critical value of 0.542. Meanwhile, the extension of large-amplitude response regarding flow velocity tends to be infinite.

Low mass ratio and high aspect ratio L/D (where L is the structural length) flexible cylinders (e.g., risers, tendons and marine cables) have been widely used in ocean and offshore engineering; consequently, much research attention has been made. Willden and Graham [15] numerically studied the CF VIV of a long (L/D = 1544), flexible pipe in uniform flow field. Its mass ratio varied from 1.0–3.0 and its effect on VIV behaviors was examined. It was shown that the long flexible pipe vibrated in a multi-modal form and the excited modes responded at the Strouhal frequency. Trim, et al. [16] conducted experiments to study the VIV response of a marine riser in uniform and linear shear currents. The riser model had a mass ratio of 1.60 and an aspect ratio of 1405. It was found that in-line (IL) fatigue damage of the marine riser was as severe as CF fatigue damage. Chaplin, et al. [17] measured the VIV response of a vertical tension riser in stepped flows by vertically towing a pipe model of length 13.12 m, diameter 2.8 cm (i.e., L/D = 468.5), and mass ratio 3.0 along a towing tank. It was observed that CF response could be excited to the 8th mode and the standard deviation of vibration amplitude ratio was close to 0.53. With the help of digital particle image velocimetry (DPIV) technique, Xu, et al. [18] investigated the velocity and vorticity fields in the vicinity of a flexible riser with a mass ratio of 1.35 and an aspect ratio of 181. The Reynolds numbers covered the range of 9400–47,000. Three vortex modes '2P', '2S', and 'P+S' were identified in the near wake of the riser. Huera-Huarte and Bearman [19,20] performed an experimental investigation on a single vertical riser of length 1.5 m, external diameter 16.0 mm (i.e., L/D = 93.75), and mass ratio 1.8 in a stepped flow. It was reported that the maximum dimensionless CF vibration amplitude and mean drag coefficient were approximately 0.7 and 3.0, respectively. Moreover, accompanied by the increasing reduced velocity, the drag effect was significantly amplified with increasing bi-directional response in the lock-in region. Song, et al. [21] experimentally studied the VIV response of a long riser model with a mass ratio of 1.0 and an aspect ratio of 1750. CF and IL amplitudes were measured to be almost $2.8D$ and $1.3D$, respectively. Huera-Huarte, et al. [22] found that, for a flexible cylinder with mass ratios of 1.1 and 2.7, the maximum VIV displacement amplitude was larger than $3.0D$. More recently, our term carried out towing tank experiments to study the streamwise VIV of a flexible cylinder with a mass ratio of 1.39 and an aspect ratio of 195.5 [23].

Several conclusions can already be drawn from the above literature: (a) the amplitude response significantly increases with the decrease of mass ratio and mass-damping parameter [7,10,13]; (b) the VIV characteristic of a low mass ratio cylinder is more complicated than that of a high mass ratio cylinder due to the influence of the added mass [11,12]; (c) the vibration feature of an elastically mounted rigid cylinder with critical mass ratio have been studied thoroughly [13,14]. However, up to now, few works are available on the hydrodynamics of flexible cylinders with very low mass ratios, despite the experimental works of Seyed-Aghazadeh and Modarres-Sadeghi [24] which investigated the VIV response of a flexible cylinder of mass ratio 0.47 and aspect ratio 67 by using a reconstruction algorithm. In this paper, a series of experimental tests were carried out to study the CF VIV responses of and hydrodynamic forces on a flexible pipe with a critical mass ratio 0.54 and an aspect ratio 181.8. The displacement amplitude, response frequency, fluid forces (including mean drag force, lift force, and added mass force) were studied and discussed. It is expected that the present results can improve the understanding of VIV characteristic of a long flexible and low mass ratio pipe. Thus, they are of high significance for the design of flexible slender structures in ocean and offshore engineering.

The organization of the rest of the paper is as follows. Section 2 describes the experimental setup. Section 3 introduces the reconstruction algorithm used for transferring the strain signals to displacement responses. Section 4 presents and analyzes the measurement results. Finally, some conclusions as well as future prospect are drawn in Section 5.

2. Experimental Setup

The experiment was performed in a 137-m-long, 7-m-wide, and 3.3-m-deep towing tank at the State Key Laboratory of Hydraulic Engineering Simulation and Safety, Tianjin University. The experimental device mainly consisted of four parts: a polypropylene random (PPR) pipe model, an axial tension adjustment device, a vertical supporting system, and a horizontal supporting frame. The pipe model was fixed on a moving carriage. During the tests, it was towed in still water along the water tank to implement a uniform fluid flow condition. A similar device has been used in our earlier experiments, in which the streamwise VIV [23] and the VIV reduction of a flexible cylinder fitted with helical strakes [25,26] were studied. Figure 1 shows the sketch of the experimental setup.

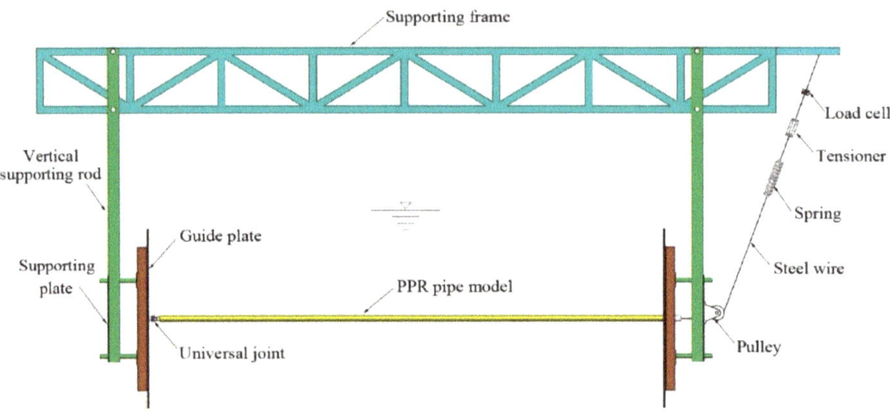

Figure 1. Sketch of the experimental setup.

The pipe was made of PPR due to its small density and strong deformation resistivity. The mass per unit length of the PPR pipe model was 0.205 kg/m, so the mass ratio is approximately equal to 0.54. The value is close to the critical mass ratio of an elastically mounted rigid circular cylinder undergoing VIV [13,14]. The pipe was 4.0 m long and had a small bending stiffness, $EI = 4.88$ Nm^2 (where E is the Young's modulus and I is the moment of inertia). Figure 2 illustrates the seven measurement positions G1–G7 that evenly distributed along the pipe. At each position, as shown in Figure 3, four resistance strain gages were pasted to the outer surface of pipe. Among them, two gages were oriented toward the x-direction (IL) while the other two gages were oriented toward the y-direction (CF). A heat shrink tube was covered on the outer surface of PPR pipe. In this way, a smooth wall boundary was implemented and the instrumentation cables and strain gages could be protected and insulated from the fluid. The final external diameter of the pipe model was 22.0 mm and the aspect ratio was equal to 181.8.

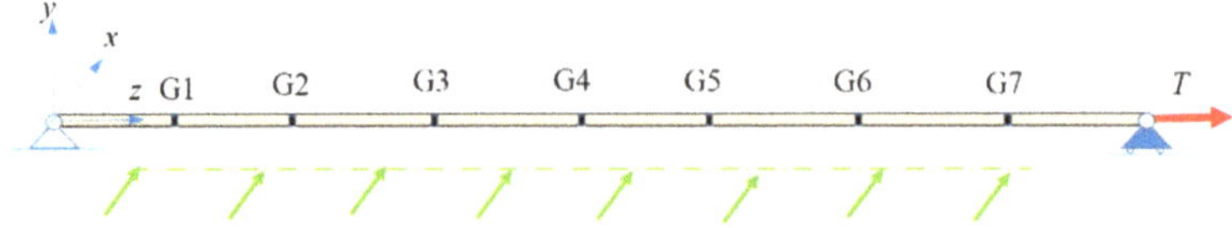

Figure 2. Strain gages arrangement along the simply supported pipe.

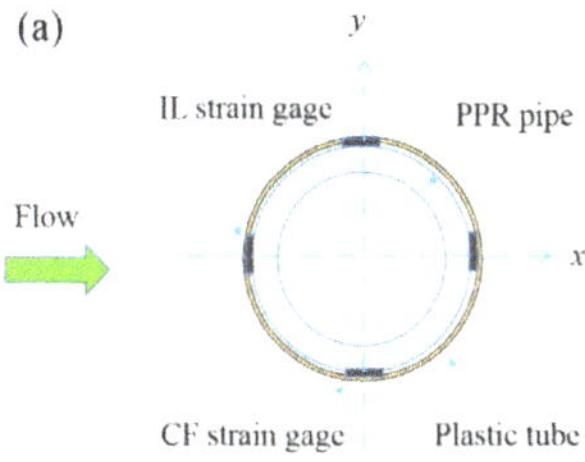

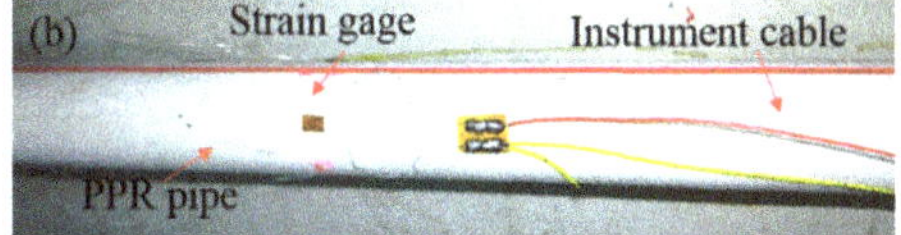

Figure 3. PPR pipe model used in the experiment; (**a**) cross-section view of the PPR pipe model; (**b**) measurement position with strain gage; and (**c**) instrument cables.

The axial pre-tension in the experiment is a parameter needs to be carefully weighed. On the one hand, there will be a considerable drag-induced IL deflection if the axial pre-tension is small. On the other hand, a large pre-tension will lead to a very stiff pipe. Thus, high mode responses of VIV can only be excited in the case of high towing velocities. After a reasonable trade off, two axial tension forces ($T = 200$ N and 300 N) were adopted in our experiment. They were exerted through adjusting the tensioner. A load cell was used to measure real-time axial tension on the vibrating pipe. Free decay tests were performed in the air and still water, respectively. It showed that the damping ratio of the pipe in the air was 0.0082 and the CF fundamental frequencies in the water were 2.32 Hz and 2.83 Hz with respect to the two axial tension forces. All main physical properties of PPR pipe are summarized in Table 1.

Table 1. Physical properties for the PPR pipe model.

Items	Values
Pipe length, L	4.0 m
Outer diameter, D	0.022 m
Bending stiffness, EI	4.88 Nm^2
Axial tension, T	200, 300 N
Fundamental frequency, f_1	2.32, 2.83 Hz
Mass per unit length, m_s	0.205 kg/m
Mass ratio, $4m_s/(\pi \rho D^2)$	0.54
Aspect ratio, L/D	181.8

The vertical supporting system was composed of two vertical supporting rods, two supporting plates and two guide plates. The vertical supporting rods were fixed to the horizontal supporting frame at its top end. On the bottom end, it was connected with the supporting plates. Parallel to the supporting plates, there mounted two guide plates. The guide plates were designed to weaken the flow disturbance caused by the supporting plates and vertical supporting rods. One end of the pipe was pinned connected with the guide plate through a universal joint, while the other end was treated as a simple support condition. Thus, the pipe could bend in both IL and CF directions, but its torsion and translation were resisted. Furthermore, the axial elongation of the pipe was free. It was controlled by a steel wire that passed through the hollow poles on the supporting plate and connected to a spring.

The horizontal supporting structure refers to the truss structure that mounted on the moving carriage. To avoid the free-surface effect, the pipe model was submerged 1.0 m below the still water level. The scheme of the experimental installation is shown in Figure 4. The towing velocity of the carriage ranged from 0.1–1.0 m/s with an increment of 0.05 m/s, yielding a maximum Reynolds number of 2.2×10^4. A sampling frequency of 100 Hz was adopted. It was in accordance with the Nyquist–Shannon sampling theorem, so the discrete strain signals were able to reflect the whole information of a continuous signal. Nearly 40 runs were performed in the experiment. Each run lasted for a duration of 50 s. Two consecutive runs were brought to at least 15 min halt to calm down the disturbing water.

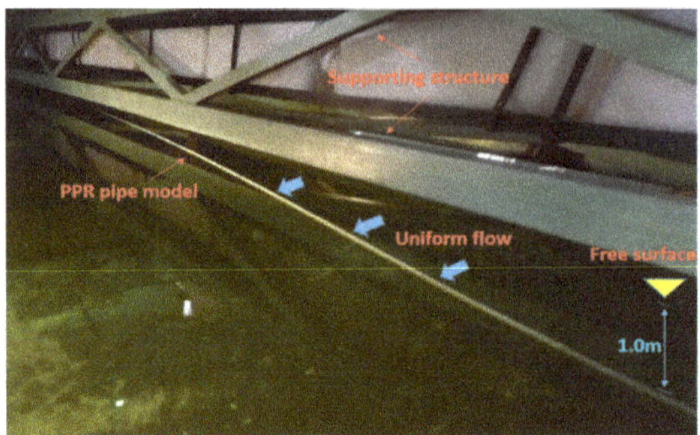

Figure 4. Scheme of the experimental installation.

3. Data Analysis

This paper was concerned with the CF displacement response of the pipe model, while the measured signals were the structural strains at positions G1–G7. Therefore, a reconstruction algorithm based on the model analysis technique was adopted. It was originally proposed by Lie and Kaasen [27] and subsequently applied in the experimental works of [16,17,21–23,25–28].

In the linear regime, the CF displacement of a flexible pipe $y(z, t)$ can be written as the sum of the products of a series of mode shapes and their corresponding modal weights:

$$y(z,t) = \sum_{n=1}^{\infty} w_n(t)\varphi_n(z),\tag{1}$$

where t is the time, z is the coordinate along the pipe axis, n is the mode number, $w_n(t)$ is the modal weight, and $\varphi_n(z)$ is the mode shape.

For a simply supported pipe as shown in Figure 2, $\varphi_n(z)$ has the following form:

$$\varphi_n(z) = \sin \frac{n\pi z}{L},\tag{2}$$

where L is the pipe length.

The curvature of a flexible pipe $\kappa(z,t)$ is defined as:

$$\kappa(z,t) = \frac{y''(z,t)}{\left[1 + y'(z,t)^2\right]^{3/2}} \approx y''(z,t),\tag{3}$$

Also, $\kappa(z,t)$ can be calculated by:

$$\kappa(z,t) = \frac{\varepsilon(z,t)}{R},\tag{4}$$

where $\varepsilon(z,t)$ is the strain of the pipe and R is the outer radius of the pipe. Hence:

$$\frac{\varepsilon(z,t)}{R} = y''(z,t),\tag{5}$$

Substituting Equation (1) into Equation (5), we have:

$$\frac{\varepsilon(z,t)}{R} = -\sum_{n=1}^{\infty}\left(\frac{n\pi}{L}\right)^2 w_n(t)\sin\left(\frac{n\pi z}{L}\right), \tag{6}$$

By solving Equation (6) using the strain signals measured at positions G1–G7, the unknown modal weights $w_n\,(t)$ can be obtained. Note that since seven measurement positions are set, the infinite expansion on the right hand side of Equation (6) is truncated after the first seven terms. That is to say, $n = 1, 2, \ldots, 7$. The truncation error of Equation (6) is negligible, because as shown in Figures 5 and 6 VIV responses of orders higher than three can hardly be observed. After obtaining w_n (t), the CF displacement of at any positions along the pipe axis can be calculated by using Equation (1). The interested reader is referred to the works of Trim, et al. [16], Chaplin, et al. [17], and Lie and Kaasen [27] for more details about the modal analysis technique.

Two points need to be mentioned before applying the above approach for displacement reconstruction. One is that the CR strains measured by strain gages are composed of two parts: the tensile strain due to the axial pretension and the tensile or compressive strains caused by VIV bending. Between them, the latter strains should be used when calculating $w_n(t)$. They can be obtained by subtracting the pretension-induced strain measured by the load cell from the composite strains. The other one is that a band pass filtering operation should be taken to remove the undesirable frequencies. Frequencies lower than 1.0 Hz are excluded from the strain signal to eliminate the interference from the carriage and supporting structure. The value is chosen to be less than half of the fundamental vibration frequency of the pipe in water. Meanwhile, frequencies higher than 40.0 Hz are cut off to avoid the 50.0-Hz-noise of alternating current (AC) signal. It is sufficiently large to cover high-order vibration frequencies concerned in this study.

4. Results and Discussion

In this section, the CF VIV features of the flexible pipe of a critical mass ratio is investigated by sequentially analyzing the strain, displacement, and fluid force. Figure 5 gives an example of time-varying strains and corresponding response frequencies in the case of towing velocity $U = 0.55$ m/s and axial tension $T = 200$ N. The left panel exhibits strain signals measured at positions G1–G7 within $t = 20$–50 s. It is inferred that those positions where strain amplitudes are large are close to the antinodes of dominant modes. Conversely, those positions with small strain amplitudes are near the mode nodes. The right panel displays the frequency spectra of strains calculated using the Fast Fourier Transform (FFT) technique. Although the strain amplitude varies with measurement positions, the response frequencies are almost identical. It is also found that the CF strain response spectra are dominated by one or two strong frequencies and accompanied by a series of weak frequencies. It is known that the odd-order harmonics, such as $3f_y \approx 13.12$ Hz, are generally related to the CF VIV. In this case, the even-order harmonics, such as $2f_y$, can also be observed in the CF vibration. A similar result was presented by Song, et al. [21], who experimentally studied the VIV response of a long flexible riser pipe with a low mass ratio of 1.0. We infer that the even-order CF VIV is induced by the IL VIV response via those strain gages that could not strictly aligned with the pipe. Specifically, the CF VIV is dominated by odd-order harmonics while the IL vibration is dominated by even-order ones. The strain gages were manually pasted to the pipe targeting at an identical direction with the pipe axis. Unfortunately, this is difficult to be realized. Thus, the micro angle between the strain gage and pipe axis transmits the even-order IL harmonics to the CF results.

Figure 6a shows the typical CF VIV displacements at positions G1–G7 when $U = 0.75$ m/s and $T = 300$ N. The response displacements are obtained based on the reconstruction algorithm introduced in Section 3. It can be seen that the maximum displacement amplitude occurs at positions G2 and G6 and the minimum one appears at position G4. The maximum displacement amplitude is nearly $0.60D$. Figure 6b gives the root-mean-square (RMS) of dimensionless displacements along the pipe axis. Obviously, the VIV response is affected most by the 2-order mode and the maximum CF RMS displacement is up to 0.45*D*. In Figure 6c, the dominant frequency is 6.50 Hz, which is corresponding to the second-order mode. Moreover, only one strong frequency peak is observed with U = 0.75 m/s and T = 300 N. This trend of displacement response is consistent with that of a flexible cylinder with relatively high mass ratio undergoing VIV [21,26,28].

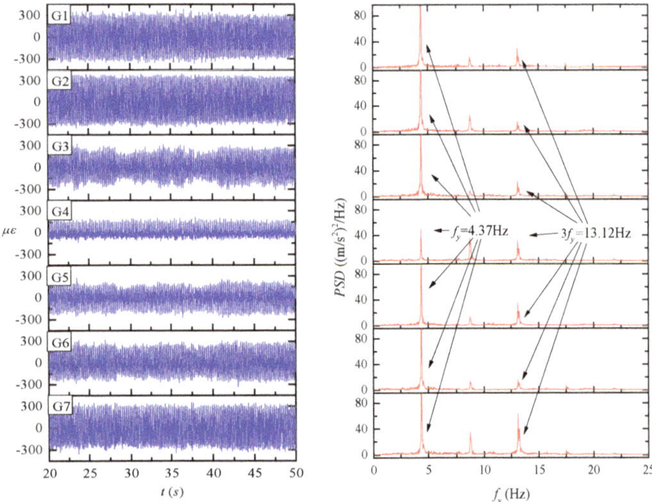

Figure 5. An example of time-varying strains and corresponding frequency spectra at measuring positions G1–G7 with U = 0.55 m/s and T = 200 N.

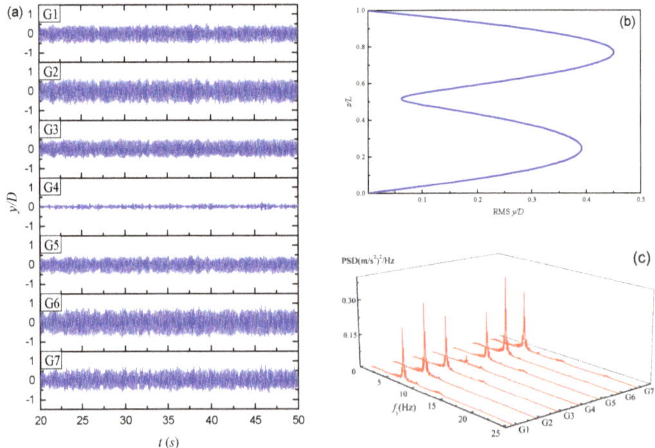

Figure 6. An example of time-varying displacements with U = 0.75 m/s and T = 300 N: (**a**) Dimensionless displacement response for the last 30 s.; (**b**) spanwise evolution of the RMS displacement; and (**c**) response frequencies at different measurement points.

Figure 7 plots the maximum CF displacement against the reduced velocity $Vr = U/f_1D$ (where f_1 is the fundamental frequency of the pipe model in water). Meanwhile, the experimental results of Song, et al. [21] and Huera-Huarte, et al. [22] for flexible cylinders with low mass ratio and that of Govardhan and Williamson [14] for an elastically-mounted rigid cylinder with the critical mass ratio are compared. It can be seen that a slightly altered pre-tension (T = 200 N and 300 N) does not make much difference in the maximum CF displacement. The present CF response amplitude has a maximum value of 2.18*D* at *Vr* = 18.63. With a larger mass ratio *m** = 1.0, the maximum CF displacement in the work of Song, et al. [21] increases to approximately 2.5*D*. Although the mass ratio is close to that of Song, et al. [21], the maximum CF displacement of Huera-Huarte, et al.'s [22] experiment can even reach up to 2.9*D*. The discrepancy is originated from the different current conditions as well as the unequal damping ratio. As for the elastically-mounted rigid cylinder with *m** = 0.542 [14], the amplitude peak value is only 1.2*D*.

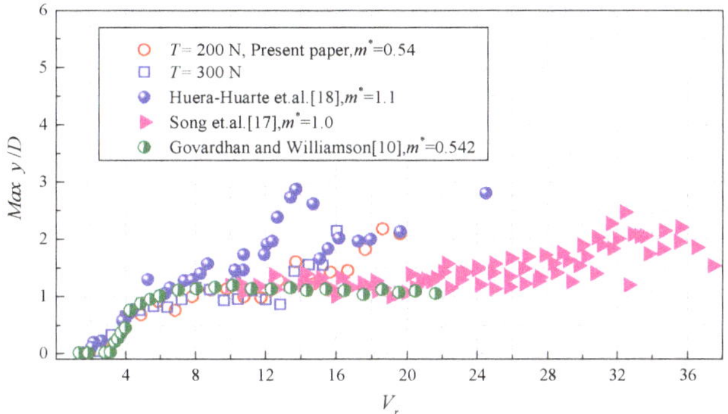

Figure 7. Maximum CF displacement versus the reduced velocity.

Figure 8 gives the CF dominant mode and dimensionless dominant frequency f_y/f_1 against the reduced velocity *Vr*. The dominant mode refers to the order of natural frequency f_n excited by f_y. It increases with *Vr* and is within the range of 1-order to 3-order. When *Vr* < 8, the CF VIV response is dominated by 1st mode. An exception occurs at *Vr* ≈ 3, where the second mode is observed. The flexible pipe in the experiment has two degrees of freedom. At low flow velocity, the CF vibration is stabilized at first mode while the IL vibration mode varies between first-order and second-order. Since the CF displacement amplitude is comparatively small then that in the IL direction, and inevitably there is micro direction error when manually pasting a strain gage to the pipe, the measured CF vibration mode is more or less affected by the high-order IF vibration. As *Vr* ranging from 8 to 14, the 2nd mode is excited. When *Vr* > 14, the third mode dominates the VIV response. The bottom panel of Figure 8 revels that f_y increases monotonically with *Vr*. This trend can be well fitted by a linear regression. Using the least square method, the slope of the fitting line (i.e., the Strouhal number $St = f_s D/U$, where f_s is the vortex-shedding frequency that is nearly the same as f_y in the lock-in region) is found to be 0.184.

A finite element technique proposed by Huera-Huarte, et al. [29] was applied to calculate the mean drag coefficient of the flexible pipe. It has been extensively used and proved effective in the VIV studies not limited to the works of [20,22,30,31]. Figure 9 shows the variation of mean drag coefficient C_{D0} with the reduced velocity *Vr*. Some typical results of VIV experiments on flexible cylinders with low mass ratios [20,22] are also plotted in the figure for comparison. It is observed Huera-Huarte and Bearman [20]'s experimental result has the same trend as that of Huera-Huarte and Bearman [22]. C_{D0} in the work of Huera-Huarte and Bearman [20] increases from 1.4 to 3.3 with the increase of *Vr* from 0 to 5.5. Then, as *Vr* exceeds 5.5 and continues to increase, C_{D0} decreases from 3.3 to 1.4 gradually. While the maximum C_{D0} in Huera-Huarte, et al. [22] can approach to 4.5. For the present work, the peak value of C_{D0} is nearly 2.0. Furthermore, the present result is not as scattered as that of Huera-Huarte, et al. [22]. The discrepancy might be attributed to the different axial tensions employed in the two experiments.

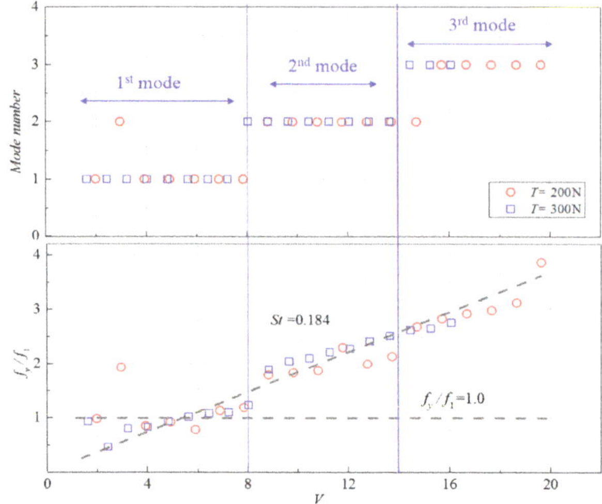

Figure 8. Dominant mode and frequency versus the reduced velocity.

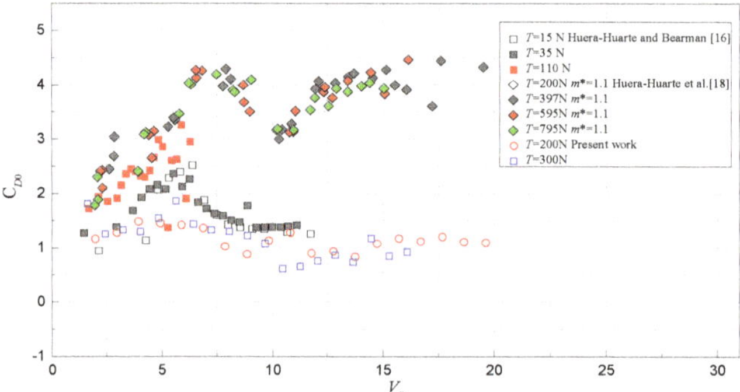

Figure 9. Mean drag coefficient as a function of the reduced velocity.

Under the current experimental techniques, it is still challenging to measure the hydrodynamic forces exerted on a flexible pipe undergoing VIV without interfering with the flow field. However, the hydrodynamic forces can be obtained with the help of the inverse analysis method. The flexible pipe is simplified as a tensioned Euler-Bernoulli beam model. Then, the following equation for CF vibration can be established [29]:

$$m_s \frac{\partial^2 y(z,t)}{\partial t^2} + c \frac{\partial y(z,t)}{\partial t} + EI \frac{\partial^4 y(z,t)}{\partial z^4} - T \frac{\partial^2 y(z,t)}{\partial z^2} = F_y(z,t), \tag{7}$$

In the above equation, F_y (z,t) is the total fluid force in the CF direction [31]:

$$F_y(z,t) = C_L \frac{\rho D}{2\sqrt{2}\dot{y}_{RMS}(z)} U^2 \dot{y}(z,t) - C_a \frac{\rho \pi D^2}{4} \ddot{y}(z,t), \tag{8}$$

where $\dot{y}_{RMS}(z)$ is the RMS value of the response velocity $\ddot{y}(z,t)$. C_L and C_a denote the lift force coefficient and added mass coefficient, respectively. They can be calculated using the following equations [31]:

$$C_L = \frac{2\sqrt{2}\,\dot{y}_{RMS}(z)}{\rho D U^2} \frac{(\lambda_2\lambda_5 - \lambda_3\lambda_4)}{\lambda_2^2 - \lambda_1\lambda_4}, \tag{9}$$

$$C_a = \frac{4}{\rho \pi D^2} \frac{(\lambda_1\lambda_5 - \lambda_2\lambda_3)}{\lambda_2^2 - \lambda_1\lambda_4}, \tag{10}$$

λ_1–λ_5 can be obtained according to following expressions:

$$\lambda_1 = \sum_{i=1}^{S} \dot{y}(z,t_i)^2, \ \lambda_2 = \sum_{i=1}^{S} \dot{y}(z,t_i)\cdot\ddot{y}(z,t_i), \ \lambda_3 = \sum_{i=1}^{S} F_y(z,t_i)\cdot\dot{y}(z,t_i),$$
$$\lambda_4 = \sum_{i=1}^{S} \ddot{y}(z,t_i)^2, \ \lambda_5 = \sum_{i=1}^{S} F_y(z,t_i)\cdot\dot{y}(z,t_i), \tag{11}$$

Figure 10 gives the variations of lift force coefficient C_L and added mass coefficient C_a of the flexible pipe against the reduced velocity Vr. It is observed that C_L under two axial tension cases ($T = 200$ N and 300 N) are nearly the same when Vr is smaller than 4.0. Moreover, in the 1st mode and 3rd mode lock-in regions, axial tension T also makes little influence on C_L. Different situations happen in the 2nd mode lock-in region, where a larger axial tension $T = 300$ N leads to an overall larger C_L. The maximum C_L covering the whole range of Vr is approximate 1.70. The corresponding Vr and T are 14.45 and 300, respectively. As for C_a, it goes up with the increase of Vr so long as CF VIV does not occur. Then it turns to decrease with the increase of Vr once lock-in appears at approximately $Vr = 4.0$ (as shown in Figure 8). The minimum C_a covering the whole range of Vr is about -0.41. The corresponding Vr and T are 15.25 and 300, respectively. This above trend can also be observed in the VIV experiment of Vikestad, et al. [32] where a lightly damped elastically-mounted rigid cylinder undergoing uniform flow was concerned.

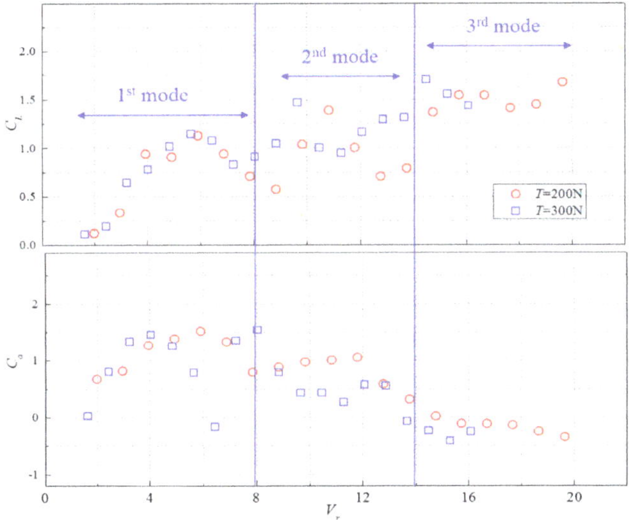

Figure 10. Lift force coefficient and added mass coefficient as functions of the reduced velocity.

5. Conclusions and Future Work

A towing tank experiment was conducted to study the CF VIV response of a flexible pipe with a critical mass ratio of 0.54 and an aspect ratio of 181.8. Two different axial pre-tensions (200 N and 300 N) were enforced and a reconstruction algorithm was applied to obtain the VIV displacement based on the measured strains discretely distributed along the pipe.

The conclusion indicates that the CF strain response of the flexible pipe near the critical mass ratio is dominated by one or two frequencies and accompanied by a series of weak frequencies. Furthermore, both odd-order and even-order harmonics can be observed in the CF frequency spectrum. The CF displacement has a maximum value of $2.18D$ and the CF dominant mode numbers are in the range of 1–3. The fitting result shows that the present flexible pipe has a Strouhal number of 0.184.

In addition, new fluid force coefficients of a flexible pipe undergoing VIV are supplemented into the existing dataset. It is hoped to provide reference to the later experimental and numerical studies. The maximum mean drag coefficient and lift force coefficient are close to 2.0 and 1.7, respectively, in the present experiment. The added mass coefficient first increases and then declines with the increase of the reduced velocity. The turning point locates at the reduced velocity at which the lock-in occurs. Moreover, the minimum added mass coefficient can be as small as −0.41.

In the future, uncertainty quantification (UQ) analysis would be carried out to strengthen the scientific rigor of the present work. Among various existing methods, a newer approach adopted by Rezaeiravesh, et al. [33] is being considered.

Author Contributions: Conceptualization, W.X. and M.H.; experiments conduction, W.X. and Z.X.; formal analysis, X.G.; writing—original draft preparation, X.G. and Z.X.; writing—review and editing, M.H.

Funding: This research was funded by National Natural Science Foundation of China [grant numbers 51679167, 51709201], and China Postdoctoral Science Foundation [grant number 2017M621074].

Conflicts of Interest: The authors declare no conflict of interest.

References

1. Meneghini, J.R.; Saltara, F.; De Andrade Fregonesi, R.; Yamamoto, C.T.; Casaprima, E.; Ferrari, J.A., Jr. Numerical simulations of VIV on long flexible cylinders immersed in complex flow fields. *Eur. J. Mech. B Fluids* **2004**, *23*, 51–63. [CrossRef]
2. Zheng, C.; Liu, Z.; Wu, T.; Wang, H.; Wu, Y.; Shi, X. Experimental investigation of vortex-induced vibration of a thousand-meter-scale mega-tall building. *J. Fluids Struct.* **2019**, *85*, 94–109. [CrossRef]
3. Matsumoto, M.; Yagi, T.; Shigemura, Y.; Tsushima, D. Vortex-induced cable vibration of cable-stayed bridges at high reduced wind velocity. *J. Wind Eng. Ind. Aerodyn.* **2001**, *89*, 633–647. [CrossRef]
4. Cachafeiro, H.; De Arevalo, L.F.; Vinuesa, R.; Goikoetxea, J.; Barriga, J. Impact of solar selective coating ageing on energy cost. *Energy Procedia* **2015**, *69*, 299–309. [CrossRef]
5. Sarpkaya, T. A critical review of the intrinsic nature of vortex induced vibrations. *J. Fluids Struct.* **2004**, *19*, 389–447. [CrossRef]
6. Gabbai, R.D.; Benaroya, H. An overview of modeling and experiments of vortex-induced vibration of circular cylinders. *J. Sound Vib.* **2005**, *282*, 575–616. [CrossRef]
7. Williamson, C.H.K.; Govardhan, R. A brief review of recent results in vortex-induced vibrations. *J. Wind Eng. Ind. Aerodyn.* **2008**, *96*, 713–735. [CrossRef]
8. Wu, X.; Ge, F.; Hong, Y. A review of recent studies on vortex-induced vibrations of long slender cylinders. *J. Fluids Struct.* **2012**, *28*, 292–308. [CrossRef]
9. Rashidi, S.; Hayatdavoodi, M.; Esfahani, J.A. Vortex shedding suppression and wake control: A review. *Ocean. Eng.* **2016**, *126*, 57–80. [CrossRef]
10. Vandiver, J.K. Dimensionless Parameters Important to the Prediction of Vortex-Induced Vibration of Long, Flexible Cylinders in Ocean Currents. *J. Fluids Struct.* **1993**, *7*, 423–455. [CrossRef]
11. Feng, C.C. The Measurement of Vortex-Induced Effects in Flow Past Stationary and Oscillating Circular and D-Section Cylinders. Master's Thesis, University of British Columbia, Vancouver, BC, Canada, 1968.
12. Khalak, A.; Williamson, C.H.K. Investigation of relative effects of mass and damping in vortex-induced vibration of a circular cylinder. *J. Wind Eng. Ind. Aerodyn.* **1997**, *69–71*, 341–350. [CrossRef]
13. Govardhan, R.; Williamson, C.H.K. Resonance forever: Existence of a critical mass and an infinite regime of resonance in vortex-induced vibration. *J. Fluid Mech.* **2002**, *473*, 147–166. [CrossRef]
14. Govardhan, R.; Williamson, C.H.K. Critical mass in vortex-induced vibration of a cylinder. *Eur. J. Mech. B Fluids* **2004**, *23*, 17–27. [CrossRef]

15. Willden, R.H.J.; Rham, J.M.R. Multi-modal Vortex-Induced Vibrations of a vertical riser pipe subject to a uniform current profile. *Eur. J. Mech. B Fluids* **2004**, *23*, 209–218. [CrossRef]
16. Trim, A.D.; Braaten, H.; Lie, H.; Tognarelli, M.A. Experimental investigation of vortex-induced vibration of long marine risers. *J. Fluids Struct.* **2005**, *21*, 335–361. [CrossRef]
17. Chaplin, J.R.; Bearman, P.W.; Huera-Huarte, F.J.; Pattenden, R.J. Laboratory measurements of vortex-induced vibrations of a vertical tension riser in a stepped current. *J. Fluids Struct.* **2005**, *21*, 3–24. [CrossRef]
18. Xu, J.; He, M.; Bose, N. Vortex modes and vortex-induced vibration of a long, flexible riser. *Ocean Eng.* **2009**, *36*, 456–467. [CrossRef]
19. Huera-Huarte, F.J.; Bearman, P.W. Wake structures and vortex-induced vibrations of a long flexible cylinder-part 1: Dynamic response. *J. Fluids Struct.* **2009**, *25*, 969–990. [CrossRef]
20. Huera-Huarte, F.J.; Bearman, P.W. Wake structures and vortex-induced vibrations of a long flexible cylinder-part 2: Drag coefficients and vortex modes. *J. Fluids Struct.* **2009**, *25*, 991–1006. [CrossRef]
21. Song, J.-N.; Lu, L.; Teng, B.; Park, H.-I.; Tang, G.-Q.; Wu, H. Laboratory tests of vortex-induced vibrations of a long flexible riser pipe subjected to uniform flow. *Ocean Eng.* **2011**, *38*, 1308–1322. [CrossRef]
22. Huera-Huarte, F.J.; Bangash, Z.A.; González, L.M. Towing tank experiments on the vortex-induced vibrations of low mass ratio long flexible cylinders. *J. Fluids Struct.* **2014**, *48*, 81–92. [CrossRef]
23. Xu, W.; Qin, W.; Gao, X. Experimental study on streamwise vortex-induced vibration of a flexible slender cylinder. *Appl. Sci.* **2018**, *8*, 311. [CrossRef]
24. Seyed-Aghazadeh, B.; Modarres-Sadeghi, Y. Reconstructing the vortex-induced-vibration response of flexible cylinders using limited localized measurement points. *J. Fluids Struct.* **2016**, *65*, 433–446. [CrossRef]
25. Xu, W.; Luan, Y.; Han, Q.; Ji, C.; Chen, A. The effect of yaw angle on VIV suppression for an inclined flexible cylinder fitted with helical strakes. *Appl. Ocean Res.* **2017**, *67*, 263–276. [CrossRef]
26. Xu, W.; Luan, Y.; Liu, L.; Wu, Y. Influences of the helical strake cross-section shape on vortex-induced vibrations suppression for a long flexible cylinder. *China Ocean Eng.* **2017**, *31*, 438–446. [CrossRef]
27. Lie, H.; Kaasen, K.E. Modal analysis of measurements from a large-scale VIV model test of a riser in linearly sheared flow. *J. Fluids Struct.* **2006**, *22*, 557–575. [CrossRef]
28. Gao, Y.; Yang, J.D.; Xiong, Y.M.; Wang, M.H.; Peng, G. Experimental investigation of the effects of the coverage of helical strakes on the vortex-induced vibration response of a flexible riser. *Appl. Ocean Res.* **2016**, *59*, 53–64. [CrossRef]
29. Huera-Huarte, F.J.; Bearman, P.W.; Chaplin, J.R. On the force distribution along the axis of a flexible circular cylinder undergoing multi-mode vortex-induced vibrations. *J. Fluids Struct.* **2006**, *22*, 897–903. [CrossRef]
30. Gu, J.; Vitola, M.; Coelho, J.; Pinto, W.; Duan, M.; Levi, C. An experimental investigation by towing tank on VIV of a long flexible cylinder for deepwater riser application. *J. Mar. Sci. Technol.* **2013**, *18*, 358–369. [CrossRef]
31. Song, L.J.; Fu, S.X.; Cao, J.; Ma, L.X.; Wu, J.Q. An investigation into the hydrodynamics of a flexible riser undergoing vortex-induced vibration. *J. Fluids Struct.* **2016**, *63*, 325–350. [CrossRef]
32. Vikestad, K.; Vandiver, J.K.; Larsen, C.M. Added mass and oscillation frequency for a circular cylinder subjected to vortex-induced vibrations and external disturbance. *J. Fluids Struct.* **2000**, *14*, 1071–1088. [CrossRef]
33. Rezaeiravesh, S.; Vinuesa, R.; Liefvendahl, M.; Schlatter, P. Assessment of uncertainties in hot-wire anemometry and oil-film interferometry measurements for wall-bounded turbulent flows. *Eur. J. Mech. B Fluids* **2018**, *72*, 57–73. [CrossRef]

Journal of
*Marine Science
and Engineering*

Article

Transmission of Low-Frequency Acoustic Waves in Seawater Piping Systems with Periodical and Adjustable Helmholtz Resonator

Boyun Liu * and Liang Yang

College of Power Engineering, Naval University of Engineering, Wuhan 430033, China; yangliang0306@163.com
* Correspondence: boyunliu@163.com; Tel.: +86-027-6546-1023

Received: 30 August 2017; Accepted: 15 November 2017; Published: 4 December 2017

Abstract: The characteristics of acoustic wave transmitting in a metamaterial-type seawater piping system are studied. The metamaterial pipe, which consists of a uniform pipe with air-water chamber Helmholtz resonators (HRs) mounted periodically along its axial direction, could generate a wide band gap in the low-frequency range, rendering the propagation of low-frequency acoustic waves in the piping system dampened spatially. Increasing the air volume in the Helmholtz chamber would result in a sharply decrease in the central frequency of the resonant gap and an extension in the bandwidth in the beginning, yet very slowly as the air volume is further augmented. Acoustic waves will experience a small amount of energy loss if the acoustic–structure interaction effect is considered. Also, the structure-borne sound will be induced because of the interaction effects. High pressure loadings on the system may bring in a shrink in the band gap; nevertheless, the features of broad band gaps of the system is still be maintained.

Keywords: seawater piping systems; low-frequency sound; acoustic metamaterials; acoustic band gap; high pressure

---

1. Introduction

The topic of low-frequency acoustic waves transmitting in seawater piping systems in ocean surface ships and under-water vehicles, of which their outlets are underneath the water surface, etc., is of special interest [1,2]. The seawater pipe is an excellent medium that can carry acoustic waves, especially the low-frequency sounds, to a distance far way and almost without energy loss. Therefore, a lot of researchers have committed themselves to controlling the noise transmission in piping systems [3,4].

It can be seen from the open public literature that there has been a considerable number of noise reduction methods available for the piping systems, e.g., (i) arrangement of elastic joints, corrugated pipe/bellows, or other components to the piping systems [5–8]; (ii) installation of a muffler elbow and micro-perforated plates [9,10]; (iii) avoidance of short-radius elbow, sharp bends, and branch pipes in pipeline [11]; (iv) laying damping materials [9]; (v) active control technologies [12–15]; and installation of pipe mufflers [16–19], etc. Particularly, installation of a pipe muffler is the most widely used means among these methods. Research has shown that installing a muffler at the broadside inlet of a seawater pipe can effectively suppress or isolate the noise radiating from ship piping system to the outboard field. Nevertheless, existing problems that may beset the current silencers may be ascribed to the limitation in low-frequency noise control capacity. Take the Helmholtz resonator (HR) as an example. Although it can achieve a suppression effect in the low-frequency range, its effective silencing band is always too narrow [4,18–20]. In contrast, the expansion-type silencer can achieve a broadband character of noise reduction, yet its silencing frequency is often too high [21,22].

Recently, the propagation of elastic or acoustic waves in artificial materials/structures called acoustic metamaterials (AMs) has received considerable attention [23–25]. The most important characteristic of AMs is their periodic structure or array of substructures. The initial attraction of using periodic structures may be due to their unusual dynamical characteristics, e.g., the existence of wave bands within which the propagation of elastic waves is forbidden over some selected frequency ranges. Bradley has examined the propagation of linear, dissipative, time-harmonic waves in a broad class of periodic waveguides and confirmed the peculiar dispersion characteristics marked by the phenomenon of stop bands or band gaps in a frequency domain [26]. Following on Bradley's work, Sugimoto and Horioka investigated dispersion characteristics of sound waves propagating in a tunnel with an array of Helmholtz resonators connected axially, as well as the effects of wall friction and the thermoviscous diffusivity of sound [27]. Advantage of the array of resonators is to increase the effective silencing frequency range for the piping system and a downshifting effect on the central of anechoic bands, rendering the bandwidth of noise elimination broadened and the location moved toward the low frequency domain. The introduction of visco-thermal losses tend to increase the damping effects on non-dispersive sound waves such that noise transmitting in the pipe system will be quickly damped spatially. However, in a plausible case, the wall friction and the diffusive effect of sound are very small as compared to that of the stopping bands. More recent studies have been on the nonlinear acoustic behavior of dispersion relation for the air-filled waveguide tube with an array of axially distributed resonators installed [28–30]. These works showed that the usually observed band gaps for the wave transmission coefficient through the system are found to be amplitude-dependent.

In the present work, a metamaterial-based periodic system, consisting of a seawater pipe upon which are mounted axially an array of HRs, is constructed. The difference between the current and foregoing works lies in the improvements of HR design. Here, the HR cavity is designed to be a gas-liquid hybrid chamber, such that the bandwidth may be notably expanded. A transfer matrix (TM) method is utilized to conduct the investigation, and an acoustic-structure interaction model is constructed to give an estimation for the sound transmission loss in a quasi-experimental case. Further, the effects of high pressure on the properties of acoustic band gaps are considered, and comparisons between the liquid-filled chamber and the gas-liquid hybrid chamber are carried out.

2. Acoustic Equations and Calculation Method

The essential model of an HR could be viewed as a spring-mass dynamical absorber with mass $M_H = \rho_w S_n l_{ne}$ and spring stiffness $K_H = \kappa_w S_n^2 / V_c$, or could be analogous to an inductor-capacitor circuit, whose acoustic impedance L_H and capacitance C_H are $L_H = \rho_w l_{ne} / S_n$ and $C_H = V_c / \kappa_w$, respectively, for the HR and the pipe, which are filled with liquid (e.g., water) only [29]. Symbol l_{ne} is the effective length of the neck, which can be calculated by $l_{ne} = l_n + 1.4 r_n$ [31,32]. l_n, r_n, and S_n represent the length, the radius, and the cross-sectional area of HR neck, respectively. ρ_w and κ_w are, respectively, the fluid density and the bulk modulus. Subscripts "w" and "a" indicate the fluid "water" and "air," respectively. Such an equivalent physical model is sufficiently accurate if the frequency range concerned was low enough. The dimension of HR will be smaller than the corresponding wavelength. In this sense, the HR can be defined as the local resonator according to the AMs theory. When the HR cavity is composed of a liquid chamber and a gas chamber, e.g., a water and air hybrid chamber, the acoustic impedance L_H and capacitance C_H should be changed to $L_H = \rho_w l_n / S_n$ and $C_H = (V_{cw} / \kappa_w^2 + V_{ca} / \kappa_a^2)$, respectively. Mounting this HR periodically into the seawater piping system with a fixed lattice space l_a that is much smaller than the acoustic wavelength λ ($l_a \sim \lambda/5$), the periodic system will behave as a homogenized effective medium where the acoustic band gaps may be expected [32]. Numerical validation will be addressed in next section. Construction of the periodic seawater piping system is sketched in Figure 1. Wave equation of the acoustic medium inside the seawater pipe can be given by the following formula [4,17,28]:

$$\nabla^2 p - \frac{1}{c_{w(a)}^2}\frac{\partial^2 p}{\partial t^2} = 0 \tag{1}$$

wherein p is the acoustic pressure. Volumes of the liquid- and gas- chambers are respectively expressed as V_{cw} and V_{ca}; S_p is the cross-sectional area of the pipe; bulk modulus κ is calculated by ρc^2, in which ρ and c denote the density and the acoustic speed of fluid inside the pipe and the HR, respectively. Based on the time-harmonic assumption, the above acoustic equations can be simplified to be the following form:

$$\frac{\partial^2 p}{\partial x^2} + k^2 p = 0 \tag{2}$$

of which the term exp $(-j\omega t)$ has been suppressed throughout. ω is the radian frequency, and k is the wave number that formulated by ω/c. Hence, the acoustic pressure within the tube can be expressed as

$$p = A_t e^{jkx} + A_r e^{-jkx} \tag{3}$$

and acoustic speed v as [1]

$$v = \rho_w^{-1} c_w^{-1}\left(A_t e^{jkx} - A_t e^{-jkx}\right) \tag{4}$$

wherein A_t and A_r indicate the amplitude coefficients of transmitted and reflected waves, respectively. Acoustic speed v is derived from pressure p through the relation between the sound pressure and the acoustic velocity: $v = -\rho_0^{-1}\int \partial p/\partial x dt$. Volume speed Q is the product of acoustic speed v and cross-sectional area of pipe S_p. Thus, acoustic states at the two ends of a uniform pipe section with length of l_a (e.g., the uniform pipe section between the $(n-1)$th and the nth periodic pipe cells as sketched in Figure 1) has the following transfer matrix relation [1,17]:

$$\left\{\begin{array}{c} p_{n-1,R} \\ Q_{n-1,R} \end{array}\right\} = \left[\begin{array}{cc} \cos\frac{\omega l_a}{c} & j\frac{\rho c}{S_p}\sin\frac{\omega l_a}{c} \\ j\frac{S_p}{\rho c}\sin\frac{\omega l_a}{c} & \cos\frac{\omega l_a}{c} \end{array}\right]\left\{\begin{array}{c} p_{nL} \\ Q_{nL} \end{array}\right\} \tag{5}$$

Letter "n" indicates the relevant variables for the nth periodic cell. When a HR is mounted to the pipe section, the corresponding transfer matrix relation in terms of the classical state variables has to modify to the following form:

$$\left\{\begin{array}{c} p_{n-1,L} \\ Q_{n-1,L} \end{array}\right\} = \left[\begin{array}{cc} \cos\frac{\omega l_a}{c} & j\frac{\rho c}{S_p}\sin\frac{\omega l_a}{c} \\ \frac{1}{Z_H}\cos\frac{\omega l_a}{c} + j\frac{S_p}{\rho c}\sin\frac{\omega l_a}{c} & \cos\frac{\omega l_a}{c} + j\frac{\rho c}{Z_H S_p}\sin\frac{\omega l_a}{c} \end{array}\right]\left\{\begin{array}{c} p_{nL} \\ Q_{nL} \end{array}\right\} \tag{6}$$

wherein Z_H is the acoustic impedance of HR. It can be formulated by $Z_H = j\omega L_H + (j\omega C_H)^{-1}$. Introducing the state vector $\mathbf{\Gamma} = \{p, Q\}'$ into the above equation, Equation (6) can be rewritten into the following abbreviated form:

$$\mathbf{\Gamma}_{n-1} = \mathbf{T}_c \cdot \mathbf{\Gamma}_n \tag{7}$$

in which $\mathbf{T}_c$ is the corresponding transfer matrix relating to the state vectors at the two ends of a periodic cell. Moreover, the two state vectors at the left and the right sides of periodic cell should also satisfy the following restriction due to the periodic boundary condition [33]:

$$\mathbf{\Gamma}_{n-1} = e^{j\mu a_p} \cdot \mathbf{\Gamma}_n \tag{8}$$

Combining Equations (7) with (8) gives rise to

$$\left|\mathbf{T}_c - e^{j\mu a_p}\mathbf{I}\right| = 0 \tag{9}$$

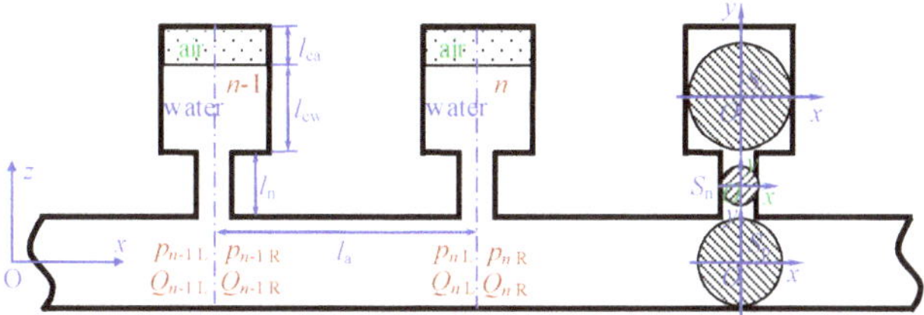

Figure 1. Sketch of the seawater pipe system with Helmholtz resonators (HRs) mounted periodically.

Solving the values of μ in Equation (9), as functions of ω, one can easily obtain the dispersive relation to describe the state of acoustic waves propagating in the infinite periodic pipe, that is, the well-known acoustic band structure. Such a band structure comprises two parts: the real and the imaginary parts. The real part is the so-called phase constant, and the imaginary part is referred to as attenuation constant. In this sense, the symbol μ can be referred as the effective wavenumber inside the metamaterial. Wave propagation is possible within frequency bands where μ is real (pass bands), whereas attenuation occurs for the frequency values that provide an imaginary part to μ, referred to as band gap. It describes the attenuation degree of amplitude coefficient of waves transmitting from one side of a periodic cell to the other side [33].

For a finite periodic pipe consisting of N Helmholtz resonators mounted equidistantly in a uniform pipe, the transmitting relationship for the acoustic states at the inlet and the outlet can be given by

$$\Gamma_i = T_c^N \cdot \Gamma_o \tag{10}$$

It can be further simplified to

$$t_p = \left| \frac{2\sin(\mu l_a)\left(\tau_p^2 - 1\right)\tau_p^N}{\left(1 - \tau_p e^{-j\mu l_a}\right)^2 - \tau_p^{2N}\left(\tau_p - e^{-j\mu l_a}\right)^2} \right| \tag{11}$$

if the inlet and the outlet of the finite periodic pipe are perfectly impedance matched and $\tau_p = \exp(\pm j k l_a)$. Accordingly, the sound transmission loss TL can be given by $10\lg(t_p^{-1})$. Up to now, propagation characteristics of acoustic waves in the infinite and the finite periodic pipes can be examined by the calculation of TL and band structure, respectively, upon which relative analysis for the periodic system could be carried out.

3. Results and Discussion

In what follows, numerical examples are addressed to illuminate the acoustic characteristics of low-frequency wave transmission in the metamaterial-type periodic pipe designed in the current work. The cavity shape of Helmholtz resonance can be spherical, cylindrical, or even an irregular cavity. Also, the cross section of the neck tube can be a regular shape, such as circular, square, oval, or other shape. In fact, the most important geometric factors that affect the performance of HR are the volume of Helmholtz chamber, the cross-sectional area, and the length of Helmholtz tube, etc. In the low frequency range, the effects of geometric shape of the neck tube and chamber on the performance of muffler may be neglected. Without loss of generality, this paper chooses the shape for the HR chamber and neck tube to be cylindrical. Radii for the pipe, HR chamber, and neck are respectively employed as $r_p = 5$ cm, $r_c = 2r_p$, and $r_n = 0.8r_p$; lengths for the periodic cell, HR chamber, and neck are chosen to be $l_a = 0.96$ m, $l_{cw} = 4.5r_p$, and $l_n = 1.5r_p$, respectively.

In the first place, the periodic seawater pipe system is assumed to be filled with water; acoustic speed and density for water are respectively 1500 m/s and 1000 kg/m^3. Figure 2 exhibits the corresponding numerical results for a seawater pipe system with four HRs installed equidistantly, as shown by the solid line; the dashed line corresponds to the simulation of Comsol commercial software for the same periodic system. The good agreement of solid line with dashed line validates the accuracy of the numerical algorithm developed in this paper. Comparing the sound transmission loss of the periodic pipe system with that of a same pipe system with a single HR mounted, as illustrated by the dash-dotted line, one can see that there are two attenuation ranges of sound transmission in the former case. They are 356.5–645.5 Hz and 781–1094 Hz. Within these two frequency zones, acoustic wave propagation in the periodic system is attenuated evidently; in contrast, sound transmitting in the latter one, i.e., the pipe system with one HR attached, is damped only in a very narrow frequency range near the resonant frequency peak f_H of HR that determined by $f_H = (2\pi)^{-1}(C_H L_H)^{-1/2}$. With regard to the infinite periodic pipe system, the acoustic propagation characteristics are captured by the band structure, as shown in Figure 3, of which the lattice constant l_a and other geometric parameters involved in the calculation are the same as those in Figure 2. On examination of Figure 3, it can be seen that the location and bandwidths of these two BGs are exactly the same as that of the sound suppression zones in the finite periodic structure, as well as the attenuation effects, as revealed by the imaginary part of μ in Figure 3 and the TL in Figure 2. In fact, both the sound transmission loss and the band structure are equivalent in describing acoustic characteristics for periodic systems. The difference of these two approaches lies in that the former is used for finite periodic systems and the latter is for ideal periodic structures, i.e., the infinite periodic systems. The damping in the band gap has nothing to do with energy loss transformed into heat, so it can occur in the lossless cases. Introduction of visco-thermal losses into the metamaterial-type seawater pipe system may strengthen the damping effects in both the stop and the pass bands, yet it tends to smooth out the sharply cusped features that occur at the boundaries of the Bragg gaps and at scatterer resonance gaps [27]. Moreover, the diffusive effect of sound in a plausible case is very small except for the stopping bands, thus the dissipation is not taken into account in the current work.

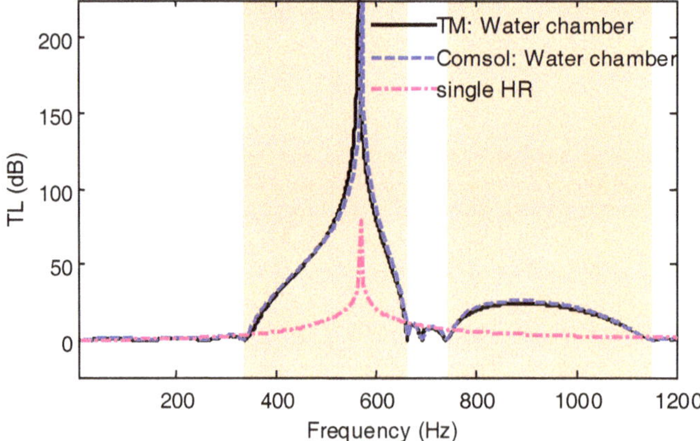

Figure 2. Sound transmission losses for the seawater pipe system with a single HR and a HR array mounted, respectively.

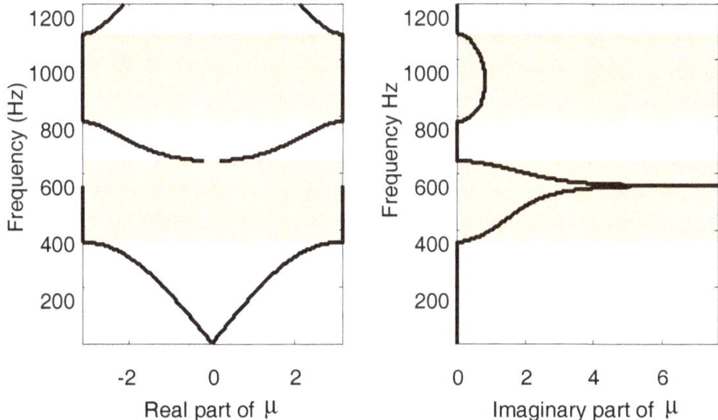

Figure 3. Acoustic band structure of an infinite periodic seawater pipe system.

Next, we consider the water-filled chamber of HR to be replaced by an air-water hybrid chamber and investigate the acoustic characteristics of the periodic pipe. As mentioned above, if the HR chamber is filled by air and water, then the acoustic capacitance C_H will be changed from $C_H = V_{cw}/\kappa_w^2$ to $C_H = (V_{cw}/\kappa_w^2 + V_{ca}/\kappa_a^2)$, such that the location of the first BG will be notably lowered, as the resonant frequency of HR f_H is decreased with the increase of C_H. Figure 4 validates such a conclusion. Lengths for the water and the air chambers employed in the calculation are $l_{cw} = 0.4r_p$ and $l_{ca} = 0.5r_p$, respectively. Other parameters are kept the same with those in Figure 3. Density and sound speed of air are 1.225 kg/m^3 and 342 m/s, respectively.

On observation of Figure 4, it is easy to see that band edge of the first band gap can be decreased as low as 9 Hz. Moreover, the bandwidth is broadened to be 444 Hz. This change in the first band gap is, of course, good for the low-frequency noise control for the seawater piping system. Probing into the formation mechanism of the first band gap, it may be ascribed to the co-resonance of HR array (scattering resonances) in the periodic system. The resonant peak with maximum attenuation coefficient in the band gap exactly corresponds to the resonant frequency f_H of HR, i.e., 13.3 Hz under the aforementioned parameters. Consequently, this band gap is categorized as resonant gap (RG). As for the second band gap in Figure 4, namely the frequency range 781–1036 Hz, one can see that it is almost unchanged as compared to that in Figure 3. Tracing it to its causation, we know that the behavior of the second band gap is dominated by the Bragg scattering mechanism. In other words, the generation of the second gap could be ascribed to the effects of interference between the incident, reflected, and transmitted acoustic waves in the system cells. Thereupon, one of the band edge f_B, is determined by $mc/2l_a$, i.e., the Bragg condition; $m = 1, 2, 3, \ldots$, denotes the mth band gap induced by this Bragg scattering mechanism. In this sense, the second band gap is defined as Bragg-type gap (BG). From the Bragg condition, it can be known that the adoption of air-water hybrid chamber in the HR medium in fact has little influence on the Bragg condition, hence the BG would experience no change in its band gap features, such as bandwidth, location, etc.

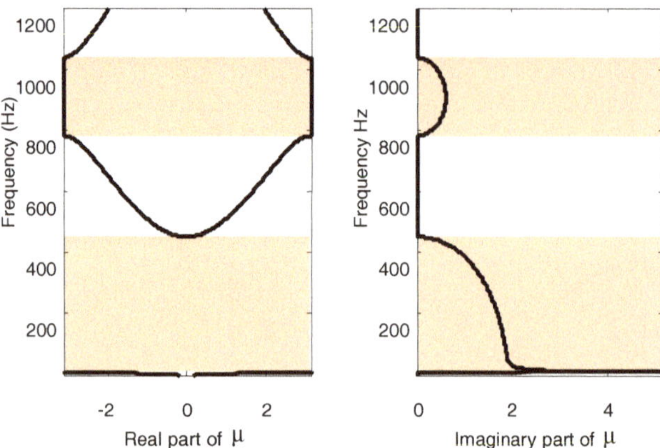

Figure 4. Acoustic band structure of a periodic pipe with its HR chamber filled with air and water.

Figure 5 shows the corresponding sound transmission loss for the same model as studied in Figure 4, except that the seawater pipe now is assumed to be a finite structure. The seawater pipe system discussed at this stage is loaded by four identical water-air chamber HRs in an axial array of lattice l_a. Material and geometric parameters involved in the calculation are kept the same as those applied in Figure 4. The solid and the dashed lines correspond to the simulation of TM method and Comsol commercial software, respectively. The TL curves predicted by these two different numerical methods agree in the low frequency range, while only small differences can be observed near the Bragg gap between the TM method and the Comsol simulation. This may due to a nonlinear conversion of propagative waves toward evanescent waves and due to the hypothesis used in the TM algorithm that the HR neck connected to pipe wall is viewed as a point such that its neck geometric dimension is neglected. The attenuation zones in this plot agree well with the two gaps shown in Figure 4, thus the effectiveness of low-frequency noise suppression capacity by changing the cavity of a water-filled HR by a mix between the liquid and gas cavity (water and air) is validated once again.

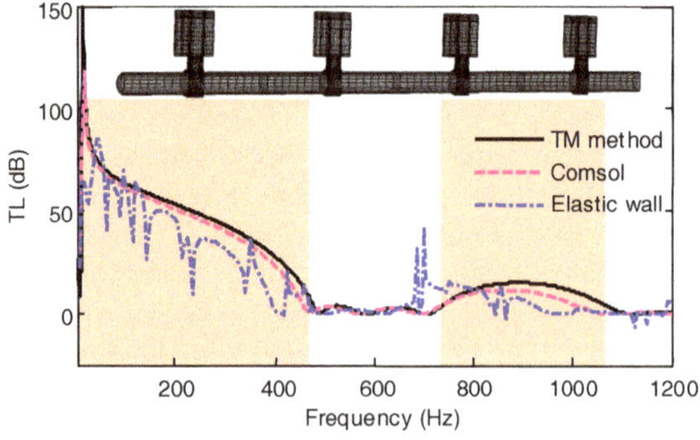

Figure 5. Sound transmission loss of a finite length pipe with four liquid-air HRs attached equidistantly.

Toward the practical consideration for engineering piping systems, there are some experimental constraints that should be discussed. One key factor is that acoustic waves will propagate not only in water, but also in the material of the pipe wall, e.g., steel, due to the low impedance contrast between media. This effect will be presented in any experiment, so here, an acoustic-structure interaction model is constructed to give an estimation for the sound transmission loss in a plausible case. The evaluation model, i.e., a finite element model (FEM), used in the investigation is presented in the inset of Figure 5.

The thicknesses for the elastic pipe and, the HR cavity, and neck walls are chosen as 4 mm, 9.5 mm, and 3.5 mm, respectively; other geometric parameters are kept unchanged. Numerical results of such an acoustic-structure model under the aforementioned parameters are shown by the dash-dotted line in Figure 5. It is seen that total amount of noise elimination in this case experiences a bit of a decrease over most of the band gap ranges. Moreover, there are numerous "resonant peaks" in the previously smooth TL curve, rendering the noise damping effects in the band gaps deteriorated at these frequencies. In fact, acoustic waves transmission in this FEM experiences a small amount of energy loss due to the sound radiation through the pipe wall and also motivates the structure vibration modes that occurred because of the acoustic-structure interaction effects, as demonstrated by the subsequent Figure 6a,c for the von misses stress at frequencies 235 Hz and 340 Hz, respectively. Thereupon, structure-borne sound is induced, as shown by the acoustic pressure fields in Figure 6b,d, respectively corresponding to these two frequency locations. The pressure excitation is loaded at the inlet (the left end) of the pipe, and signals are picked up in the outlet (the right end). For better comparison, the von misses stress distribution and the acoustic pressure field that located outside the band gap (for example, at frequency 500 Hz) are given in Figure 6e,f. Apparently, sound transmission in the pass band could propagate freely through the pipe system without any attenuation, whereas in the stop gap, wave propagation is blocked forward, even in a worse case wherein the structure-borne sound is excited.

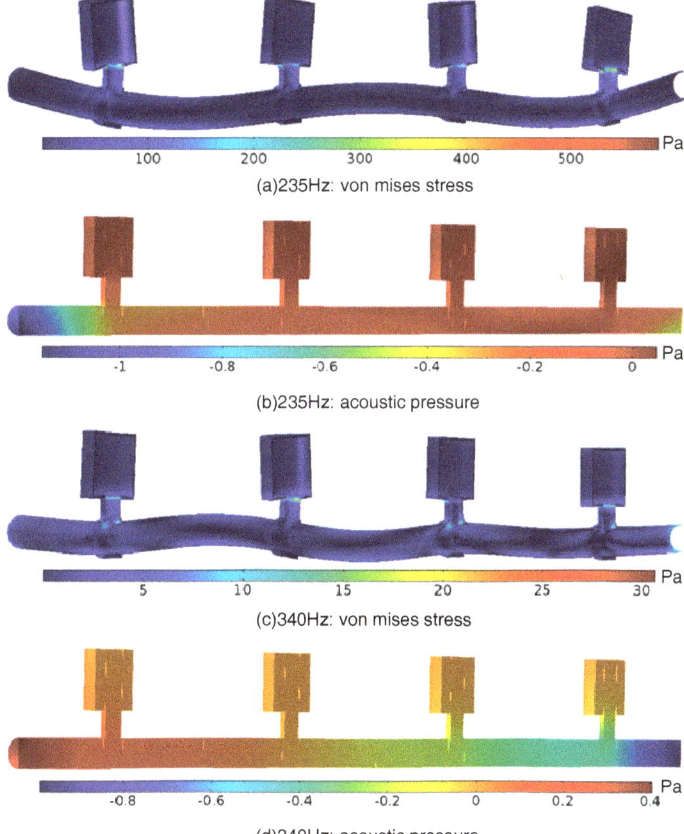

Figure 6. *Cont.*

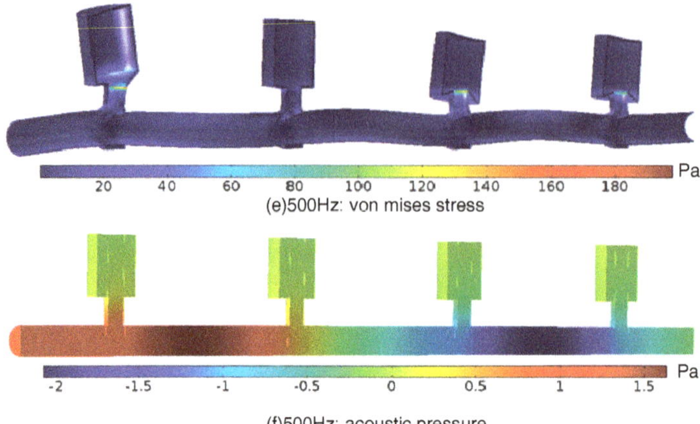

Figure 6. Von misses stress and acoustic pressure field for the seawater pipe system: (**a**,**c**,**e**) are respectively corresponding to the von mises stress fields at 235 Hz, 340 Hz and 500 Hz; (**b**,**d**,**f**) correspond to the acoustic pressure fields at 235 Hz, 340 Hz and 500 Hz, respectively.

In what follows, some key factors that may modulate the band gap features will be discussed. As revealed in the foregoing, the design technology of using an air cavity could greatly increase the acoustic capacitance of HR C_H, therein resulting in a low-frequency and broad RG. Hence, it should be noted that the filling ratio of air chamber plays an important part in generating this low-frequency and wide RG. RG with broader width and lower central frequency may be achieved if the volume of HR air cavity is increased. Figure 7 presents a complete surface of imaginary parts of μ, as functions of frequency and volume of air chamber (i.e., length of the air chamber l_{ca}), through which detailed information on the behavior of the band gap location, width, and attenuation coefficient can be roundly known. According to the planform view of this attenuation constant surface, a conclusion can be reached as follows: a little bit of air chamber added to the water chamber of HR will lead to a sharply decrease in the central frequency of the resonant gap, as well as an extension in the bandwidth; however, a further increase of air volume cannot bring in such a remarkable change in the RG as that in the beginning. Still, it does broaden the bandwidth and lower the band gap location to some extent. Anyway, revelation of the RG and BG behavior against these key parameters will eventually prove to be useful in obtaining a broad, low-frequency band gap for noise transmitting suppression, mainly to form experience for reference.

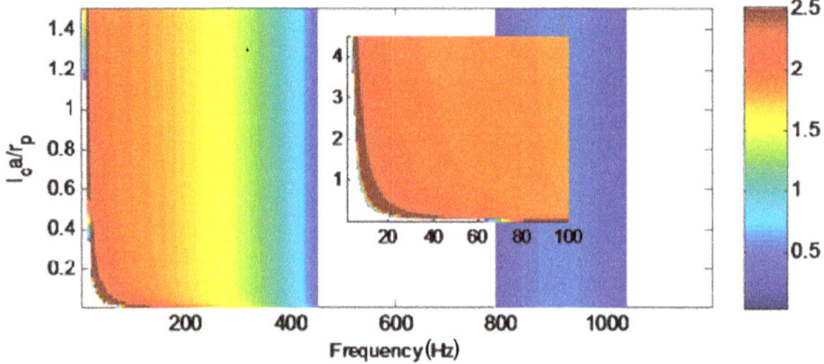

Figure 7. Imaginary parts of μ, as functions of l_{ca}, for the periodic pipe of which its HR chambers are filled with air and water.

In terms of practical application, especially for under-water vehicles with outlets of seawater piping systems that are underneath the water surface, the influence of high pressure on the transmission of seawater pipe noise to the outlets might attract much more attention. Hence, at this stage, focus is put on the effects of high pressure on the band gaps of the periodic seawater pipe constructed in the current work. The most directly influence of high pressure on the seawater pipe system is the variation of acoustic characteristics of the media inside both of the HR chamber and the tube. That is to say the external pressure on the piping system would lead to a change to either the acoustic speed or the density, or both, for the filled water and air. The gas density ρ relating with pressure p could be written as $\rho = pM/RT$, which is deduced from the classical ideal gas law that may be given by $pV = nRT$, wherein V and n correspond to volume and number of moles of a substance, respectively; T, M, and R are respectively corresponding to absolute temperature, molar mass, and ideal gas constant, approximately 8.3144621 J/(mol·K). For air, the average molar mass M is approximately 0.029 kg/mol. Sound speed in an ideal gas depends only on its temperature and composition. The speed has a weak dependence on frequency and pressure in ordinary air, deviating slightly from ideal behavior. In general, the speed of sound c is given by the Newton-Laplace equation: $c = (K_f/\rho)^{1/2}$, in which the bulk modulus K_f is simply the gas pressure p multiplied by the dimensionless adiabatic index γ, which is about 1.4 for air. Effects of external pressure on the sound transmission properties of the seawater pipe system with air-water chamber HRs attached periodically are illustrated in Figure 8. Geometric parameters are chosen to be the same as those in Figure 4. Temperature of the surroundings and the internal media of the pipe system is assumed to be 25 degrees Celsius. The dash-dotted, the dashed, and the solid lines correspond to pressure values of 1 bar, 5 bars, and 15 bars, respectively.

On examination of Figure 8, some useful information can be obtained: (i) the augment of external pressure on the periodic system will bring in an increase in the lower band edge of the RG, yet has no change in the higher band edge, thus causing a reduction to the RG bandwidth. (ii) The variation of pressure does not change the second band gap induced by the Bragg scattering mechanism. This is expectable, since such a gap is generated by the interference of reflected waves by the periodic geometric cells of the system. (iii) The features of broad band gaps of the periodic system are still unchanged, although high pressure is loaded on the pipe system. The good news is that such a noise control strategy of periodic design may be applicable in the area of noise control for underwater vehicles where high pressure may be subjected. It is important to mention that the acoustic nonlinearities due to the intrinsic air behavior in the HR cavity and due to the high amplitude waves were not taken into account here. The presence of nonlinearities may introduce an interplay between nonlinear effects and spatial periodicity. As a consequence, the HR band gap width tends to increase with the acoustic amplitude in some extent, as revealed in Ref. [28]. In this study, our analytical model is simplified (i.e., based on a linear model). However, this simplification allowed us to provide a full understand of the band gap characteristics of the metamaterial-type pipe and to guide the structure design for a seawater pipe system to avoid the low-frequency noise control problem as mentioned in the introduction. A further consideration of the present work may be the nonlinear effects, so as to capture more detailed behavior of the acoustic waves in the periodic system subjected to high pressure loadings. Nevertheless, as a first step toward the study of nonlinear effects, the investigation of linear system is reasonable.

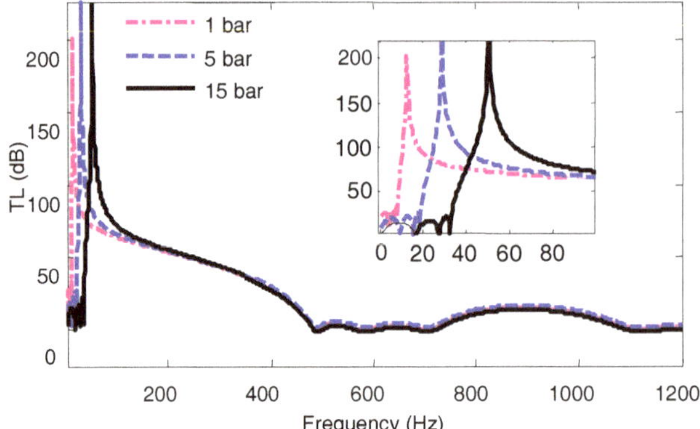

Figure 8. Sound transmission loss of the periodic pipe system under various pressure loadings.

4. Conclusions

In conclusion, the current work attempts to design a metamaterial-based periodic structure to solve the low-frequency noise control problem in the seawater pipe systems of ships. By installing the HR with its chamber consisting of a water chamber and an air chamber periodically into the pipe system, a low-frequency and broad band gap can be generated. Within the band gap, the propagation of low-frequency acoustic waves in the seawater pipe system is suppressed effectively. This means that the transmission of acoustic noise through such a periodic pipe structure will be effectively attenuated before it arrives at the outlet underneath the water surface. The damping in the band gap has nothing to do with energy loss transformed into heat, so it can occur in the lossless cases. Changing the cavity of a water-filled HR by a mix between the liquid and gas cavity (water and air) would bring in a decrease in the effective bulk modulus of the cavity, thus lowering the resonance frequency of the HR and therefore expanding the low frequency band gap. However, a large filling ratio of air chamber to water chamber is not necessary, since a small rate of filling air in HR chamber is enough to lower and broaden the resonant-type band gap. A further increase of the air volume cannot make any more notable change to the RG. In terms of a practical case, acoustic waves will propagate in water and also in the material of the pipe wall, e.g., steel, due to the low impedance contrast between media. As a result, acoustic waves will experience a small amount of energy loss due to the sound radiation through the pipe wall. Also, the structure-borne sound will be induced because of the acoustic-structure interaction effects. Besides, the characteristics of the low-frequency band gap of the periodic system can be maintained even when subjected to high pressure loadings. Nevertheless, high pressure will bring in some sort of increase in the lower band edge of the RG, thus causing a decrease in the RG bandwidth.

A further extension of this work is the nonlinear effects on the dispersion relation and an experimental investigation, and this is now under consideration. Hopefully, this research will be helpful and interesting to the research being conducted on noise control for piping systems.

Acknowledgments: This work was funded by the National Natural Science Foundation of China (Grant No. 51705529).

Author Contributions: Boyun Liu and Liang Yang conceived and designed the metamaterial-type seawater pipe; Boyun Liu performed the numerical calculation and Liang Yang analyzed the data; Boyun Liu wrote the paper.

Conflicts of Interest: The authors declare no conflict of interest.

References

1. Li, Y.; Shen, H.; Zhang, L.; Su, Y.; Yu, D. Control of low-frequency noise for piping systems via the design of coupled gap of acoustic metamaterials. *Phys. Lett. A* **2016**, *380*, 2322–2328. [CrossRef]
2. Liu, B.; Pan, J.; Li, X.; Tian, J. Sound radiation from a water-filled pipe, radiation into light fluid. *J. Acoust. Soc. Am.* **2002**, *112*, 2814–2824. [CrossRef] [PubMed]
3. Zhu, Y.-W.; Zhu, F.-W.; Zhang, Y.-S.; Wei, Q.-G. The research on semi-active muffler device of controlling the exhaust pipe's low-frequency noise. *Appl. Acoust.* **2017**, *116*, 9–13. [CrossRef]
4. Chiu, M.-C. Numerical assessment of two-chamber mufflers hybridized with multiple parallel perforated plug tubes using simulated annealing method. *J. Low Freq. Noise Vib. Act. Control* **2017**, *36*, 3–26. [CrossRef]
5. Golliard, J.; Belfroid, S.; Vijlbrief, O.; Lunde, K. Direct measurements of acoustic damping and sound amplification in corrugated pipes with flow. In Proceedings of the ASME Pressure Vessels and Piping Conference, Boston, MA, USA, 19–23 July 2015; pp. 19–23.
6. Amielh, M.; Anselmet, F.; Jiang, Y.; Kristiansen, U.; Mattei, P.-O.; Mazzoni, D.; Pinhede, C. Aeroacoustic source analysis in a corrugated flow pipe using low-frequency mitigation. *J. Turbul.* **2014**, *15*, 650–676. [CrossRef]
7. Zhang, T.; Zhang, Y.O.; Ouyang, H. Structural vibration and fluid-borne noise induced by turbulent flow through a 90° piping elbow with/without a guide vane. *Int. J. Press. Ves. Pip.* **2015**, *125*, 66–77. [CrossRef]
8. Chiang, Y.K.; Choy, Y.S.; Tang, S.K. Vortex sound radiation in a flow duct with a dipole source and a flexible wall of finite length. *J. Acoust. Soc. Am.* **2017**, *141*, 1999–2010. [CrossRef] [PubMed]
9. Bravo, T.; Maury, C.; Pinhède, C. Absorption and transmission of boundary layer noise through flexible multi-layer micro-perforated structures. *J. Sound Vib.* **2017**, *395*, 201–223. [CrossRef]
10. Xiang, L.; Zuo, S.; Wu, X.; Liu, J. Study of multi-chamber micro-perforated muffler with adjustable transmission loss. *Appl. Acoust.* **2017**, *122*, 35–40. [CrossRef]
11. Koh, J.P.; Lyu, S.J.; Lee, T.B. Analysis of pipeline noise and vibrations using FEM. In Proceedings of the 22nd International Congress on Sound and Vibration, Florence, Italy, 12–16 July 2015; pp. 12–16.
12. Bhagwat, S.; Creasy, M.A. Adjustable pipes and adaptive passive damping. *Eur. J. Phys.* **2017**, *38*, 035003. [CrossRef]
13. Park, J.H.; Lee, S.K. A novel adaptive algorithm with an IIR filter and a variable step size for active noise control in a short duct. *Int. J. Automot. Technol.* **2012**, *13*, 223–229. [CrossRef]
14. Hwang, Y.; Lee, J.M.; Kim, S.-J. New active muffler system utilizing destructive interference by difference of transmission paths. *J. Sound Vib.* **2003**, *262*, 175–186. [CrossRef]
15. Guan, C.; Jiao, Z.; Wu, S.; Shang, Y.; Zheng, F. Active control of fluid pressure pulsation in hydraulic pipe system by bilateral-overflow of piezoelectric direct-drive slide valve. *J. Dyn. Sys. Meas. Control.* **2014**, *136*, 031025.
16. Cheng, H.-C.; Chiu, M.-C. Shape optimization of two-chamber mufflers hybridized with multiple parallel dissipative tubes using simulated annealing method. *J. Mar. Sci. Technol.* **2016**, *24*, 744–758.
17. Guo, R.; Zhu, W.W. Acoustic attenuation performance of a perforated resonator with a multi-chamber and its optimal design. *Proc. Inst. Mech Eng. Part D J. Automob. Eng.* **2014**, *228*, 1051–1060. [CrossRef]
18. Yasuda, T.; Wu, C.Q.; Nakagawa, N.; Nagamura, K. Studies on an automobile muffler with the acoustic characteristic of low-pass filter and Helmholtz resonator. *Appl. Acoust.* **2013**, *74*, 49–57. [CrossRef]
19. Oh, S.; Wang, S.; Cho, S. Development of energy efficiency design map based on acoustic resonance frequency of suction muffler in compressor. *Appl. Energy* **2015**, *150*, 233–244. [CrossRef]
20. Xu, M.B.; Selamet, A.; Kim, H. Dual Helmholtz resonator. *Appl. Acoust.* **2010**, *71*, 822–829. [CrossRef]
21. Singh, N.K.; Rubini, P.A. Large eddy simulation of acoustic pulse propagation and turbulent flow interaction in expansion mufflers. *Appl. Acoust.* **2015**, *98*, 6–19. [CrossRef]
22. Mimani, A.; Munjal, M.L. 3-D acoustic analysis of elliptical chamber mufflers having an end-inletand a side-outlet: An impedance matrix approach. *Wave Motion* **2012**, *49*, 271–295. [CrossRef]
23. Jiménez, N.; Romero-García, V.; Pagneux, V.; Groby, J.-P. Quasiperfect absorption by subwavelength acoustic panels in transmission using accumulation of resonances due to slow sound. *Phys. Rev. B* **2017**, *95*, 014205. [CrossRef]
24. Lee, S.H.; Park, C.M.; Yong, M.S.; Wang, Z.G.; Kim, C.K. Composite Acoustic Medium with Simultaneously Negative Density and Modulus. *Phys. Rev. Lett.* **2010**, *104*, 054301. [CrossRef] [PubMed]

25. Shi, X.F.; Mak, C.M. Sound attenuation of a periodic array of micro-perforated tube mufflers. *Appl. Acoust.* **2017**, *115*, 15–22. [CrossRef]
26. Bradley, C.E. Time harmonic acoustic Bloch wave propagation in periodic waveguides. *J. Acoust. Soc. Am.* **1994**, *96*, 1844. [CrossRef]
27. Sugimoto, N.; Horioka, T. Dispersion characteristics of sound waves in a tunnel with an array of Helmholtz resonators. *J. Acoust. Soc. Am.* **1995**, *97*, 1446. [CrossRef]
28. Richoux, O.; Tournat, V.; Le Van Suu, T. Acoustic wave dispersion in a one-dimensional lattice of nonlinear resonant scatterers. *Phys. Rev. E* **2007**, *75*, 026615. [CrossRef] [PubMed]
29. Achilleos, V.; Richoux, O.; Theocharis, G.; Frantzeskakis, D.J. Acoustic solitons in waveguides with Helmholtz resonators: Transmission line approach. *Phys. Rev. E* **2015**, *91*, 023204. [CrossRef] [PubMed]
30. Richoux, O.; Lombard, B.; Mercier, J.-F. Generation of acoustic solitary waves in a lattice of Helmholtz resonators. *Wave Motion* **2015**, *56*, 85–99. [CrossRef]
31. Hao, L.-M.; Ding, C.-L.; Zhao, X.-P. Tunable acoustic metamaterial with negative modulus. *Appl. Phys.* **2012**, *106*, 807–811. [CrossRef]
32. Fang, N.; Xi, D.; Xu, J.; Ambati, M.; Srituravanich, W.; Sun, C.; Zhang, X. Ultrasonic metamaterials with negative modulus. *Nat. Mater.* **2006**, *5*, 452–456. [CrossRef] [PubMed]
33. Shen, H.; Wen, J.; Yu, D.; Asgari, M.; Wen, X. Control of sound and vibration of fluid-filled cylindrical shells via periodic design and active control. *J. Sound Vib.* **2013**, *332*, 4193–4209. [CrossRef]

Journal of
*Marine Science
and Engineering*

MDPI

Article

An Integrated Numerical Model for the Design of Coastal Protection Structures

Theophanis V. Karambas [1] and Achilleas G. Samaras [1,2,*]

[1] Department of Civil Engineering, Aristotle University of Thessaloniki, University Campus,
    PC 54 124 Thessaloniki, Greece; karambas@civil.auth.gr
[2] THALASSA Coastal Engineering Modelling, Greece
*   Correspondence: asamaras@civil.auth.gr; Tel.: +30-6972-728807

Received: 7 August 2017; Accepted: 13 October 2017; Published: 24 October 2017

Abstract: In the present work, an integrated coastal engineering numerical model is presented. The model simulates the linear wave propagation, wave-induced circulation, and sediment transport and bed morphology evolution. It consists of three main modules: WAVE_L, WICIR, and SEDTR. The nearshore wave transformation module WAVE_L (WAVE_Linear) is based on the hyperbolic-type mild slope equation and is valid for a compound linear wave field near coastal structures where the waves are subjected to the combined effects of shoaling, refraction, diffraction, reflection (total and partial), and breaking. Radiation stress components (calculated from WAVE_L) drive the depth averaged circulation module WICIR (Wave Induced CIRculation) for the description of the nearshore wave-induced currents. Sediment transport and bed morphology evolution in the nearshore, surf, and swash zone are simulated by the SEDTR (SEDiment TRansport) module. The model is tested against experimental data to study the effect of representative coastal protection structures and is applied to a real case study of a coastal engineering project in North Greece, producing accurate and consistent results for a versatile range of layouts.

Keywords: coastal protection structures; integrated numerical model; waves; hydrodynamics; sediment transport; morphology evolution

---

1. Introduction

Nowadays, numerical models are the main tool for engineers involved in the design of coastal and marine structures. There are numerous examples in relevant literature of more or less advanced models, covering various aspects of wave-, hydro-, and morpho-dynamics from deep water to the nearshore, at different scales and at varying levels of detail [1–8]. However, coastal engineering practice requires robust integrated models that are able to represent in a reliable way the full range of processes governing coastal dynamics, including the effect of the presence of structures, while maintaining the computational effort needed at reasonable levels. Model versatility should also be considered as an essential requirement, since such models should be able to be adapted to a wide range of design layouts and perform satisfactorily for an equally wide range of field conditions.

For a long time, the design of coastal protection structures was essentially based on engineering experience and empirical rules. Such approaches were gradually replaced by models of varying complexity, focusing on the structures' effects on wave dynamics, circulation patterns, and morphological evolution in coastal areas. Morphodynamic processes are among the most complex ones to accurately reproduce, since they depend on the combined effect of waves and currents, whose interaction becomes increasingly complicated when moving within the breaker zone and towards the swash. Morphological evolution modelling has indeed come a long way throughout the years, from simple conceptual models to fully 3D ones, currently encompassing a series of improvements in our understanding of the involved processes. Regarding relevant literature (on the more advanced 2D horizontal, quasi-3D, and 3D models), one may indicatively refer to: pioneering [9] and more recent works [10–12] on bed morphology evolution due to the presence of detached breakwaters, studies on the morphological effects of groins and groin systems [13,14], and more complete/inclusive works on modelling coastal morphodynamics in the presence of coastal structures [15–18]. Focusing on the main features that separate similar integrated modelling attempts, one could refer to: (a) Model capabilities to represent the effects of various types of structures on coastal dynamics; (b) the representation of swash zone dynamics; and (c) the approaches and formulae used to calculate bed load and suspended load sediment transport.

In this work, an integrated coastal engineering numerical model—developed by the authors—is presented and described in detail. The model consists of three main modules that simulate: linear wave propagation (i.e., WAVE_L), wave-induced circulation (i.e., WICIR), and sediment transport along with bed morphology evolution (i.e., SEDTR). The model is capable of simulating the presence of various types of structures (vertical structures; groins and groin systems; emerged, submerged, and floating breakwaters) and includes a novel approach for the representation of swash zone hydrodynamics, while sediment transport is modelled based on the formula proposed by Camenen and Larson [19,20]. The model is tested against experimental data to study the effect of representative coastal protection structures [21,22], and—given the good agreement between calculated results and measured data—is afterwards applied to a real case study of a coastal engineering project in North Greece (combination of submerged breakwaters and beach nourishment).

2. Model Description

2.1. Nearshore Wave Transformation Module–WAVE_L

Linear wave propagation is simulated by applying a mild-slope model [23,24], derived without the assumption of progressive waves. The module WAVE_L is based on the hyperbolic-type mild slope equation and is valid for a compound wave field near coastal structures where waves are subjected to the combined effects of shoaling, refraction, diffraction, reflection (total and partial), and breaking. The module consists of the following pair of equations [23,24]:

$$\frac{\partial \eta}{\partial t} + \frac{c}{c_g} \nabla \frac{c_g}{c} \mathbf{Q}_w = 0 \tag{1}$$

$$\frac{\partial \mathbf{U}_w}{\partial t} + \frac{c^2}{d} \nabla \eta = \nu_h \nabla^2 \mathbf{U}_w \tag{2}$$

where η is the surface elevation; $\mathbf{U_w}$ is the mean velocity vector $\mathbf{U_w} = (U_w, V_w)$; d is the depth, $\mathbf{Q_w} = \mathbf{U_w} h_w = (Q_w, P_w)$; h_w is the total depth ($h_w = d + \eta$); c is the celerity; and c_g is the group velocity ($c_g = (gd)^{0.5}$). The term ν_h is a horizontal eddy viscosity coefficient introduced in order to include breaking effects based on the formulation of [25]:

$$\nu_h = 2d \left(\frac{D}{\rho}\right)^{1/3} \tag{3}$$

In Equation (3), D is the dissipation of wave energy expressed as:

$$D = \frac{1}{4} Q_b f \rho g H_m^2 \tag{4}$$

where H_m is the maximum wave height; ρ is the water density; f is the wave frequency; and Q_b is the probability of a wave breaking at a certain depth, expressed as $(1 - Q_b)/(\ln Q_b) = (H_{rms}/H_m)^2$ according to [26]. The mean square wave height H_{rms} is calculated from $H_{rms} = 2(<2\eta^2>)^{1/2}$, with the brackets denoting a time-mean quantity. It should be noted that—since linear wave models are not capable of describing waves in the swash zone—in WAVE_L, the water depth from the rundown point (i.e., depth equal to $R/4$; R is the runup height) and up to the runup point (i.e., depth equal to $-R$) is considered to be constant and equal to $R/4$.

WAVE_L is adapted for engineering applications based on the following:

1. The input wave is introduced at a line inside the computational domain according to [27,28].
2. A sponge layer boundary condition is used to absorb the outgoing waves at the four sides of the domain [27].
3. The presence of vertical structures is incorporated by introducing a total reflection boundary condition ($\mathbf{U_w} = (U_w. V_w) = 0$ normal to the boundary, where U_w is the mean velocity vector; for a rectilinear grid, the above is equivalent to $U_w = 0$ or $V_w = 0$).
4. Partial reflection is also simulated, by introducing an artificial eddy viscosity coefficient ν_h. The values of ν_h are estimated from the method developed by Karambas and Bowers [29], using the reflection coefficient values proposed by Bruun [30].
5. The presence of submerged structures is incorporated as in [31].
6. The presence of floating structures is incorporated as in [32].

The numerical solution is based on the well-documented explicit second order finite difference staggered scheme using a mid-time method [24].

2.2. Wave-Induced Circulation Module—WICIR

The depth and shortwave-averaged 2D continuity and momentum equations are used for simulating nearshore currents in the coastal zone. They are expressed as:

$$\frac{\partial \zeta}{\partial t} + \frac{\partial (Uh)}{\partial x} + \frac{\partial (Vh)}{\partial y} = 0 \tag{5}$$

$$\frac{\partial U}{\partial t} + U\frac{\partial U}{\partial x} + V\frac{\partial U}{\partial y} + g\frac{\partial \zeta}{\partial x} = -\frac{1}{\rho h}\left(\frac{\partial S_{xx}}{\partial x} + \frac{\partial S_{xy}}{\partial y}\right) + \frac{1}{h}\frac{\partial}{\partial x}\left(\nu_h h\frac{\partial U}{\partial x}\right) + \frac{1}{h}\frac{\partial}{\partial y}\left(\nu_h h\frac{\partial U}{\partial y}\right) - \frac{\tau_{bx}}{\rho h} \tag{6}$$

$$\frac{\partial V}{\partial t} + U\frac{\partial V}{\partial x} + V\frac{\partial V}{\partial y} + g\frac{\partial \zeta}{\partial y} = -\frac{1}{\rho h}\left(\frac{\partial S_{xy}}{\partial x} + \frac{\partial S_{yy}}{\partial y}\right) + \frac{1}{h}\frac{\partial}{\partial x}\left(\nu_h h\frac{\partial V}{\partial x}\right) + \frac{1}{h}\frac{\partial}{\partial y}\left(\nu_h h\frac{\partial V}{\partial y}\right) - \frac{\tau_{by}}{\rho h} \tag{7}$$

where S_{xx}, S_{yy}, and S_{xy} are the radiation stresses; $h = d + \zeta$ (ζ being the mean water elevation); U and V are the depth-averaged current velocities; and τ_{bx} and τ_{by} are the bottom shear stresses. Based on linear wave theory, Copeland [33] derived the equations for radiation stresses (S_{ij}) without the typical assumption of progressive waves, expressed as:

$$\begin{aligned} \frac{S_{xx}}{\rho} = \ & d^2 < U_w^2 > A_r - d^2 < \left(\frac{\partial U_w}{\partial x} + \frac{\partial V_w}{\partial y}\right)^2 > B_r + \frac{\partial}{\partial x} < \left[U_w\left(\frac{\partial U_w}{\partial x} + \frac{\partial V_w}{\partial y}\right)\right] > D_r + \\ & \frac{\partial}{\partial y} < \left[V_w\left(\frac{\partial U_w}{\partial x} + \frac{\partial V_w}{\partial y}\right)\right] > D_r + \frac{1}{2}g < \eta^2 > \end{aligned} \tag{8}$$

$$\begin{aligned} \frac{S_{yy}}{\rho} = \ & d^2 < V_w^2 > A_r - d^2 < \left(\frac{\partial U_w}{\partial x} + \frac{\partial V_w}{\partial y}\right)^2 > B_r + d^2\frac{\partial}{\partial y} < \left[V_w\left(\frac{\partial U_w}{\partial x} + \frac{\partial V_w}{\partial y}\right)\right] > D_r + \\ & d^2\frac{\partial}{\partial x} < \left[U_w\left(\frac{\partial U_w}{\partial x} + \frac{\partial V_w}{\partial y}\right)\right] > D_r + \frac{1}{2}g < \eta^2 > \end{aligned} \tag{9}$$

$$\frac{S_{xy}}{\rho} = d^2 < U_w V_w > A_r \tag{10}$$

$$A_r = \frac{k}{4\sinh^2 kd}(\sinh 2kd + 2kd) \tag{11}$$

$$B_r = \frac{1}{4k\sinh^2 kd}(\sinh 2kd - 2kd) \tag{12}$$

$$D_r = \frac{d}{4\sinh^2 kd}\left(\frac{1}{2kd}\sinh 2kd - \cosh 2kd\right) \tag{13}$$

where k is the wave number and the brackets denote time-mean quantities.

In nearshore circulation models, the treatment of the bottom stress is critical. The bottom shear stresses $\boldsymbol{\tau}_b = (\tau_{bx}, \tau_{by})$ in WICIR are calculated based on the formulae proposed by Kobayashi et al. [34]:

$$\tau_{bx} = \frac{1}{2}\rho f_b \sigma_T^2 G_{bx} \tag{14}$$

$$\tau_{by} = \frac{1}{2}\rho f_b \sigma_T^2 G_{by} \tag{15}$$

$$G_{bx} = \frac{U}{\sigma_T}\left[1.16^2 + \left(\frac{|U|}{\sigma_T}\right)^2\right]^{0.5} \tag{16}$$

$$G_{by} = \frac{V}{\sigma_T}\left[1.16^2 + \left(\frac{|U|}{\sigma_T}\right)^2\right]^{0.5} \tag{17}$$

where f_b is the bottom friction factor, σ_T is the standard deviation of the oscillatory horizontal velocity, and $|U| = (U^2 + V^2)^{0.5}$.

Since it acts as the "link" between WAVE_L and SEDTR in the framework of the proposed integrated model, it is important for WICIR to be able to reproduce a number of processes that are essential for the realistic description of sediment transport.

Regarding the surf zone, it should be noted that the existence of the undertow (i.e., the current directed offshore) cannot be directly predicted by depth-averaged models; nonetheless, its representation is essential in the aforementioned context. In WICIR, quasi-3D effects are introduced by adopting the analytical expression for the vertical distribution of the cross-shore flow below the wave trough level proposed by Stive and Wind [35], expressed as:

$$v_u = \frac{1}{2}\left[(\xi-1)^2 - \frac{1}{3}\right]\frac{h-\zeta_t}{\rho\,\nu_\tau}\frac{dR}{dy} + \left(\xi - \frac{1}{2}\right)\frac{(h-\zeta_t)\tau_s}{\rho\nu_\tau} - \frac{M\cos\Theta}{h-\zeta_t} \tag{18}$$

where v_u is the undertow velocity in the direction normal to the shore, $\xi = z/(h-\zeta_t)$, $h = d + \zeta$ (ζ being the mean water elevation), ζ_t is the wave trough level, $dR/dy = 0.14\rho gdh/dy$, τ_s is the shear stress at the wave trough level, M is the wave mass flux above trough level (including surface roller effects), Θ is the direction of wave propagation ($\Theta = \arctan[(<Q_w{}^2>/<P_w{}^2>)^{1/2}]$), and ν_τ is the eddy viscosity coefficient according to De Vriend and Stive [36]:

$$\nu_\tau = 0.025h\left(\frac{D}{\rho}\right)^{1/3} \tag{19}$$

Regarding the swash zone, an essential process for shoreline evolution is longshore sediment transport. For obliquely incident waves, the trajectory of the bore-front follows a parabolic movement in the swash, in the direction of the net longshore flow per wave period. The mean longshore transport velocity V_R at the shoreline is determined according to Baba and Camenen [37] as:

$$V_R = \sqrt{2gR}\sin\Theta \tag{20}$$

where R is the runup height ($R = 1.6H_0\xi_0$, where H_0 is the deep water wave height and ξ_0 is the Iribarren number) and Θ is the wave direction near the rundown point at depth $d = R/4$. The longshore velocity V_R is presumed constant within the swash zone, the width of which is considered as extending from $d = R/4$ (i.e., the rundown point) to $d = -R$. The above velocity is indirectly introduced in the model by increasing the radiation stresses in the swash zone, based on the rationale described in the following. Longshore velocity can be expressed analytically by:

$$\overline{U} = 2.7\frac{\gamma}{2}\sqrt{gd_b}\sin\alpha_b\cos\alpha_b \tag{21}$$

where γ is the breaking index, and d_b and α_b are the water depth and incident wave angle at the breaking point, respectively. Assuming a linear variation of $\overline{U}$, the velocity at the shoreline can be approximated as:

$$\overline{U} = 2.7\gamma\sqrt{gd_b}\sin\alpha_b\cos\alpha_b\frac{d_s}{d_b} \tag{22}$$

where d_s is the water depth at the shoreline. A comparison of Equation (20) to Equation (22) shows that the square of the ratio does not deviate significantly from an empirical factor, a_s, expressed as:

$$a_s = 16\sqrt{\gamma\xi^{1.8}(H_0/L_0)^{0.2}} \tag{23}$$

where ξ_0 is the Iribarren number, and H_0 and L_0 are the wave height and wavelength, respectively, for deep water conditions. Accordingly, the aforementioned increase in radiation stresses in the swash zone is achieved by multiplying them by the factor a_s.

Finally, regarding flooding due to wave setup, in WICIR, this process is simulated using the "dry bed" boundary condition which, according to Militello et al. [38], can be written as the following set of pairs of conditions for any given grin point (i,j):

if $(d + \zeta)_{i,j} > h_{cr}$ and $(d + \zeta)_{i-1,j} \leq h_{cr}$ and $U_{i,j} > 0 \rightarrow U_{i,j} = 0$
if $(d + \zeta)_{i,j} > h_{cr}$ and $(d + \zeta)_{i,j-1} \leq h_{cr}$ and $V_{i,j} > 0 \rightarrow V_{i,j} = 0$

if $(d + \zeta)_{i,j} \leq h_{cr}$ and $(d + \zeta)_{i-1,j} \leq h_{cr} \rightarrow U_{i,j} = 0$
if $(d + \zeta)_{i,j} \leq h_{cr}$ and $(d + \zeta)_{i,j-1} \leq h_{cr} \rightarrow V_{i,j} = 0$

if $(d + \zeta)_{i,j} \leq h_{cr}$ and $(d + \zeta)_{i-1,j} > h_{cr}$ and $U_{i,j} < 0 \rightarrow U_{i,j} = 0$
if $(d + \zeta)_{i,j} \leq h_{cr}$ and $(d + \zeta)_{i,j-1} > h_{cr}$ and $V_{i,j} < 0 \rightarrow V_{i,j} = 0$

where ζ is the mean water surface elevation and h_{cr} is a terminal depth below which drying is assumed to occur (e.g., in WICIR this depth is set to $h_{cr} = 0.001$ m).

The numerical solution in WICIR is also (as in WAVE_L) based on the explicit second order finite difference staggered scheme using a mid-time method [24].

2.3. Sediment Transport Module—SEDTR

The mode of sediment movement on the coast is usually divided into bed load, suspended load, and sheet flow transport. Different model concepts are being presently used for the prediction of each one, which range from empirical transport formulae to more sophisticated bottom boundary layer models. In the present work, bed load transport (q_b) is estimated with a quasi-steady, semi-empirical formulation, developed by Camenen, and Larson [19,20] for an oscillatory flow combined with a superimposed current under an arbitrary angle:

$$\Phi_b = \begin{cases} \dfrac{q_{b,w}}{\sqrt{(s-1)gd_{50}^3}} = a_w\sqrt{\theta_{cw,net}}\theta_{cw,m}\exp\left(-b\frac{\theta_{cr}}{\theta_{cw}}\right) \\[2ex] \dfrac{q_{b,n}}{\sqrt{(s-1)gd_{50}^3}} = a_n\sqrt{\theta_{cn}}\theta_{cw,m}\exp\left(-b\frac{\theta_{cr}}{\theta_{cw}}\right) \end{cases} \tag{24}$$

where the subscripts w and n correspond, respectively, to the wave direction and the direction normal to the wave direction; s (= ρ_s/ρ) is the relative density between the sediment (ρ_s) and water (ρ); g is the acceleration due to gravity; d_{50} is the median grain size; a_w, a_n, and b are empirical coefficients; $\theta_{cw,m}$ and θ_{cw} are the mean and maximum Shields parameters due to the wave-current interaction, respectively; θ_{cn} is the current-related Shields parameter in the direction normal to the wave direction, and θ_{cr} is the critical Shields parameter for the inception of transport. The net Shields parameter $\theta_{cw,net}$ in Equation (24) is given by:

$$\theta_{cw,net} = \left(1 - \alpha_{pl,b}\right)\left(1 + \alpha_\alpha\right)\theta_{cw,on} - \left(1 + \alpha_{pl,b}\right)\left(1 - \alpha_\alpha\right)\theta_{cw,off} \tag{25}$$

where $\theta_{cw,on}$ and $\theta_{cw,off}$ are the mean values of the instantaneous Shields parameter over the two half "periods" T_{wc} (crest-onshore) and T_{wt} (trough-offshore), $\alpha_{pl,b}$ is a coefficient for the phase-lag effects [19], and α_α is a coefficient for the acceleration effects [39]. The Shields parameter θ_{cw} is defined by:

$$\theta_{cw,j} = \frac{1}{2}f_{cw}U^2_{cw,j}/[(s-1)gd_{50}] \tag{26}$$

with U_{cw} being the wave and current velocity and f_{cw} the friction coefficient taking into account the wave and current interaction, while the subscript j should be replaced either by *onshore* or *offshore*. In the above formulation (since linear wave theory cannot be used), the estimation of nonlinear time-varying near-bottom wave velocities is also needed. For the incorporation of nonlinear velocity characteristics (i.e., skewness and asymmetry) in SEDTR, the parameterisation proposed by Isobe and Horikawa [40] is adopted.

The incorporation of the suspended sediment transport rate in SEDTR is done by solving the depth-integrated transport equation for suspended sediment [41,42]:

$$\frac{\partial(hC)}{\partial t} + \frac{\partial(hCU)}{\partial x} + \frac{\partial(hCV)}{\partial x} = c_R w_s - w_s\frac{C}{\beta_d} \tag{27}$$

where h is the total mean depth, C is the depth-averaged volumetric sediment concentration, c_R is the reference concentration at the bottom [19], w_s is the sediment fall velocity, and β_d is a coefficient calculated based on [20] by:

$$\beta_d = \frac{\varepsilon}{w_s}\left[1 - \exp\left(-\frac{w_s h}{\varepsilon}\right)\right] \tag{28}$$

with ε being the sediment diffusivity (related to the eddy viscosity coefficient), estimated by [36]:

$$\varepsilon = 0.025\,h\,(D/\rho)^{1/3} \tag{29}$$

Cross-shore sediment transport in the swash zone in SEDTR is calculated according to [43], while for the longshore sediment transport, only the increased mean velocity is taken into account, as described in Section 2.2.

These sediment transport rates are then used for the simulation of the coastal bathymetry changes by the module SEDTR. The methodology adopted for the series of model applications can be encoded into the steps described in the following. First, the initial bathymetry is inserted into the wave and wave-induced circulation modules (WAVE_L and WICIR, respectively) in order to estimate the wave and current fields. These fields are afterwards used by the sediment transport module SEDTR to calculate the sediment transport rates. Finally, bathymetry is updated by SEDTR solving the equation of the conservation of sediment transport (for the previously calculated transport rates; [44]). The procedure is repeated for a user-specified time period or until a state of morphologic equilibrium is reached. The aforementioned repetitions take place after bottom change in the order of 10–15% is observed in the field, so that the changes in wave and wave-induced current fields calculated by WAVE_L and WICIR for the updated bathymetry are significant.

3. Model Applications

3.1. Comparison with Experimental Data

The integrated numerical model was set-up and applied in order to reproduce the small-scale laboratory experiments of: (a) Ming and Chiew [21], who studied the shoreline changes caused by the presence of a detached breakwater under the influence of pure wave action; and (b) Badiei et al. [22], who studied the morphological effects of groins on an initially straight beach exposed to oblique irregular waves.

The experiments of Ming and Chiew [21] were conducted in a 10 m long, 5 m wide, and 0.7 m high wave basin. A plunger-type wavemaker was used to generate monochromatic waves and sponge was placed behind the wavemaker in order to minimize wave reflection. The 6 m long beach consisted of uniformly distributed sand with a median grain size of d_{50} = 0.25 mm. The duration of the tests was approximately 15 h (which was the duration needed for the beach to reach an equilibrium state). Three different cases were reproduced numerically and are presented in the following, for normally incident waves of H_0 = 0.05 m deep water wave height and T = 0.85 s wave period. The test cases, presented in Table 1, differed in breakwater length (B) and breakwater distance from the initial shoreline (X), in order to cover a wide range of B/X ratios resulting in both tombolo and salient formation behind the breakwaters.

Table 1. Test conditions for the numerically reproduced experiments of Ming and Chiew [21].

Test	B = Breakwater Length (m)	X = Distance from the Initial Shoreline (m)	B/X	Formation of Salient/Tombolo
3	1.5	0.6	2.50	tombolo
10	1.2	1.2	1.00	salient
11	1.5	1.2	1.25	tombolo

Baidei et al. [22] employed a series of mobile bed process models (according to [45]) in order to investigate the impact of groins on nearshore morphology under the attack of obliquely incident random waves. Two series of tests were carried out at the Queen's University Coastal Engineering Laboratory (QUCERL) and the Hydraulic Laboratory of the National Research Council of Canada (NRCC). The physical model regarded an initially plane sloping beach (1:10 slope), composed of D_{50} = 0.12 mm sand grains. The beach—without the presence of the groins—was exposed to wave action for a duration of 4 h until the formation of a nearly stable bathymetry (clear offshore bar trough/step formation). The installation of the groins followed, and the tests continued thereafter in 2 h cycles. In this work, the case of a single groin was modelled, exposed to waves of H_{s0} = 0.08 m deep water significant wave height, T_p = 1.15 s peak period, and θ_0 = 11.6° deep water incident wave angle, for a total duration of 12 h after groin installation (Test NT2, NRCC test series).

3.2. Application to Paralia Katerinis Beach (Greece)—Coastal Protection with Submerged Breakwaters

Following its validation for laboratory experiments, the presented integrated model was applied to a real case study of a coastal engineering project regarding the protection of a beach by using detached submerged breakwaters. The study area is located in the Region of Central Macedonia, Greece, at a sandy beach north of the fishing port of Katerini (Figure 1). The area has been facing coastal erosion problems for well over 30 years, which started after the construction of the harbour seen in the left part of Figure 1 (1980–1984). As a result, and due to the prevailing SE winds, the coastal zone south of the fishing port showed strong accretion, while the coast north from the port was eroded, with a shoreline retreat in the order of 20 m; the erosive phenomena stretched over a zone of 800 m north from the port. In order to reverse erosion, a groin field consisting of 13 rubble mount groins was constructed (1990–1997). The project not only failed to further protect the beach—since additional erosion occurred in between the constructed groins—but it also transferred coastal retreat northwards.

Furthermore, the semi-closed basins that formed between the constructed groins (being in the non-tidal Mediterranean Sea) caused significant environmental problems regarding water quality due to the limited renewal rates. In 2010, a new coastal protection project was designed and constructed; the groin field was replaced by a set of three 200 m long submerged breakwaters placed at a distance of approximately 200 m from the coast, and the gaps between the structures were approximately 110 m. The breakwaters were designed to have transmission coefficients in the order of 0.4 ($K_t \approx 0.4$). In addition, a beach nourishment project was also designed and applied to restore the beach to its previous condition. Bathymetry measurement data for the area are available for the period right after the completion of coastal works (beach nourishment and construction of the submerged breakwaters), as well as for three years later [46].

Figure 1. Location and satellite image of the study area at Paralia Katerinis, Greece ([47]; privately processed).

The integrated model was applied to simulate coastal morphodynamics after the realization of the coastal protection project. Since wave data were not available for the area, hindcast data were used. The main incident wind directions are: NE, E, and SE-S. The model was run by applying three representative waves (i.e., three equivalent wave heights on an annual basis; see Table 2). The workflow for the coupled module runs can be summed-up in the following. Starting with the initial bathymetry, WAVE_L, WICIR and SEDTR modules were run in sequence for the characteristics of the first representative wave and taking into account its annual frequency of occurrence for the simulation of morphology evolution. The updated bed morphology was then used to run the integrated model for the second representative wave (in the same way) and the morphology at the end of this second run for the third representative wave. The aforementioned simulation steps were repeated until the total duration of the wave action was reached.

Table 2. Characteristics of the three representative waves used for the Paralia Katerinis beach runs.

Wave Direction	Significant Wave Height H_s (m)	Peak Wave Period T_p (s)	Annual Frequency of Occurence f (%)
SE-S	1.34	5.4	12.70
E	1.09	4.4	1.12
NE	0.94	4.6	1.17

4. Results and Discussion

4.1. Comparison with Experimental Data

Figure 2 shows the initial wave-induced current velocity field and the comparison between the computed and measured shoreline evolution data for Test 11 of Ming and Chiew [21]. The presence of the breakwater leads to the formation of two opposing eddies in the area behind it, as currents move towards the sheltered area along the foreshore from both sides of the structure. A secondary cause of the observed circulation pattern is the mean sea level gradient between the illuminated and sheltered areas due to diffraction effects, while the representation of swash zone hydrodynamics by the model should also be highlighted (see also Section 2.2). Regarding morphology evolution, the model results are in good agreement with the measured data of Ming and Chiew [21], satisfactorily representing the formation of the tombolo in Test 11 (Figure 2b), as well as the formation of the tombolo and salient in Tests 3 (Figure 3a) and 10 (Figure 3b), respectively.

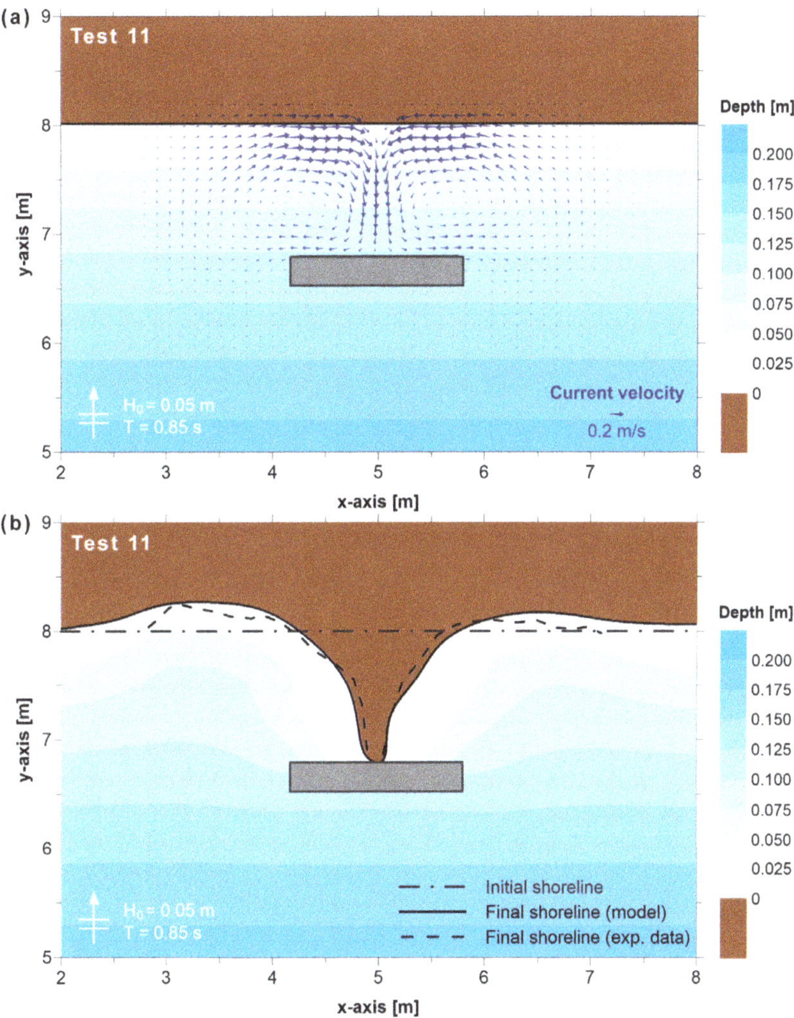

Figure 2. (a) Initial wave-induced current velocity field (contours represent the initial bathymetry) and (b) comparison between the computed and measured shoreline evolution (contours represent the final computed bathymetry) for Test 11 of Ming and Chiew [21].

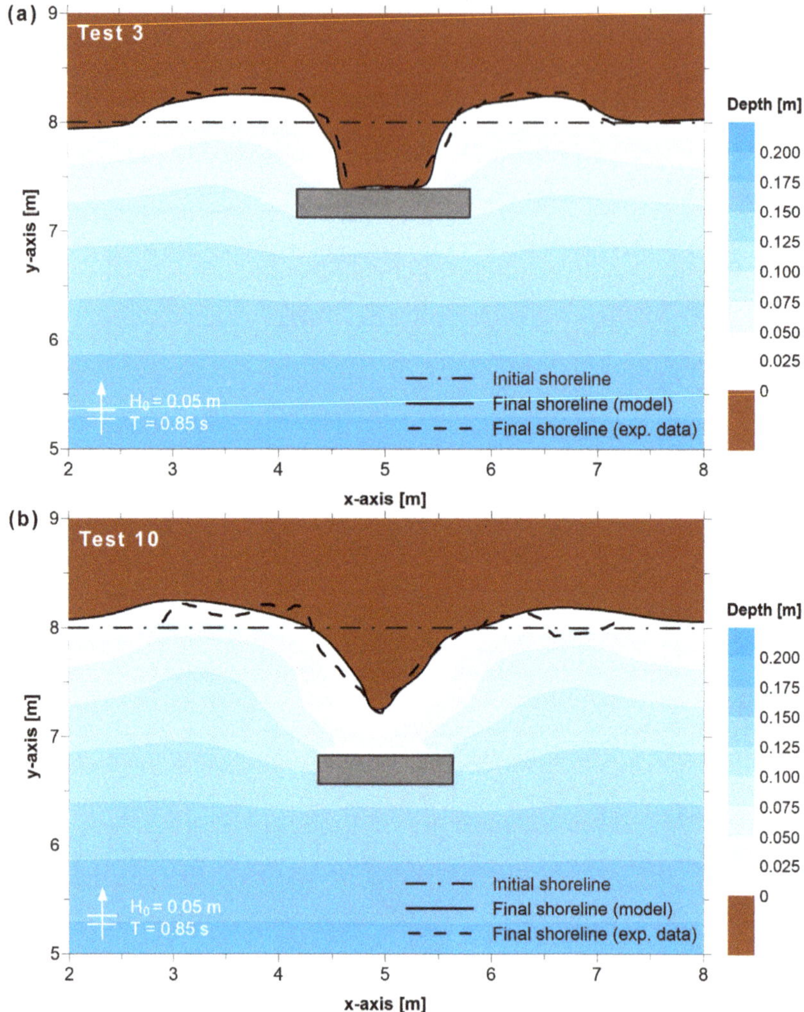

Figure 3. Comparison between the computed and measured shorelines evolution (contours represent the final computed bathymetry) for: (**a**) Test 3 and (**b**) Test 10 of Ming and Chiew [21].

Figure 4 shows the initial breaking wave-induced current velocity field and the comparison between the computed and measured shoreline evolution for the single groin test (Test NT2) of Baidei et al. [22]. The sediment accretion updrift of the groin-type structure results in an advance of the shore, while the lack of sediments at the lee of the groin leads to a retreat of the shore; shoreline evolution is well-reproduced, with model results and measurement lines practically overlapping.

In general, the integrated model results in smooth but consistent (considering also test runs that are not presented in this work) bathymetries behind detached breakwaters and in the vicinity of groin-type structures, while it appears to be smoothing in a close-to-natural way eventual local shoreline irregularities which, on the other hand, are present in the experimental results.

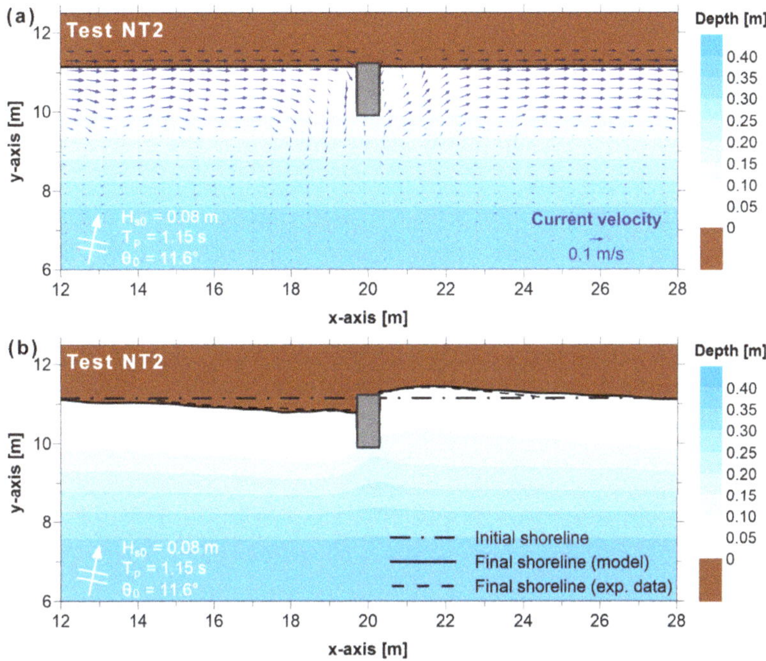

Figure 4. (**a**) Initial breaking wave-induced current velocity field (contours represent the initial bathymetry) and (**b**) comparison between the computed and measured shoreline evolution (contours represent the final computed bathymetry) for Test NT2 of Baidei et al. [22].

4.2. Application to Paralia Katerinis Beach (Greece)—Coastal Protection with Submerged Breakwaters

Figure 5 shows the breaking wave-induced current field for the prevailing SE-S waves. The submerged breakwaters allow some wave transmission and overtopping that cause an additional supply of water behind the structures, which is taken into account by the circulation module and affects the current field. This net transport of water into the lee zone causes a water level rise and is balanced mainly by outgoing currents at the heads of the structures. Consequently, the main flow pattern is characterized by an onshore flow over the submerged breakwaters, an offshore flow at the gaps between them (eroding rip currents), and nearshore eddies similar to those formed in the case of emerged breakwaters (although with lower relative intensity). The first two flow patterns do not exist in the latter case, while the third flow pattern is not extended up to the structures, where onshore flow leads to the formation of—more intense—opposite direction eddies.

Figure 6 shows the initial bed morphology of Paralia Katerinis beach and the comparison between the computed and measured bed morphology evolution, along with the satellite image of the same area (2016). The workflow for the coupled module runs is described in Section 3.2. The model results are in close agreement with the measurements (respective lines are practically overlapping), with the integrated model succeeding in reproducing all morphological patterns behind the breakwaters and up to the shoreline, under the presence of both permanent structures (emerged groins) and the realized beach nourishment project there.

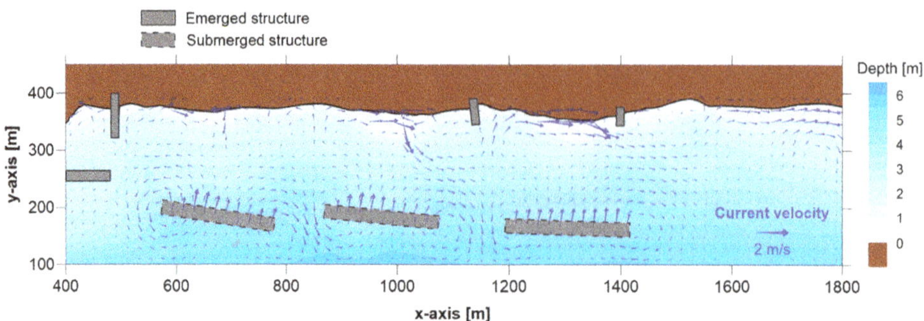

Figure 5. Breaking wave-induced current field for the prevailing SE waves at Paralia Katerinis beach (Greece) after the construction of the detached submerged breakwaters.

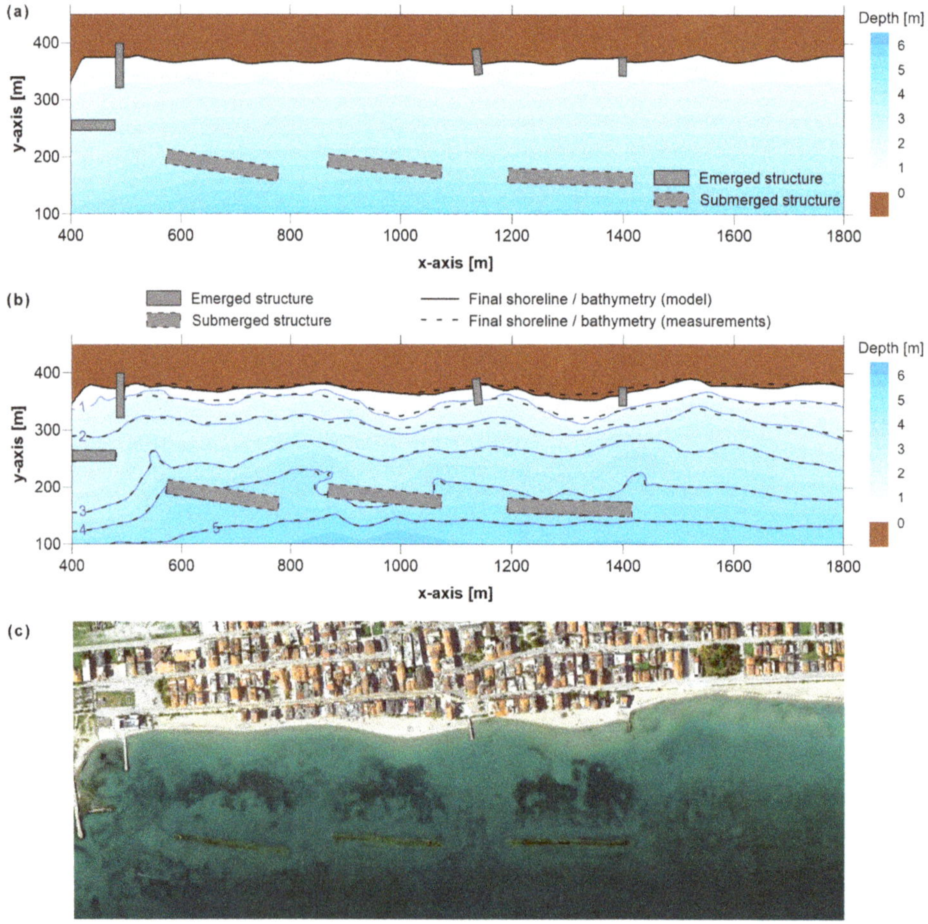

Figure 6. (**a**) Initial bed morphology of Paralia Katerinis beach, (**b**) comparison between the computed and measured bed morphology evolution, and (**c**) satellite image of the same area ([47]; privately processed).

4.3. General Discussion

Elaborating further on the integrated model's performance in simulating the morphological effects of the presence of coastal protection structures, particular insights can be drawn by the discussion in the following.

Model setup was based on the successful model calibration for the experimental data presented in this work (see Sections 3.1 and 4.1), along with specific modelling choices based on extensive experience in morphological modelling for both research and engineering applications. The calibration process of the presented model mainly refers to: (a) swash zone incorporation, i.e., the introduction of coefficient a_s (see Equation (23)) and the region of its application (i.e., the swash zone as defined in Section 2.2); (b) the use of the bottom friction formulae proposed by Kobayashi et al. [34]; and (c) the use of the Camenen and Larson [19,20] formulae for the calculation of bed and suspended load (instead of other approaches).

The satisfactory agreement between model predictions and experimental/field data should be mainly attributed to the above choices, which are deemed to distinguish the presented model from relevant work. The novelty of incorporating swash zone hydrodynamics using a linear wave model and a nonlinear wave-induced circulation model particularly allows for the simulation of swash zone morphodynamics which play an essential role in nearshore morphology evolution. The above, keeping in mind that the objective of this work was to present an integrated model that could be adapted to a wide range of design layouts and perform satisfactorily for an equally wide range of field conditions, while maintaining the computational effort needed at reasonable levels, an aspect that usually limits the applicability of more complex models in engineering applications.

5. Conclusions

This work presents an integrated coastal engineering numerical model that simulates linear wave propagation, wave-induced circulation, sediment transport, and bed morphology evolution. The model consists of three main modules: The nearshore wave transformation model WAVE_L, properly adapted for the simulation of compound linear wave fields near coastal structures; the wave-induced circulation module WICIR, which includes a novel approach for the representation of swash zone hydrodynamics; and the sediment transport and bed morphology evolution module SEDTR.

The model is tested against experimental data to study the effect of representative coastal protection structures, such as detached breakwaters (data from [21]) and groins (data from [22]). Given the good agreement between model results and laboratory measurements, the model was also successfully applied to a real case study of a coastal engineering project in North Greece (combination of submerged breakwaters and beach nourishment). The model is deemed to constitute a suitable tool for the design and evaluation of the morphological influence of harbour and coastal protection works, being able to deliver results in a fast and seamless way at all times for a wide range of design layouts.

Acknowledgments: The authors would like to thank the anonymous reviewers for their constructive comments and suggestions.

Author Contributions: T.V.K. and A.G.S. collaboratively developed the presented integrated numerical model, conceived and designed this work, performed the numerical simulations for the model runs, and wrote the paper.

Conflicts of Interest: The authors declare no conflict of interest.

References and Note

1. Rakha, K.A. A Quasi-3D phase-resolving hydrodynamic and sediment transport model. *Coast. Eng.* **1998**, *34*, 277–311. [CrossRef]
2. Karambas, T.V.; Koutitas, C. Surf and swash zone morphology evolution induced by nonlinear waves. *J. Waterw. Port Coast. Ocean Eng.* **2002**, *128*, 102–113. [CrossRef]
3. Karambas, T.V.; Karathanassi, E.K. Longshore sediment transport by nonlinear waves and currents. *J. Waterw. Port Coast. Ocean Eng.* **2004**, *130*, 277–286. [CrossRef]

4.   Karambas, T.V. Prediction of sediment transport in the swash-zone by using a nonlinear wave model. *Cont. Shelf Res.* **2006**, *26*, 599–609. [CrossRef]

5.   Karambas, T.V.; Samaras, A.G. Soft shore protection methods: The use of advanced numerical models in the evaluation of beach nourishment. *Ocean Eng.* **2014**, *92*, 129–136. [CrossRef]

6.   Gallerano, F.; Cannata, G.; De Gaudenzi, O.; Scarpone, S. Modeling bed evolution using weakly coupled phase-resolving wave model and wave-averaged sediment transport model. *Coast. Eng. J.* **2016**, *58*, 58. [CrossRef]

7.   Gallerano, F.; Cannata, G.; Lasaponara, F. A new numerical model for simulations of wave transformation, breaking and long-shore currents in complex coastal regions. *Int. J. Numer. Methods Fluids* **2016**, *80*, 571–613. [CrossRef]

8.   Klonaris, G.T.; Memos, C.D.; Drønen, N.K.; Deigaard, R. Boussinesq-type modeling of sediment transport and coastal morphology. *Coast. Eng. J.* **2017**, *59*, 59. [CrossRef]

9.   Watanabe, A.; Maruyama, K.; Shimizu, T.; Sakakiyama, T. Numerical prediction model of three-dimensional beach deformation around a structure. *Coast. Eng. J.* **1986**, *29*, 179–194.

10. Nicholson, J.; Broker, I.; Roelvink, J.A.; Price, D.; Tanguy, J.M.; Moreno, L. Intercomparison of coastal area morphodynamic models. *Coast. Eng.* **1997**, *31*, 97–123. [CrossRef]

11. Zyserman, J.A.; Johnson, H.K. Modelling morphological processes in the vicinity of shore-parallel breakwaters. *Coast. Eng.* **2002**, *45*, 261–284. [CrossRef]

12. Postacchini, M.; Russo, A.; Carniel, S.; Brocchini, M. Assessing the hydro-morphodynamic response of a beach protected by detached, impermeable, submerged breakwaters: A numerical approach. *J. Coast. Res.* **2016**, *32*, 590–602. [CrossRef]

13. Roelvink, J.A.; Uittenbogaard, R.E.; Liek, G.J. Morphological modelling of the impact of coastal structures. In *Coastal Structures '99, Proceedings of the International Conference Coastal Structures '99, Santander, Spain, June 1999*; Losada, I.J., Balkema, A.A., Eds.; CRC Press: Boca Raton, FL, USA, 2000; pp. 865–871.

14. Johnson, H.K. Coastal area morphological modelling in the vicinity of groins. In Proceedings of the 29th International Conference on Coastal Engineering, Lisbon, Portugal, 19–24 September 2004; pp. 2646–2658.

15. Ding, Y.; Wang, S.Y.; Jia, Y. Development and validation of a quasi-three-dimensional coastal area morphological model. *J. Waterw. Port Coast. Ocean Eng.* **2006**, *132*, 462–476. [CrossRef]

16. Ding, Y.; Wang, S.S.Y. Development and application of a coastal and estuarine morphological process modeling system. *J. Coast. Res.* **2008**, 127–140. [CrossRef]

17. Leont'yev, I.O. Modelling of morphological changes due to coastal structures. *Coast. Eng.* **1999**, *38*, 143–166. [CrossRef]

18. Nam, P.T.; Larson, M.; Hanson, H.; Hoan, L.X. A numerical model of beach morphological evolution due to waves and currents in the vicinity of coastal structures. *Coast. Eng.* **2011**, *58*, 863–876. [CrossRef]

19. Camenen, B.; Larson, M. *A Unified Sediment Transport Formulation for Coastal Inlet Application*; ERDC/CH: CR-07-1; US Army Corps of Engineers, Engineering Research and Development Center: Vicksburg, MS, USA, 2007; p. 247.

20. Camenen, B.; Larson, M. A general formula for noncohesive suspended sediment transport. *J. Coast. Res.* **2008**, *24*, 615–627. [CrossRef]

21. Ming, D.; Chiew, Y. Shoreline changes behind detached breakwater. *J. Waterw. Port Coast. Ocean Eng.* **2000**, *126*, 63–70. [CrossRef]

22. Badiei, P.; Kamphuis, J.; Hamilton, D. Physical experiments on the effects of groins on shore morphology. In Proceedings of the 24th International Conference on Coastal Engineering, Kobe, Japan, 23–28 October 1994; pp. 1782–1796.

23. Copeland, G.J.M. A practical alternative to the "mild-slope" wave equation. *Coast. Eng.* **1985**, *9*, 125–149. [CrossRef]

24. Watanabe, A.; Maruyama, K. Numerical modelling of nearshore wave field under combined refraction, diffraction and breaking. *Coast. Eng. Japan* **1986**, *29*, 19–39.

25. Battjes, J.A. Modelling of turbulence in the surf zone. In Proceedings of the Symposium on Modelling Techniques, San Francisco, CA, USA, 3–5 September 1975; ASCE: San Francisco, CA, USA, 1975; pp. 1050–1061.

26. Battjes, J.A.; Janssen, J.P.F.M. Energy loss and set-up due to breaking of random waves. In Proceedings of the 16th International Conference on Coastal Engineering, Hamburg, Germany, 27 August–3 September 1978; pp. 569–587.

27. Larsen, J.; Dancy, H. Open boundaries in short wave simulations—A new approach. *Coast. Eng.* **1983**, *7*, 285–297. [CrossRef]

28. Lee, C.; Suh, K.D. Internal generation of waves for time-dependent mild-slope equations. *Coast. Eng.* **1998**, *34*, 35–57. [CrossRef]

29. Karambas, T.V.; Bowers, E.C. Representation of partial wave reflection and transmission for rubble mound coastal structures. *WIT Trans. Ecol. Environ.* **1996**, *12*, 415–423.

30. Bruun, P. *Design and Construction of Mounds for Breakwaters and Coastal Protection*; Elsevier: Amsterdam, The Netherland, 1985; Volume 37.

31. Kriezi, E.; Karambas, T. Modelling wave deformation due to submerged breakwaters. *Marit. Eng. Proc. Inst. Civ. Eng.* **2010**, *163*, 19–29. [CrossRef]

32. Koutandos, E.V.; Karambas, T.V.; Koutitas, C.G. Floating breakwater response to waves action using a Boussinesq model coupled with a 2DV elliptic solver. *J. Waterw. Port Coast. Ocean Eng.* **2004**, *130*, 243–255. [CrossRef]

33. Copeland, G.J.M. Practical radiation stress calculations connected with equations of wave propagation. *Coast. Eng.* **1985**, *9*, 195–219. [CrossRef]

34. Kobayashi, N.; Agarwal, A.; Johnson, B.D. Longshore current and sediment transport on beaches. *J. Waterw. Port Coast. Ocean Eng.* **2007**, *133*, 296–304. [CrossRef]

35. Stive, M.J.F.; Wind, H.G. Cross-shore mean flow in the surf zone. *Coast. Eng.* **1986**, *10*, 325–340. [CrossRef]

36. De Vriend, H.J.; Stive, M.J.F. Quasi-3D modelling of nearshore currents. *Coast. Eng.* **1987**, *11*, 565–601. [CrossRef]

37. Baba, Y.; Camenen, B. Importance of the Swash Longshore Sediment Transport in Morphodynamic Models. In Proceedings of the Coastal Sediments '07 6th International Symposium on Coastal Engineering and Science of Coastal Sediment Process, Louisiana, LA, USA, 13–17 May 2007; Kraus, N.C., Rosati, J.D., Eds.; ASCE: San Francisco, CA, USA, 2012.

38. Militello, A.; Reed, C.W.; Zundel, A.K.; Kraus, N.C. *Two-Dimensional Depth-Averaged Circulation Model M2D: Version 2.0, Report 1, Technical Documentation and User's Guide*; ERDC/CHL TR-04-2; US Army Corps of Engineers, Engineering Research and Development Center: Vicksburg, MS, USA, 2004; p. 134.

39. Camenen, B.; Larson, M. Phase-lag effects in sheet flow transport. *Coast. Eng.* **2006**, *53*, 531–542. [CrossRef]

40. Isobe, M.; Horikawa, K. Study on water particle velocities of shoaling and breaking waves. *Coast. Eng. Japan* **1982**, *25*, 109–123.

41. Kobayashi, N.; Tega, Y. Sand Suspension and Transport on Equilibrium Beach. *J. Waterw. Port Coast. Ocean Eng.* **2002**, *128*, 238–248. [CrossRef]

42. Nam, P.T.; Larson, M.; Hanson, H.; Hoan, L.X. A numerical model of nearshore waves, currents, and sediment transport. *Coast. Eng.* **2009**, *56*, 1084–1096. [CrossRef]

43. Larson, M.; Wamsley, T.V. A formula for longshore sediment transport in the swash. In Proceedings of the Coastal Sediments '07 6th International Symposium on Coastal Engineering and Science of Coastal Sediment Process, Louisiana, LA, USA, 13–17 May 2007; Kraus, N.C., Rosati, J.D., Eds.; ASCE: Reston, VA, USA, 2007; pp. 1924–1937.

44. Leont'yev, I.O. Numerical modelling of beach erosion during storm event. *Coast. Eng.* **1996**, *29*, 187–200. [CrossRef]

45. Kamphuis, J.W. Physical Modeling. In *Handbook of Coastal and Ocean Engineering*; Herbich, J., Ed.; Gulf Publishing Company: Huston, TX, USA, 1991; Volume 2, pp. 1049–1066.

46. Christopoulos, S. *Coastal Protection Works of Paralia Katerinis*; Technical Report; Municipality of Katerini: Katerini, Greece, 2014.

47. Google Earth, Image ©2017 TerraMetrics, Data SIO, NOAA, U.S. Navy, NGA, GEBCO. 2017.

Article

Stochastic Modeling of Forces on Jacket-Type Offshore Structures Colonized by Marine Growth

Hamed Ameryoun [1], Franck Schoefs [2,*], Laurent Barillé [3] and Yoann Thomas [4]

[1] IXEAD/CAPACITES Society, Université de Nantes, 44322 Nantes, France; hamed.ameryoun@capacites.fr
[2] Research Institute in Civil Engineering and Mechanics-(GeM)-UMR CNRS 6183, Sea and Littoral Research Institute, Université de Nantes, IUML-FR CNRS 3473, 44322 Nantes, France
[3] Mer Molécules Santé-(MMS)-EA 2160, Sea and Littoral Research Institute, Université de Nantes, IUML-FR CNRS 3473, 44322 Nantes, France; laurent.barille@univ-nantes.fr
[4] IRD LEMAR-IUEM, rue Dumont d'Urville, 29280 Plouzané, France; yoann.thomas@ird.fr
* Correspondence: franck.schoefs@univ-nantes.fr; Tel.: +33-2-98-49-86-43

Received: 14 April 2019; Accepted: 13 May 2019; Published: 22 May 2019

Abstract: The present paper deals with the stochastic modeling of bio-colonization for the computation of stochastic hydrodynamic loading on jacket-type offshore structures. It relies on a multidisciplinary study gathering biological and physical research fields that accounts for uncertainties at all the levels. Indeed, bio-colonization of offshore structures is a complex phenomenon with two major but distinct domains: (i) marine biology, whose processes are modeled with biomathematics methods, and (ii) hydrodynamic processes. This paper aims to connect these two domains. It proposes a stochastic model for the marine organism's growth and then continues with transfers for the assessment of drag coefficient and forces probability density functions that account for marine growth evolution. A case study relies on the characteristics (growth and shape) of the blue mussel (*Mytilus edulis*) in the northeastern Atlantic.

Keywords: marine growth; biofouling; wave loading; stochastic modeling; reliability; jacket structures

1. Introduction

Actual challenges for requalification of existing offshore structures through the reassessment process emphasize the importance of updating information about the structural condition state. One of the most important phases during the design or re-assessment level is a re-evaluation of environmental loads and updating knowledge concerning the state of biocolonization, structural damage, and corrosion. The random nature of biofouling and the uncertainty inherent to biological processes make modeling of environmental loading complicated. Biofouling is a complex phenomenon involving a diversity of marine species, which constitute communities whose dynamic is driven by physical and biological processes. It has many negative impacts on offshore structures such as loading excess, structures occlusion, increase in drag coefficient, and corrosion [1,2]. Therefore, it represents a challenge for engineers with respect to design and maintenance programs. Several standardized methods of inspections and in-situ measurements of the marine growth have been developed to obtain relevant information about species composition, percent cover, weight, thickness, and roughness, allowing the determination of structural design, cleaning, and maintenance strategies. Biocolonization processes show spatial and temporal variations related to several environmental factors (water temperature, hydrodynamics, turbidity, distance from the shore, bottom characteristics) acting at regional and local scales. However, the results are often more qualitative than quantitative and suffer from a lack of consistent modeling for structural engineers, except when a big database is available. Cost-effective, safety management of offshore structures involves allocating the optimal amount of resources to periodical inspections and maintenance activities in order to control risks (expected life of the structure).

The growth of marine organisms on offshore structures has long been a significant issue for the oil and gas industry [1,2]. In the 70's and 80's, studies focused on the effect of biofouling on hydrodynamic forces acting on offshore structures. Numerous experimental studies were carried out with different types of marine growth, cylinder diameter, and hydrodynamic conditions to provide a better understanding of their interactions with hydrodynamic forces and to highlight the key relationships. Despite the great variability due to the complexity and instabilities of the flow regime around structures, abacuses were built and are still recommended by offshore standards such as American Petroleum Institute (API) [3] and Det Norske Veritas (DNV) [4]. Only a few studies considered the global modeling of the loading in a probabilistic context [5,6] and none of them consider the modeling of the organismal growth itself. In fact, there are few available databases containing on-site measurements with time [7,8]. This paper proposes a modeling of characteristics of the external marine growth layer consistent with structural engineering needs. As the first year is crucial for future colonization patterns, we focused here on the building of the first layer of biofouling by a macro-fouler. The blue mussel *Mytilus edulis* was considered for the modeling, as it is an ubiquist bio-fouler in European waters [9,10].

The present paper considers the biocolonization as a stochastic process. Biocolonization is represented as cumulative deterioration process and this study defines two phases for it: an initiation phase and a propagation phase. The paper reviews meta-models and it describes database construction, which consists of the influencing factors. It proposes a stochastic modeling of biofouling based on a non-stationary, state-dependent Gamma process for the blue mussel *Mytilus edulis*. The developed Gamma process [11,12] provides individual shell length time series for blue mussels in the first year of colonization. Its parameters are identified from simulations carried out by a biological model. To this aim, a biological model based on the Dynamic Energy Budget (DEB) theory [13,14] was used to simulate the variations of individual mussel shell size depending on environmental data. Thereafter, the study focuses on the drag term of Morison's equation. It reviews a response surface method to model the drag force as well as the effect of physical characteristics of structural members, such as surface roughness (k) and the average thickness of marine growth (Th). Moreover, the drag force exerted by extreme waves for colonized structural members during the typical macro-colonization years is determined. The probabilistic macro-colonization, shell length time-series considering the occurrence probability of typical macro-colonization years are provided. The evolution of the drag coefficient with regard to the probabilistic shell length time series is evaluated and the results are discussed.

A case-study site was chosen offshore the Loire Estuary (France) corresponding to a future offshore wind farm site, in order to illustrate the role of biofouling on the computation of hydrodynamic forces (drag force). In order to model the colonization during the first year, two main phases of bio-colonization were considered: (1) an initiation phase without any macro-fouling on the structure, and (2) a propagation phase or macro-colonization phase, corresponding to the growth of mussels. The key influencing factors affecting these two stages were hydrological data (water temperature and chlorophyll-a concentration, as proxy of mussel food). Mussel growth was used to derive two geometrical characteristics Th and k describing marine growth. It should be noted that the added mass and inertia forces are beyond the scope of this paper.

Hereafter, the objective of this work is to propose a meta-model, which combines different disciplinary approaches accounting for several types of uncertainty and variability among (a) the temporal variability of the main influencing environmental factors; (b) the biological uncertainty of the individual's growth; (c) the uncertainty, due to the modeling of geometrical parameters of structural components caused by biofouling and needed for structural computations; and (d) the uncertainty of wave characteristics to compute the loading on structural components. To propagate the uncertainty of biological and physical marine environment (marine growth and wave), a physical matrix response surface was used in view to provide a probabilistic model of the environmental loading on jacket type offshore structures based on Schoefs & Boukinda (2010) [1]. This method was applied for quasi-static calculations of wave forces in the presence of marine growth.

2. Materials and Methods

2.1. Requirements for a Meta-Model

To develop the macro-colonization model for structural computation, several properties should be considered. The main trends of growth with time should be captured using a time step compatible with sensitivity to input parameters; it should be sensitive to the environmental parameters that govern the ecophysiology of the biofouling: temperature and food availability; it should provide intermediate parameters (shell size) from which required outputs can be easily computed (roughness and thickness) to perform a reliability analysis (stochastic processes); it should be versatile to modify the trends depending on site specificity. The next sections detail the way these requirements have been taken into account in this work.

2.2. Description of Bio-Colonization Temporal Dynamic

Bio-colonization is a complex process depending on biotic and abiotic variables with many interactions [15,16]. Indeed, it would be unrealistic to envisage a complete model involving a multilayer of various marine organisms that have complex interactions for survival, growth, and reproduction. We propose here a model that accounts for the temporal variability of the main influencing factors in a simplified but realistic case. It focuses on the growth of a single species, the blue mussel *Mytilus edulis*.

The bio-colonization process depends on two early stages: (1) the reproduction of adults, which spawn in the water column and produce larvae that will become part of the plankton transported by currents; (2) larval survival and development in the water column. The bio-colonization itself starts with the larval settlement on a structure (micro-colonization) and corresponds mainly to the macro-colonization step, i.e., growth of individuals up to the adult state. It is important to estimate the spawning date(s) and to assess the conditions allowing larval survival. This is a prerequisite before modeling macro-colonization. Consequently, the model needs to take into account (1) an initiation phase (no macro-organism present on the structures) with no significant effect on structural reliability, and (2) a propagation phase corresponding to the growth of macro-organisms (Figure 1). With these two phases, an analogy can be made with the dynamic of degradation processes like corrosion of steel rebars in reinforced concrete [17]. In this study, we considered that the larval settlement corresponds to the beginning of the propagation phase.

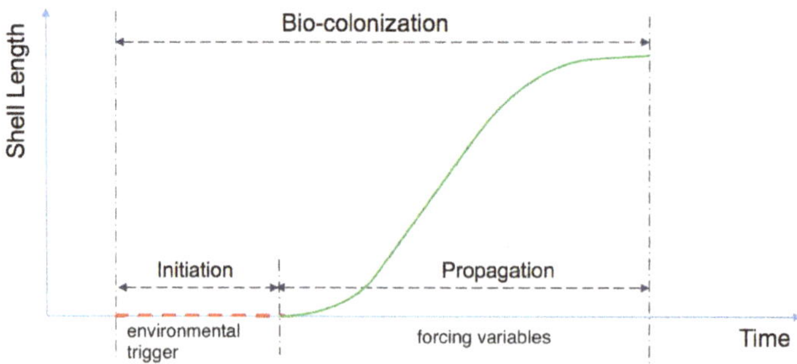

Figure 1. Schematic diagram of bio-colonization phases and their influencing parameters.

The model should be able to capture the initiation phase and then simulate a propagation phase (macro-colonization). The latter allows for obtaining the individual size and, accordingly, the physical characteristics of the colonized surface needed for the hydrodynamic calculations [6]. The initiation phase includes spawning date, larval survival, development, and settlement. We considered that this phase was mainly driven by temperature while the propagation phase (macro-colonization) corresponding to the juvenile growth was driven by both, the temperature and the concentration of chlorophyll-a, a proxy of the food available in the water column for mussels. These drivers are related to the bivalve ecophysiology, which is detailed in the following paragraph.

2.3. Description of Bio-Colonization Temporal Dynamic

The blue mussel *Mytilus edulis* (*M. edulis*) was chosen to develop a simplified (single organism colonization) but realistic bio-colonization model for the North-Atlantic coasts. *M. edulis* is a ubiquitous and abundant species in the coastal waters of the North and Mid-Atlantic Regions [18], and has been reported as a main macro-colonizer of offshore structures [19,20]. When found as a dominant hard fouler, it has an influence on the composition of the external layer of marine growth [15]. It is a suspension-feeding bivalve that attaches to substrata by byssal threads and is traditionally cultivated on ropes or wooden poles on the Western Atlantic coasts [21]. *M. edulis* is eurythermal (adaptable to a wide range of temperatures) and, under the latitude of our case-study site, is well acclimated to a 5 to 20 °C temperature range [22]. It is very common in the intertidal area forming beds on rocky and hard substrates but can be found in subtidal environment down to −10 m. Mussels feed on suspended particulate matter and their main food resource is phytoplankton cells [23–25]. Phytoplankton is also considered as the dominant food source for all life stages of *M. edulis* since larvae also rely on phytoplankton for their development. The concentration of chlorophyll-a is a widely used proxy of phytoplankton biomass, and this variable was used in this study to assess the food available for the mussel's growth. For more details on *M. edulis* morphology, physiology and ecology, the reader is referred to Gosling (2003) [21].

2.4. Initiation Phase and Propagation Phases

The spawning date, larval survival, and development are the most important stages for the initiation phase modeling. Blue mussels, like the majority of shallow water bivalves, produce large numbers of pelagic planktotrophic larvae that spend several weeks in the surface waters [25]. *M. edulis* sexes are separated, and gametes are shed into the water where fecundation occurs. At the latitude of the study site, mussels can spawn up to three times a year from April to September successively, depending on environmental factors. In bivalves, an essential condition related to spawning is a thermic threshold corresponding to a minimum water temperature [26]. Indeed, the temperature is the strongest exogenous factor controlling *M. edulis* reproduction [21]. In this work, we considered only the spring period when the mussel producers submerge ropes to collect planktonic larvae. In Pertuis Breton, which is the closest area to our study site, Barillé-Boyer (1996) [22] found a threshold of 10.5 °C, and it has been considered that spawning was not triggered below this temperature. Above this temperature, the spawning dates are not modeled but forced with observed datasets [24]. In this study, observations from mussel producers are used. A 30 day interval between two spawning was adopted in relation to the mussel gametogenesis dynamic. This delay is linked to the time necessary to reconstitute reproductive tissues [22]. Other exogenous factors, such as storms, shock, rain, etc., which can randomly trigger bivalve's spawning, were not considered. The second important step following spawning is the larval survival and development. For mussels, Bayne (1965) [27] observed that *M. edulis* larvae could reach its development within 20 to 40 days, depending on the temperature. A slower (S) larval growth and metamorphosis can take 40 days if spawning happened in early spring with a water temperature of around 10 °C, while a faster (F) larval development of around 20 days is possible at a higher temperature of 14 °C [22]. Therefore, it was considered that if during the next 20 days after spawning water temperature was >14 °C, larvae survival and development was completed in 20 days, otherwise in 40 days. For each year in our database, the spawning occurrence times and initiation phase typologies are determined by post-processing the temperature time-series.

As mentioned before, geometrical parameters (thickness and roughness) are required for load computation. They depend on geometrical specifications (shape) of organisms colonizing the structure. In this study, these parameters are linked to the shell length of blue mussel individuals. The shell growth of the blue mussel has an asymmetric sigmoid shape curve [28,29]. The growth rate of blue mussel individuals is, therefore, neither monotonic nor stationary, and the growth curve can be described by the acceleration and deceleration phases (Figure A1, in Appendix A). No clear relationship between the individual shell growth and the water temperature has been observed, while the concentration of chlorophyll-a appeared to be the main driver. This observation is consistent with several studies showing that the food supply was the most important variable explaining mussel growth [30–32].

2.5. Database Post-Treatment, Virtual Database, and Aggregation of Influencing Factors

2.5.1. Environmental Data at the Case-Study Site

In order to model the initiation and propagation phases, water temperature and chlorophyll-a (*Chl. a*) concentrations were obtained for the site of Le Croisic (47°17′33″ N, 2°31′15″ W) on the western Atlantic coast of France. This location was chosen for its proximity to the future offshore wind farm site of Banc de Guérande (47°19′41″ N, 2°25′46″ W). Data were collected by the French Observation and Monitoring program for Phytoplankton and Hydrology in coastal waters (REPHY, [33]), and implemented and managed by the French Research Institute for the Exploitation of the Sea (Ifremer). Bimonthly samples were collected at a sub-surface depth (between 0 and 1 m) during high tides between 1996 and 2012 (Figure 2). Chlorophyll-a display higher concentration between March to June corresponding to the spring phytoplanktonic bloom characteristic of northern hemisphere temperate waters. In 2004, a single spring peak was observed, while in 1996, three peaks of lower concentrations were detected. It should be noted that the water temperature cannot change abruptly in a short time. For an overview of the REPHY network, the reader is directed to Hernández Fariñas et al. (2013) [34].

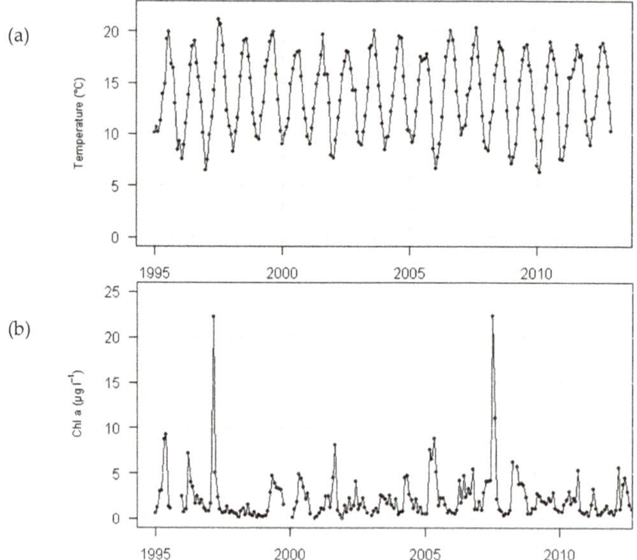

Figure 2. Inter-annual variations (1996–2012) of water temperature (°C) (**a**), and chlorophyll-a (*Chl. a*, µg·L⁻¹) (**b**), at Le Croisic sampling station (Loire-Atlantique, France). Data from Ifremer/Quadrige/Rephy©.

In order to model bio-colonization and to standardize the time intervals of data acquisition, each month has been divided into three 10 day periods. The database used hereafter has been therefore constructed from periodic observations at established time intervals τ equal to 10 days. The average value of all temperatures and *Chl. a* values measured during each decade has been assigned as the decade temperature and *Chl. a* values. It should be noted that the water temperature cannot change abruptly in each decade. On the other hand, *Chl. a* will not interfere directly in the model but rather will form the growth potential parameter that is explained in next section. That is why if there were no available measurements available for some decades, a linear interpolation from adjacent measurements was carried out. The number of the database time-series [Year Time-step] is defined by N, representing the number of years for which the database base has been prepared, and t, which represents the number of observations each year depending on the data acquisition time intervals (in days) τ; in our case $N = 17$, $t = 37$ and $\tau = 10$. The long-term time-variant modeling of input factors being out of the scope of this work, we assumed that N is statistically sufficient for computing the frequency of each macro-colonization scenario. Therefore, the database has been constituted from the regular measurements of water temperature (T) and *Chl. a* (C), and can be denoted as:

$$\left\{ \left(T^{i}_{t,\tau}, C^{i}_{t,\tau} \right); t > 0, i \in 1, N \right\}. \tag{1}$$

Hereafter, $T^{i}_{t,\tau}$, is used for the initiation phase determination and $C^{i}_{t,\tau}$ for the modeling of the propagation phase. Four types of larval development combining the slow (S) and fast (F) growth possibilities are presented in the Table 1 for the three initiation times (corresponding to the three spawning periods) obtained from the database considering key factors and thresholds described in the previous section. The first larval development is always slow because the water temperature is below 14 °C during early spring, and the third one can be slow only if the second one is also slow (because the water temperature cannot fluctuate abruptly). These results come from the natural seasonal variations of temperature during one year. These frequencies will be considered as discrete probabilities for the modeling. At the end of this larval growth period, we considered that larvae settled on the structures, and that was the start of the propagation phase (macro-colonization) described in Table 1 for the 17 annual chronicles.

Table 1. Inter-annual development types for three main spawning events (S: slow initiation phase, F: fast initiation phase).

Development Type	Occurrence	Probability
SSS	2	0.12
SSF	10	0.59
SFS	0	0.00
SFF	5	0.29

Table 2 shows the date of start of macro-colonization, expressed in 10 day periods (1 = first 10 days of January), for three main spawning events of blue mussel between March and June. Occurrence and probability were calculated from the 17 year time-series of temperature data at the study site. Calculations revealed that macro-colonization starting date spanned from the 11th to the 20th 10 day period. The most probable macro-colonization inception times for the three spawning events corresponded to the combination of 10 days periods of 12-15-16 and 13-16-17 with 18% probability. The first macro-colonization inception occurred between the 11th and the 16th 10 day periods with the highest probability of 29% for 11th period and the lowest probability of 6% for the 16th period. The second macro-colonization inception occurred between the 14th and 17th periods with the lowest probability of 12% for the latter. The third macro-colonization inception occurred between the 17th and the 20th periods with the highest probability of 35% for the 17th period and the lowest (6%) for the 20th.

Table 2. Date of start of macro-colonization, expressed in 10 day periods, for three main spawning events of blue mussel.

Start of Macro-Colonization			Occurrence	Probability
11	14	15	2	0.12
11	14	17	2	0.12
11	15	16	1	0.06
12	15	16	3	0.18
13	14	17	1	0.06
13	16	17	3	0.18
14	15	18	1	0.06
14	17	18	1	0.06
15	16	19	2	0.12
16	17	20	1	0.06

2.5.2. Environmental Data at the Case-Study Site

There was no observation available for blue mussel shell lengths close to our study site and more generally, no database for the annual growth of mussels during the considered 17 year period. To fill this gap, we applied a bioenergetics growth model to simulate the individual shell length using environmental time-series data available at the study site. This biological model was calibrated by a site with similar environmental characteristics and for the same species of mussels. Different bioenergetics models have been developed to model the growth of bivalves depending on the environmental conditions, and among them, Dynamic Energy Budget (DEB) models [13] have been successfully applied to several bivalve species [24,35–38]. DEB models do not use empirical allometric relationships, but simply state that feeding is proportional to surface area, whereas maintenance is scaled according to structural body volume [13]. DEB theory proposes a generic energy budget approach that assumes common physiological processes among species and life stages via a set of parameters, the only difference among species lying in the values of those parameters.

In this study, we used the DEB model developed by Thomas et al. (2011) [31] to simulate the growth of *Mytilus edulis* in the Mont Saint-Michel Bay. A single parameter, the half-saturation coefficient of the food ingestion function term (XK), had to be adjusted to local hydrologic and trophic conditions. For our study, the half-saturation coefficient was calibrated at 2.9 μg·L^{-1} from growth data by Garen et al. (2004). Simulations started for 1 mm individuals (0.02 g of Dry Flesh Mass—DFM), a biometry corresponding to post-settled organisms. Results of the calibration are presented in Figure A2 in Appendix A. A good level of agreement between observations and simulations was obtained for shell length and dry flesh mass, a biological variable often used in bioenergetics models to assess the consistency of the simulations.

The model was then used to obtain individual growth trajectories with the 17 year time-series of the *Chl. a* concentration measured at Le Croisic (Loire-Atlantique, Le Croisic, France). Three starting dates were chosen, corresponding to the three spawning events and related macro-colonization starting dates (Figure 3). Note that the initiation phase and beginning of propagation phase plotted in Figure 1 cannot be measured and are not reported in Figure 3 top.

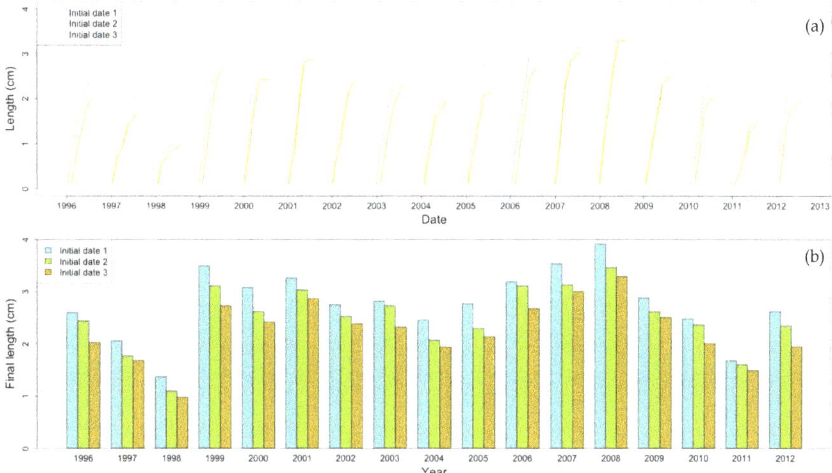

Figure 3. (**a**) Individual annual shell length trajectories simulated by a mussel Dynamic Energy Budget (DEB) model and (**b**) corresponding final length.

2.6. The Relation between Environmental Factors, Growth, and the Start of Macro-Colonization

The Gamma process simulates increments for each time interval of τ which correspond here to variations in mussel shell length ($\Delta S_{t,\tau}$). The parameterization of the function can integrate the environmental variables. Temperature is a variable of the DEB model, but *Chl. a* concentration is the main driver of growth. It was therefore decided to parameterize the Gamma process only with *Chl. a*. However, due to potential coupled effects between temperature and *Chl. a*, we analyzed the correlation between temperature and growth over the time-series. From ΔS obtained from DEB simulations, the scatter diagram of ΔS vs. temperature showed that there was no significant correlation between these two variables with a Pearson correlation coefficient $\rho = 0.21$ (Figure 4). That means that temperature is not a key driver of shell growth. On the contrary, there was a structured relationship between growth and *Chl. a* (Figure 5). It can be noted that uncertainty increases when *Chl. a* increases. Moreover, there is a ΔS plateau showing that the capability of an individual to grow is limited by the additional food supply: a concentration higher than 8 μg·L^{-1} does not lead to a larger ΔS. This is due to a well-described physiological phenomenon of maximum somatic growth in bivalves [39,40].

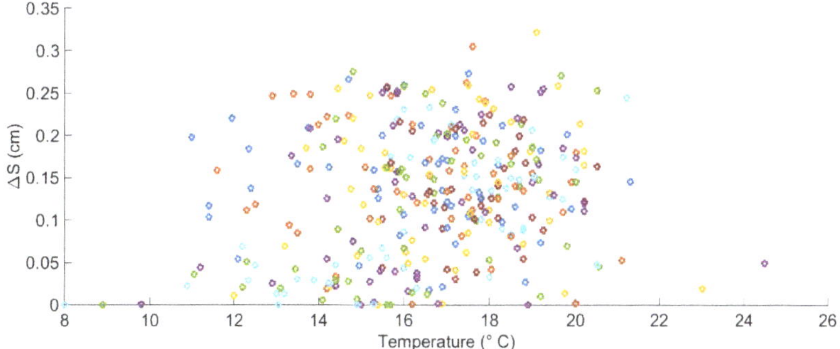

Figure 4. Scatter diagram of the variations of mussel shell length (ΔS) vs. temperature for 10 days periods during the 17 year time series (each color represents a year of the 1996–2012 time-series).

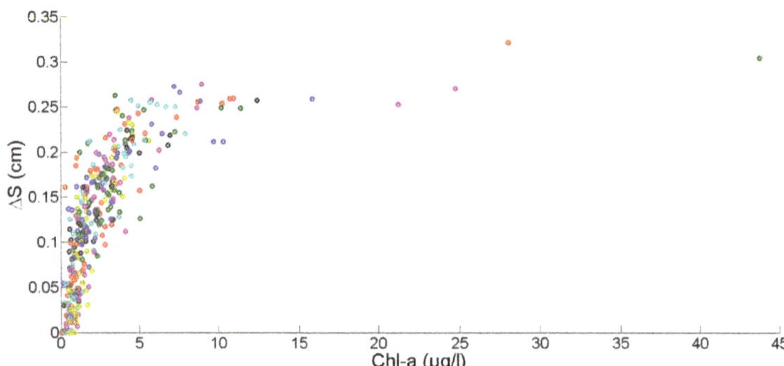

Figure 5. Scatter diagram of the variations of mussel shell length (ΔS) vs. Chlorophyll a for 10 days periods during the 17 year time series (each color represents a year of the 1996–2012 time-series).

The relationship between the start of macro-colonization and the concentration of *Chl. a* is presented in Figure A3 in Appendix A. There is no significant correlation between these two variables with a Pearson correlation coefficient ρ = −0.02. This property is of first importance, as it will govern the simulation strategy. Simulation of inception times (start of macro-colonization) requires temperature time-series only, and the modeling of mussel growth will be carried out independently using *Chl. a* time-series.

2.7. Chlorophyll Data Aggregation for Growth Computation

In order to improve the biological consistency of our simulations, we tested the possibility to link the individual growth of blue mussels to *Chl. a* concentration aggregated over a time-step instead of using instantaneous values. The integrated value of *Chl. a* was simply defined as:

$$C_{(T_{(i)}:T_{(i+n)})} = \frac{1}{n} \int_{T_{(i)}}^{T_{(i+n)}} Chl(t)dt, \; i = 1:36 - n \tag{2}$$

where $C_{(.)}$ is the aggregated *Chl. a*; $T_{(i)}$, is a 10 day period, and $Chl(t)$, is the linear equation of *Chl. a* obtained from linear interpolation between adjacent measured values for a colonization period, and n is the number of 10 day time intervals after (i), in which the data aggregation is performed. The best correlation between ΔS and *Chl. a* has been obtained for a monthly aggregation (3 time intervals, n = 2 in (2)). This time-step preserved the spring bloom typical of the seasonal dynamic of phytoplankton at the study site latitude.

In order to identify the non-linear relationship between growth and *Chl. a*, the non-linear regression (3) has been fitted with an R^2 of 0.74 (Figure 6):

$$\Delta S = \frac{0.235}{1 + 6.94e^{-1.005(Chl.-a)}} \tag{3}$$

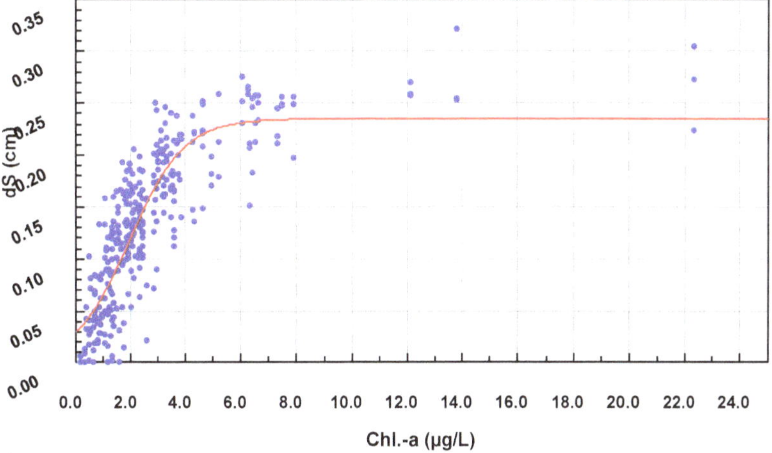

Figure 6. Identification of the *Chl. a* threshold beyond which growth saturation is observed.

The ratio of (*dΔS/dChl. a*) ≤2% has been chosen as a criterion for stabilization based on slope variations; this ratio is consistent with the accuracy of the DEB model. A concentration of *Chl. a* of 8 µg· L^{-1} was identified as the threshold beyond which ΔS remained constant at 0.235 cm/time interval. The *Chl. a* time-series was then truncated with the mentioned threshold to significantly improve the convergence of the Gamma process parameterization, without an important degradation of the database. Note that this threshold depends on the metabolism of the organism and is, therefore, species-specific.

2.8. Non-Stationary Modeling of Shell Growth through Stochastic Gamma Process

2.8.1. Growth Approximation through Gamma Processes Meta-Models

Considering the aforementioned characteristics of the mussel's colonization, and in order to model the temporal dynamic of structure deterioration, a stochastic approach based on Gamma processes has been selected [41]. Since the introduction of the Gamma process in reliability [42], it has been used commonly to model stochastic cumulative and uncertain deterioration phenomena for the maintenance optimization of various industrial systems. Indeed, the Gamma process is an analytically tractable stochastic process accumulating over time in a sequence of positive increments. Recently, it has been widely used to model cumulative degradation processes, such as corrosion, fatigue, crack growth, creep, degrading health, erosion, and wear in engineering systems and structures [11,41,43,44].

The Gamma process is a special case of a non-decreasing jump stochastic process that properly captures the temporal variability associated with the deterioration dynamic. This justified the choice of the non-stationary state-dependent Gamma process. The non-stationary Gamma process is a widely used mathematical model to describe a degradation process whose growth rate at time *t* depends only on the current state of the parameters and not on the accumulated damage up to *t* [45]. The complete Gamma process function is defined by two parameters: a shape function α_S and a scale function β_S (4). We discredited time horizon into equal intervals of length τ = 10 days. Then, the state-dependent non-stationary and bivariate Gamma process was represented as a series of state-stationary Gamma processes in each time interval. The rate of the deterioration process can thus be considered as the process resulting from the Gamma process variations from one time-interval to another. The deterioration increment in a given time interval $\Delta S_{t,\tau}$ has been considered to be a random variable with a shape function (α_S) dependent of the present deterioration state $S_{t,\tau}$ and a second variable, the state of chlorophyll-a concentration $C_{t,\tau}$. Thus, for each time step τ, we have:

$$\forall S_{t,\tau} > 0 : S_{t,\tau}, \Delta S_{t,\tau}, \tau : \Delta S(\tau; S_t, C_t) : \Gamma(\alpha_S(S_t, S_t).\tau, \beta_S) \tag{4}$$

where $S_{t,\tau}$ is the shell length for each time interval of τ and α_S and β_S are the shape and scale functions of the Gamma process, respectively. To simplify the modeling of this process, it has been assumed that the scale function β_S was constant and Gamma process was only governed by the shape function [11].

2.8.2. Parameter Estimation of the Gamma Process (Learning Phase)

In order to simulate the growth of blue mussel submitted to fluctuations of *Chl. a* in each time interval τ, the parameters of the developed Gamma process have to be estimated. The deterioration increments have been calculated by the simple subtraction of consecutive individual shell lengths and the resulting database used for deterioration density estimation is denoted as:

$$\left\{ \left(C^i_{t,\tau}, S^i_{t,\tau}, \Delta S^i_{t,\tau} \right); t > 0, \ i \in 1, N-1 \right\} \tag{5}$$

where N is the number of years (in this study N = 17). In order to estimate the parameters of the Gamma process, the Expectation-Maximization (EM) method has been employed. The program starts by scanning the database and indexes the values of $C(t)$ and $S(t)$ time-series; then using the observed data, initial parameters are estimated and used to start an iterative EM algorithm. The Gamma process parameters have been estimated and determined as:

$$\Delta S(\tau; S_t, C_t) : \Gamma\left(\alpha_S = (0.198 + 1.68 C_t)\exp\left(\frac{-(S_t - 0.44)^2}{6.512}\right), \beta_S = 0.039\right). \tag{6}$$

2.8.3. Stochastic Simulation from Gamma Process (Propagation Phase)

Once the Gamma process has been estimated, the DEB data are not needed anymore and we can use the *Chl. a* database to predict the growth rate of mussels. The macro-colonization can be then simulated with the Gamma process function (6) and consider the *Chl. a* database for 10 typical macro-colonization years from the 17 year time-series. We considered this database to be representative of the dispersion from a richer database in respect to the frequency of the main phenomena and their consequences. Thus, each environmental input (*Chl. a* time-series $C(t)$) is considered with the same weight as the growth uncertainty (biological process). The 5000 simulated growth curves obtained by the Gamma process (50 realizations of $C(t)$ for each of the 10 *Chl. a* time-series) and individual growth curves simulated with the DEB model are compared in Figure 7. Each growth curve is simulated from one realization of the Gamma process with *Chl. a* randomly sampled from the 17 year time-series database: uncertainties for *Chl. a.* and the simulation of S at a given time have herein the same weight in statistical terms. After one year of growth, the mussel shell length time-series $S(t)$ simulated with the Gamma process encompassing the extreme values obtained with the DEB model. Moreover, stabilization is reached (asymptotic behavior) after the 30th time periods when the growth is stabilized at a mature age. At the beginning of the simulations (first 25 weeks), a higher variability was observed with the Gamma process; this would lead to conservative estimations of marine growth characteristics colonizing the structure. From an engineering perspective, it is essential to reach a good representation of the distribution of maximum values of S, which is the case here.

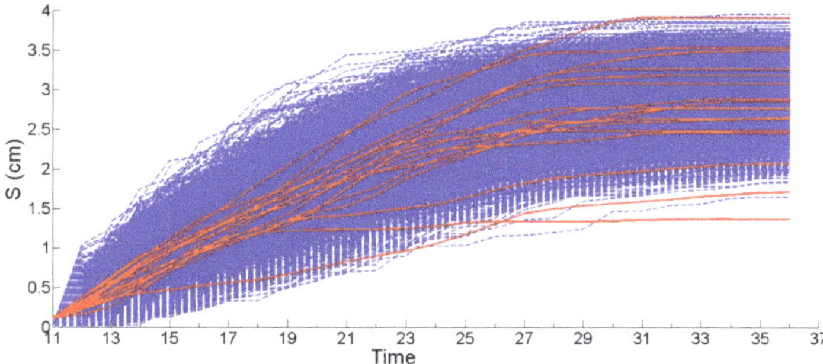

Figure 7. Individual growth trajectories obtained with the DEB model for each year of the 17 year time series (solid red lines) compared with simulated individual growth obtained with the Gamma process approach (dotted blue lines). The time unit represents 10 day periods.

Note that the Gamma process model accounts for the stabilization of shell growth during one or several time steps: that phenomenon is observed in the DEB simulation and actually represents a lack of available food. To complete this statistical analysis, the average and the standard deviation of individual shell length curves are presented in Figure 8. These curves were similar for both methods throughout the simulation period. The Gamma process simulations were a bit conservative in terms of shell length overestimation. The standard deviation curves showed differences between the two approaches. This may be due to the choice of the constant scale parameter β_S, which controls the response dispersion of the Gamma process. Note that there is also a statistical bias when estimating standard deviation from the DEB time-series due to the limited amount of data (17 trajectories).

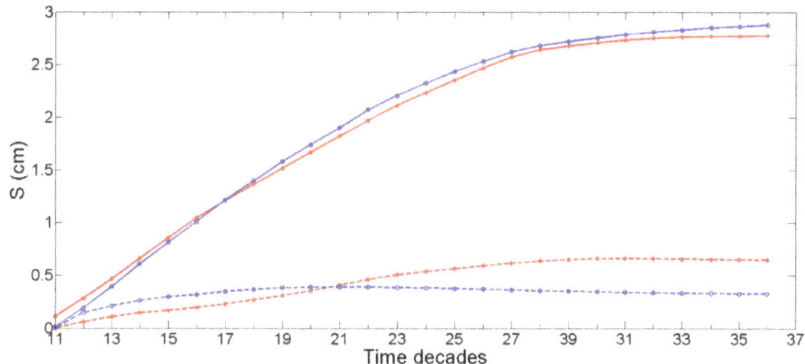

Figure 8. Comparison of average (solid lines) and standard deviation (dashed lines) of shell length from Gamma process (blue lines) and DEB model (red lines). Time unit represents 10 day periods.

2.9. Effect of Marine Growth and Hydrodynamic Forces on Jackets

From the structural point of view, marine growth may affect dynamical behavior, resistance to fatigue or extreme loading. We focused here on the latter. Offshore platforms are generally gathered in two families: bottom fixed and floating. Many works studied the effect of marine growth and hydrodynamic forces on components (cylindrical beams) of fixed steel framed offshore structures called jackets for which the component diameter (1 m) is small in comparison with wavelength during storms (100–400 m). This type being, on the one hand, the most popular in oil and gas industry and also for the offshore wind turbines substation, and on the other hand very sensitive to marine growth (fatigue and extreme loading). The analysis of marine growth effect on hydrodynamic forces could be categorized into two groups as follows:

(i)   experimental modeling of hydrodynamic forces for cylinders with different roughness conditions [46,47]. Recent studies are mostly concentrated on the water particle velocity and acceleration measurement techniques.

(ii) evaluation of hydrodynamic forces by the physical modeling of marine growth characteristics obtained from in-situ measurements [6,48]. These studies were based on inspections carried out during survey campaigns. They advocate guidelines for the probabilistic modeling of hydrodynamic forces at a given time. The biofouling database has been analyzed to propose a model of marine growth evolution and to update the design criterion. A physical response surface matrix has been proposed in order to provide a probabilistic modeling of the environmental loading on jacket type offshore structures. The key parameter is the increase of the structural diameter due to the marine growth thickness.

In the present study, we are considering the non-linear effect of the roughness of marine growth on the loading during a yearly growth. The results from laboratory studies focused mainly (i) on the regular shape of marine growth and homogeneous colonization around the cylinder and (ii) on mean thickness. Moreover, for time computation constraints (stochastic simulations of wave and marine growth) and because it allows explicitly introducing the role of marine growth, we used Morison modeling [49] for which the link between homogeneous roughness and loading is available. Because roughness is non-homogeneous and random, an uncertainty was added.

2.9.1. Effect of Marine Growth on Morison's Equation

Usually, Morison's model [49] is used to estimate hydrodynamic forces on tubular offshore structures like jackets, using the particle kinematics obtained from the wave heights and periods. It should be noted that for the jacket structures Morison's equation is valid because the structural diameters (D) are small compared to wavelengths λ ($D/\lambda < 0.2$). This equation can be employed from medium to deep-water depth [49]. It has been shown to be very appropriate for an expansion in the stochastic domain [48]. This equation is denoted as:

$$F_{Morison} = F_D + F_I = \frac{1}{2}\rho C_D D u|u| + C_M \frac{\rho \pi D^2}{4}\dot{u} \tag{7}$$

where $F_{Morison}$ is the hydrodynamic force per unit length of the member (N/m), F_D is the drag force per unit length of the member (N/m), F_I is inertia force per unit length of the member (N/m), C_D is the drag coefficient, C_M is inertia coefficient, ρ is the density of water, D is member diameter (m), and u is velocity of wave's water particles (m/s), $\dot{u}$ is the acceleration of wave's water particles (m/s^2). u and $\dot{u}$ are computed by Stoke's model [50] from the knowledge of metocean data: wave height H and period T.

Generally, the inertia term of the mentioned equation becomes important for small waves or for members with large diameters [48], otherwise the drag term will be dominant. Marine growth increases the surface roughness and hence changes both the drag and inertia forces. The variations induced by the presence of marine growth impress the hydrodynamic forces in a non-linear way.

Considering the effect of biofouling on hydrodynamic coefficients in Morison's equation, some researchers have proposed a model for the drag coefficient as a linear regression function of the thickness and roughness [51,52]. According to the recommended practice of [3], an additional parameter that affects the drag coefficient of elements with circular cross-sections is the relative roughness, $e = k/D_e$. The surface Roughness k is the average peak-to-valley height of hard growth organisms and the effective member diameter De can be obtained as:

$$D_e = D_c + 2\,Th \tag{8}$$

where D_c is the outer diameter of the clean member and Th is the biocolonization thickness (i.e., the mean of distributed thickness around the diameter) obtained by circumferential measurements [3]. API [3] gives the relationship between De and the steady-flow drag coefficient (C_{DS}) (9).

$$C_{DS} = a + \frac{b}{\frac{k}{De} + c}; \ a = 0.07152, \ b = -2.9 \times 10^{-4}, \ c = 4.12 \times 10^{-4}. \tag{9}$$

C_D is then computed from the knowledge of C_{DS} and the Keulegan-Carpenter number KC_{mg} according to [3,6].

This approach allows measuring the influence of roughness on the drag coefficient and therefore the drag force as well as their evolution with time. This leads us to choose an uncertainty model for the relationship between the size of the shell and the roughness in Morison's equation.

Coefficients of fluid-structure interactions are modeled from the knowledge of the hydraulic flow regime around the structural components [47]. Reynolds R_e and Keulegan-Carpenter KC numbers are essential for characterizing the flow regime [6]. For most offshore jacket structures in extreme conditions, Reynolds numbers are put into the post-critical flow regime, where the steady-flow drag coefficient C_{DS} for circular cylinders is independent of Reynolds number [3,53].

2.9.2. Stochastic Modeling of Marine Growth and Hydrodynamic Parameters

There is not enough knowledge, nor enough observations, about the settlement of blue mussels on offshore structures and a 100% cover was considered on the component. In order to account for the diameter of the colonized structural member, the marine growth thickness time-series $Th(t)$ should be modeled. Marine growth thickness is modeled as a Gamma process $Th_{t,\tau}$ deduced from the simulated individual shell length time-series $S_{t,\tau}$ for blue mussels in each time interval. For simplicity at this step of modeling, it has been assumed that the individual shell length time-series $S(t)$ gives the average size time-series $Th_{t,\tau}$ with a multiplying uncertain factor (10): it follows a uniform distribution with support [0.3; 0.6] at each of the i 10 day periods.

$$Th(t) = \{(Th_1, \ldots, Th_n); \ 0.3 \ S_i \leq Th_i \leq 0.6 \ S_i, \ i \in 1, 37\}. \tag{10}$$

This uncertainty accounts for the geometrical arrangement of the shells (Figure 9).

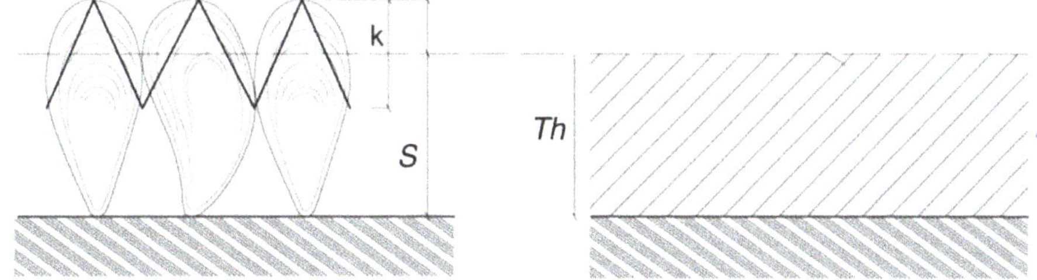

Figure 9. From biological reality to mechanical abstraction.

For roughness, on the one hand, there is a lack of on-site measurements and on the other, the available relationship between roughness and hydrodynamic forces (9) relies on a uniform roughness around the component [3]. Hence, roughness is also modeled as a Gamma process $k_{t,\tau}$ based on individual shell length time-series $S_{t,\tau}$ (11), with a random factor following a uniform distribution with support [0.2; 1]. The latter is a model error for modeling the uncertainty when quantifying the real effect of a randomly distributed roughness around the component. Note that intensive developments on underwater image processing are emerging [54–56], enabling one to envisage progress in on-site measurements. Recent works investigate the relationship between non-homogenous roughness and loading [57,58]. The wide range of uncertainty will, therefore, decrease in the next decade. Consequently, the error of computation of equivalent roughness is significant and the interval in (11) is large: it includes the stochastic distribution of shells around a tubular component and the error of model for computing the equivalent roughness.

Finally, the time-series of surface roughness $k(t)$ and marine growth average thickness $Th(t)$ have been considered independently as the random value uniformly distributed in an interval bounded to a ratio of individual shell length $S(t)$:

$$k(t) = \{(k_1, \ldots, k_n);\ 0.2\, S_n \le k_n \le S_n,\ n \in 1, 37\}. \tag{11}$$

The relative surface roughness $e(t)$ time-series are deduced ($e(t) = k(t)/De(t)$) and $C_{DS}(t)$ time-series have been simulated according to (9).

This study deals with two major time variant random variables in the hydrodynamic calculations, the meteocean data including a couple of wave height and period (H, T) and the stochastic process generating the individual shell length $S_{t,\tau}$ in each time interval τ from (4). Parameters of Gamma processes $Th_{t,\tau}$, and $k_{t,\tau}$ are dependent of individual shell length St,τ and the hydraulic parameters (Re_{mg}, KC_{mg}), and therefore drag coefficients C_D, depend on individual shell length $S_{t,\tau}$ and the couple of wave height and period (H, T).

Thus, parameters of $Th_{t,\tau}$ affect the hydrodynamic coefficients through the relationships between the hydraulic parameters (Re, KC) and the diameter of the elements, which is dependent of the coefficient of $Th_{t,\tau}$ itself [6]. The next section will explain how these cross-effects are accounted for.

2.9.3. The Stochastic Modeling Wave Loading in the Presence of Marine Growth

Figure 10 summarizes the steps of drag forces computation in a flowchart. It should be noted that the steps of hydrodynamic coefficients calculation are based on an interpolation of experimental curves [6] presented by [3].

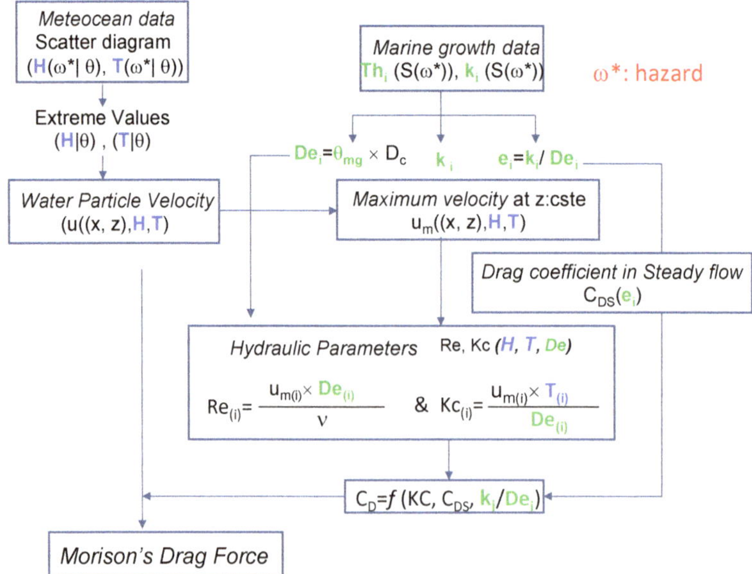

Figure 10. Schematic flowchart of drag force calculation.

The main steps of this flowchart are detailed below ((#X) means the step in circle X in the figure):

- (#1) Statistical Identification: the employed parameters are the heights of extreme waves H and associated periods T. They are modeled with a random variable, whose probability is conditioned by the wave direction θ;
- (#2) A kinematic model for the fluid for computation of water particle velocity: the Stokes model [50] is used. It assumes that the fluid is Newtonian and irrotational and the trajectory of the fluid particles is elliptical. The kinematics field deduced from the velocity potential can be defined at any point M of coordinates x and z. The maximum velocity u_m (#5) is deduced and is used in the computation of KC and Re (#6).
- (#3) The fluid-structure Interaction model: this level is involved in the hydrodynamic coefficients determined by using the recommendation of [3].
- For the probabilistic modeling of CD, in order to avoid multiplying the case studies, only vertical elements under the wave crest are analyzed. This implies high horizontal speeds and accelerations that generate very small forces, which means that the inertia forces in (7) are very low and will be neglected in the following.
- (#4) The colonized diameter $D_e(t)$ is a stochastic process that results from the increase $Th(t)$ of the initial radius of the clean component. Starting from (8), the diameter is computed by multiplying D_c by the factor θ_{mg}. The latter is computed from the thickness $Th(t)$ (12):

$$D_e = D_c + 2\,\overline{th} = \theta_{mg}\,D_c \text{ with } \theta_{mg} = \left(1 + \frac{2\,Th_{t,\tau}}{D_c}\right). \tag{12}$$

Random or stochastic nature of variables or processes is reminded at the beginning of the flowchart by writing as a function of the hazard ω.

According to API RP 2A WSD [3] and DNV-RP-C20 [4], since the flow regime is post-critical (R_e > 5 × 105) by using 100 year-return wave characteristics, the drag coefficient does not depend on R_e but rather on KC_{mg} and C_{DS}. Note that API ([3], section C2.3.1b7, p. 143 and p. 145) provides, in fact, a piecewise model on two intervals depending on KC or KC/C_{DS} and the scales of these models are different. It results in two effects on the evolution of the drag force (C_D): first for some values of C_{DS} it is the cause of discontinuity of the model at $KC = 12$ and second, it is very difficult to analyze directly the effect of C_{DS}. This is visible in Section 3.2.

Extreme wave characteristics (H, T) of the Gulf of Guinea have been considered for the hydrodynamic calculation, which is a specific site with low KC_{mg} values. Moreover, it gathers wave and wind-sea values and the spectrum is very similar to the one in French Atlantic offshore sites. Using meteocean data from this region allowed us to cover a large range of KC_{mg} to better illustrate the non-linear effects of marine growth on the drag coefficient evolution and hence on the load probabilistic distribution. This covers almost all configurations of Atlantic French offshore sites. Joint distribution of the extreme height and period for a return period of 100 years for the Gulf of Guinea are simulated based on [7]. It has been provided by recombination of sea states from the knowledge of the *H-T* scatter diagram. Representation of the joint distribution for wave height and the 100 year return period is presented in [59]. Note that breaking waves are not considered here.

We focused on drag forces acting on vertical cylindrical components under the wave crest with a diameter of 0.762 m (corresponding to the diameter of a Φ30″ leg). Note that, for simplicity, the probability of storm occurrence is independent of time and can happen in every 10 day periods of the macro-colonization period. This assumption is conservative. The time-series of surface roughness $k(t)$ and marine growth thickness $Th(t)$ obtained from the individual shell length time-series $S(t)$ have been considered for the determination of the C_{DS} (C_D in steady flow) time-series ($C_{DS}(t) = f(kt,\tau/D_e)$). In this paper, knowing C_{DS}, a numerical fitting of the curve of $C_D = f(KC_{mg})$ given in [3] is used and is plotted in Section 3.2 (lower multi-linear curve for the smooth cylinder).

3. Results

3.1. Simulation of the Drag Force Evolution from the Stochastic Time-Series of blue Mussels

This section aims to assess the evolution of the drag coefficient (CD) by mixing all of the typical macro-colonization time-series according to their occurrence probabilities by considering the macro-colonization inception times in the initiation phase (Sections 2.2–2.4). The individual shell length time-series for all the typical macro-colonization years are necessary to provide the probabilistic matrix of individual shell length. This matrix consists of the individual shell length time-series for all typical macro-colonization years, which are weighted by the occurrence probability of each typical macro-colonization year. Therefore, 30,000 simulations (10,000 simulations for each macro-colonization inception time in one year) have been performed to provide the individual shell length of blue mussels for each typical macro-colonization year.

The individual shell length time-series $S(t)$ of the blue mussels are simulated from the developed Gamma process (Section 2.8) from the inception times for typical macro-colonization years. No correlation between macro-colonization inception time conditioned by the temperature and the aggregate *Chl. a* (*Ct*) levels are observed (Section 2.6). Therefore, the levels are simulated independently. Hence, all the time-series of aggregate $C(t)$ could be used for the simulation of the individual shell length time-series $S(t)$) for each typical macro-colonization year.

The individual shell length time-series simulation procedure is as follows: the typical macro-colonization year determined by the temperature is first selected. Then, the individual shell length time-series $S(t)$ are simulated from the Gamma process. This simulation is performed according to the aggregated *Chl. a* time-series, which have been selected randomly, generating one $S(t)$ from one $C(t)$. Thus, we obtain the same statistical weight for the inception and growth by choosing an aggregated *Chl. a* time-series randomly for each simulation.

Figure 11 illustrates the estimated individual shell length time-series for the 2nd typical macro-colonization (starting dates at 11, 14, and 17 10 day periods) and 200 simulations, as an example. The highest jumps are observed for the 18th to 22nd 10 day periods, because of the important peak occurrence in the aggregated *Chl. a* time-series in 2001, 2007, and 2008 (see Figure 12).

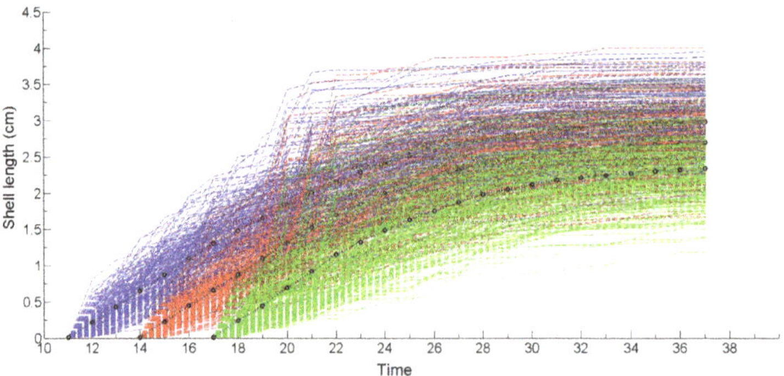

Figure 11. Simulated individual shell length of blue mussels for the 2nd typical macro-colonization (11-12-17). 200 simulations are presented.

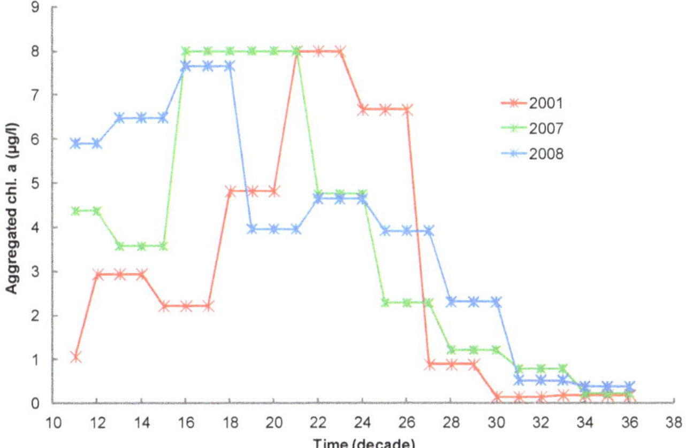

Figure 12. The aggregated *Chl. a* time-series for the years 2001, 2007 and 2008.

The simulation allows the individual shell length matrix to represent all of the typical macro-colonization years. The contribution of the individual shell length time-series could be obtained as:

$$N_t = N_S \times P_t \tag{13}$$

where, N_t is the numbers of time-series for the typical macro-colonization year of $S(t)$, which should be selected randomly, N_s is the sample size (here equal to 30,000), and P_t is the occurrence probability of the typical macro-colonization year (Table 2). The simulation procedure is illustrated in Figure 13. The *KC*, C_D, and drag forces are then computed according to the flowchart reported in Figure 10.

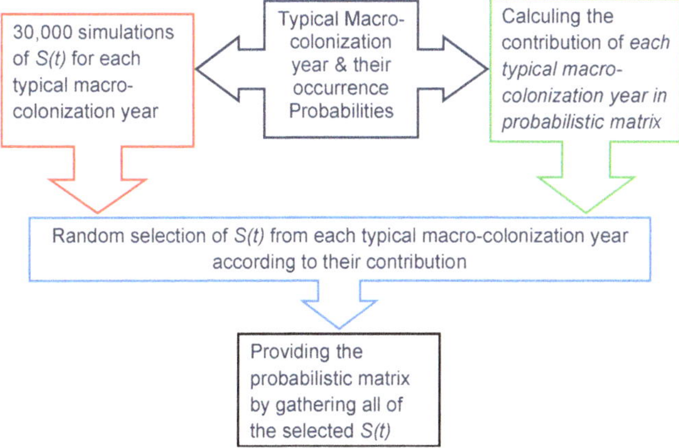

Figure 13. Schematic procedure of probabilistic individual shell length time-series from the typical macro-colonization year.

3.2. Statistical Analysis of the Transfer of Distributions

Quality of distribution transfer or uncertainty propagation is a well-known criterion for the analysis of the change of the distribution (its parameters or probabilistic law), especially for matrix response surfaces [48]. We focus first on the evolution of shell length distribution. Figure 14 illustrates the three most interesting 10 day periods representing insignificant (the 11th 10 day period), intermediate (the 18th 10 day period), and extreme (the 37th 10 day period) roughness values. The distribution of shell length changes from bimodal (the 11th and 18th 10 day periods) to normal (the 37th 10 day period), depending on time. It should be noted that the shape of the shell length distribution evolves strongly with time, which will lead to significant variations in the distribution of C_D along with its support due to the dependence of C_{DS} to k/D_e in (9). The mixing of sources of uncertainties due to independent macro-colonization inception time and independent growth builds finally a normal distribution as expected from the Central Limit Theorem.

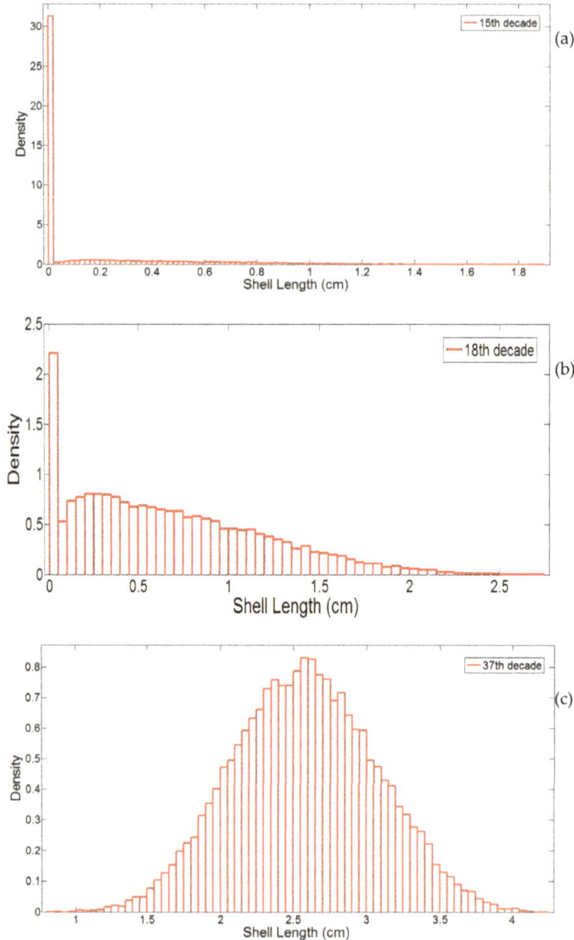

Figure 14. Evolution of shell length distribution as a function of time for three selected 10 day periods: (**a**) 15th, (**b**) 18th, (**c**) 37th.

Figure 15 illustrates the distributions of drag coefficients C_D as a function of KC for the three above-mentioned 10 day periods. First, we plot the bounds of the relationship (CD)–(KC) with lower and upper lines that depict, respectively, the smooth and roughened cylinders' drag coefficients. The discontinuity comes from the discontinuity of curves in the standards generated by the various scales $(CD/CDS, CD, KC, KC/CDS)$ used around $KC = 12$. Note that this discontinuity for the smooth and roughed cylinders follows, respectively, a potential positive and a negative jump of the C_D. Second, the scatter plots are reported in red, moving from the lower part to the upper part from 11th to 37th decade 10 day periods. Consequently, the distribution of C_D is affected. An important point is that the distribution maintains two modes, the uppermost being around 1.2 and the lowermost following the shift of the non-linear transfer function, from 0.2 to 0.6 (see the 37th 10 day period). It demonstrates the evolution of the drag coefficients C_D for the individual shell length from the non-linear transfer of the distribution of *KC* and during the probabilistic macro-colonization year (from the 11th to the 37th 10 day period). Finally, the probability of the highest values (typically 1.8) increases with time, which is a key result because it will potentially affect the distribution tail of the corresponding loading and decrease structural reliability. There is not a clear distinction between the macro-colonization inception times because of the mixing of all typical macro-colonization years. Indeed, the mixing of a large amount of potential macro-colonization inception times does not allow one to distinguish between the contribution of each year in terms of the mode in the distribution.

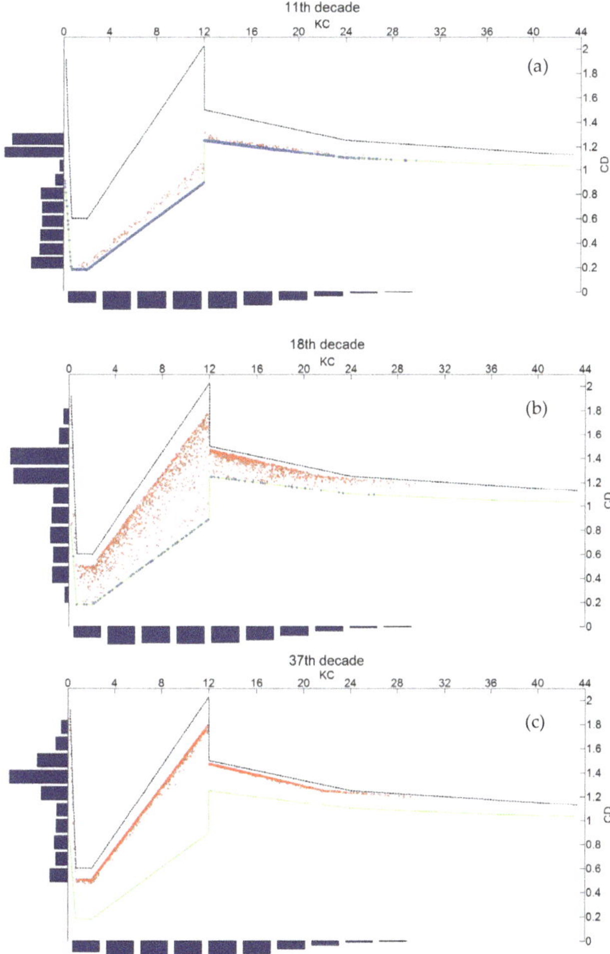

Figure 15. Distributions of the drag coefficients (C_D) as a function of (*KC*) values for three selected 10 day periods and Monte-Carlo simulations (cloud of red points): (**a**) 11th, (**b**) 18th, (**c**) 37th.

4. Discussion

Previous results give the opportunity for discussing the effect of our modeling on post-treated results, such as wave loading in the presence of marine growth. We now compare the distribution of D_e and C_{DS} (Figure 16). The distributions of D_e are mono-modal because of the combination of all typical macro-colonization years. The distributions of C_{DS} are bimodal and become mono-modal from the smooth to the ultra-roughened condition at the end of the macro-colonization period.

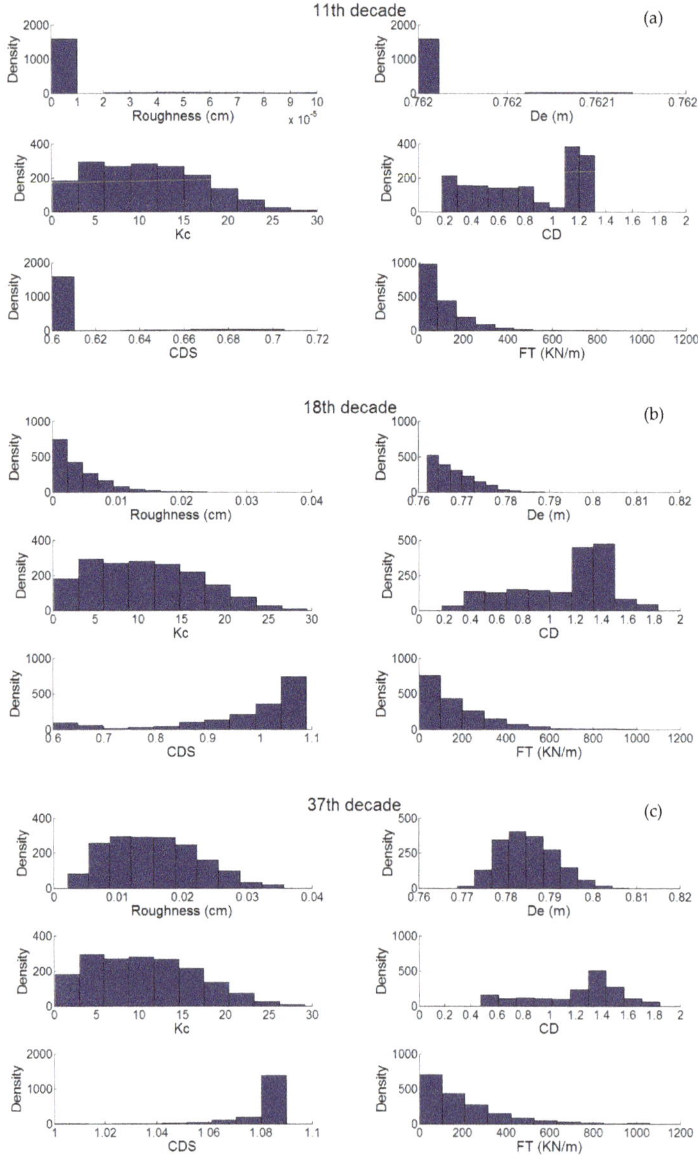

Figure 16. Comparison of the distribution of hydrodynamic parameters for the selected 10 day periods: (**a**) 11th, (**b**) 18th, (**c**) 37th.

Distribution of the drag force F_T is plotted on the same Figure 16 to better illustrate differences in distribution (mode and tails) and the transfer of these distributions. The drag force is exponentially distributed. The right distribution tail moves to higher values according to time, thereby decreasing the reliability. We analyze this distribution tail after the computation of F_{T_MAX} (note that distributions are bounded) and the fractiles $F_{T(90\%)}$, $F_{T(95\%)}$. Figure 17 shows the evolution of these statistics after each 10 day period. The latter increase smoothly with time except for the increase during one month and a half (from 10 day period 11 to 18). Finally, there is a great difference between the extreme values (F_{T_MAX}) and the fractiles ($F_{T(90\%)}$ and $F_{T(95\%)}$), confirming a long distribution tail that was observed already in Figure 16.

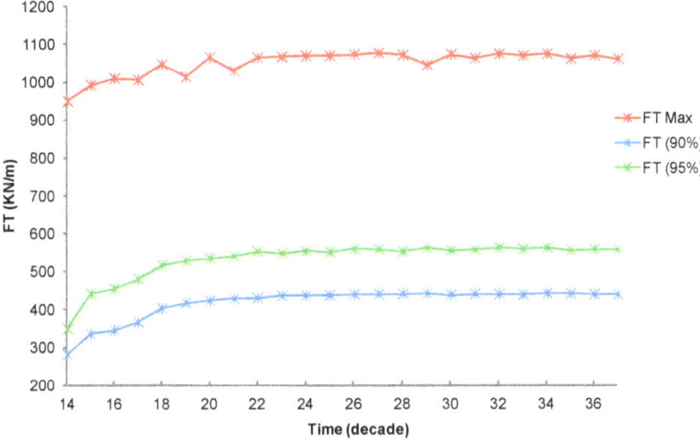

Figure 17. Evolution of F_{Tmax}, and 90%, 95% fractiles with time.

5. Conclusions

The originality of this work lies in the choice to consider biocolonization as a cumulative deterioration phenomenon and to simulate trajectories relying on individuals' characteristics through a state-dependent model to compute the probabilistic distribution of loading. The latter is a key input for structural reliability. As far as the authors know, this is the first time that bio-colonization has been considered as consecutive stochastic jumps governed by a gamma process. The developed non-stationary, state-dependent Gamma process was selected as a flexible and simple to perform methodology, which was used to generate an individual shell length time-series for blue mussels. Results of the simulation reveal that the method can capture the distribution and especially the extreme values of the observed shell length. The macro-colonization inception times determined in the initiation phase may be considered as one of the criteria for the installation or cleaning time of the structures through the maintenance programs strategy. One of its advantages is that it can be extended to other organisms, such as oysters, with the possibility of adding or modifying the parameters that influence individual growth and shape.

A model was used to investigate the drag coefficient evolution exerted by extreme waves during the mussel's growth. Three types of uncertainties have thus been considered:

- Environmental: due both to the physics of waves (height, period) and water parameters (temperature and chlorophyll-a).
- Modeling: with an uncertainty of modeling from the shell size to the thickness and the roughness in the sense of API regulation.
- Biological: accounting for the inter-individual variability.
- Moreover, calculation of hydrodynamic forces due to the biocolonization using meteo-ocean data as well as biological data is a complex task and generates two types of difficulties.

- First, the distribution of input variables that can be multi-modal (e.g., individual shell length) due to the various macro-colonization inception times.
- Second, the nonlinear transfer from the Keulegan Carpenter number to drag coefficient generates bimodal distributions from mono-modal ones.

A full probabilistic simulation that allows predicting the evolution of drag forces in a reliability context has been developed. The evolution of physical parameters due to individual growth has been presented in a time-series form. Using the empirical curves recommended by API standards to obtain wake amplification factors in a probabilistic context resulted in an abnormal discontinuity when passing the critical value $KC = 12$. Thus, these curves may not perfectly explain the evolution of the drag coefficient in a probabilistic context.

This study highlights the site-specific property of biofouling and, therefore, constructs a condition-based methodology for the modeling of biocolonization. Considering the site-specific property of biofouling, it is not logical to define a similar strategy for the maintenance and periodical cleaning programs of offshore structures without consideration of the specifications of each site. Therefore, periodical monitoring campaigns could be very useful in understanding the reaction of biofouling to environmental parameters, especially after installation or cleaning programs, and to establish the adequate maintenance strategy for each site. It allows the model to be updated as well and hence increases prediction accuracy. For some structures, it may not be necessary to clean all members completely to allow a macro-fouling community to develop and create artificial reefs that would be useful for fisheries and biodiversity. Some limitations discussed in the paper highlight that further research is requested:

- A single species was studied in a place where we can find barnacles and even algae. For the latter, relationships for the computation of drag coefficients are less developed and research is required.
- There is uncertainty in the definition of roughness and its use by engineers, which is the reason why an uncertainty of modeling is added in this paper. Recent works [60] have proposed some improvements, but this is still an open area of study. Quantification from on site inspections is possible [54], thereby opening a new area for more representative tests in laboratories.
- The probability of the occurrence of storms depends on seasons and could be introduced to reduce the conservatism.
- Effects of the Cd variations on dynamics should be introduced to expand the method to fatigue assessment.
- In the same manner, inertia forces and current could be added to get a more global influence of marine growth.

This work can be extended to floating structures once the correlation between thickness and weight is known.

6. Patents

A patent was developed in view to measure on site the marine growth and update the model: Schoefs F., Ameryoun H. (2013) «Biocolmar: Offshore Station for Measuring and Collecting Data in an Underwater Environment», 21 October 2013, N° 1360256.

Author Contributions: F.S.: methodology for coupling the models and developing gamma processes—stochastic computation; H.A.: methodology numerical implementation; L.B.: environmental modeling and metabolism of mussels; Y.T.: simulation of MEB model.

Funding: This research was funded by Université de Nantes (PhD grant from the Sea and Littoral Research Institute (www.iuml.fr)) and Région Pays de la Loire, projects COSELMAR (2012–2016) and SI3M (2008–2012).

Conflicts of Interest: The authors declare no conflict of interest.

Appendix A Additional Information about the Growth of Blue Mussels

This appendix details the intrinsic characteristics of shell growth of blue mussels (Figure A1), the comparison between the calibrated DEB model and the database in terms of shell growth, and the weight increase (Figure A2). Figure A3 illustrates the fair strait correlation between the inception date and chlorophyll-a.

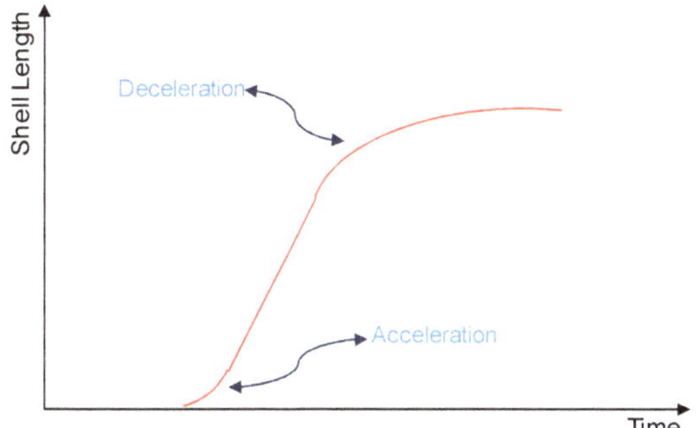

Figure A1. Schematic annual growth curve of individual blue mussels illustrating the acceleration and deceleration in the growth rate.

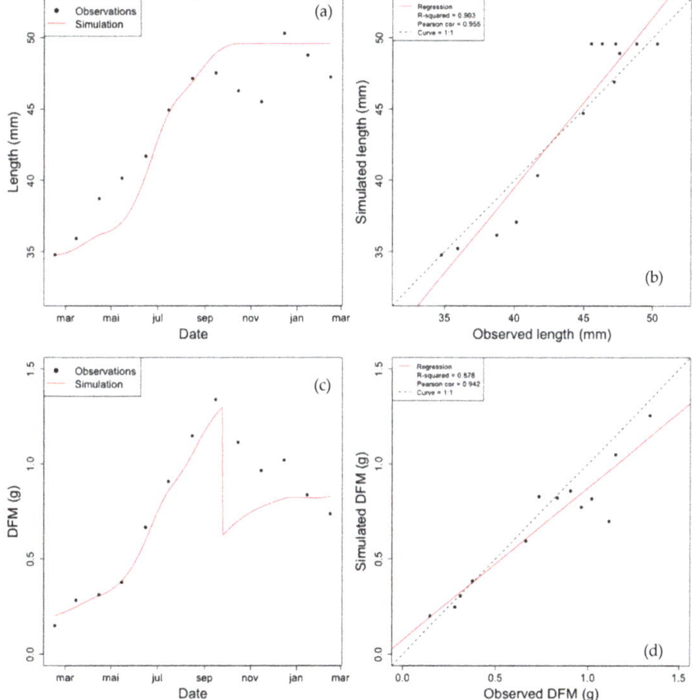

Figure A2. Calibration of the mussel DEB model used in this study to simulate shell length. (**a**) Mussel growth in length (mm), (**b**) Observed vs. simulated length; the dashed line corresponds to the 1:1 line, (**c**) Mussel growth in dry flesh mass (DFM, g); Note the strong decrease of DFM corresponding to spawning in September, (**d**) Observed vs. simulated DFM; the dashed line corresponds to the 1: 1 line.

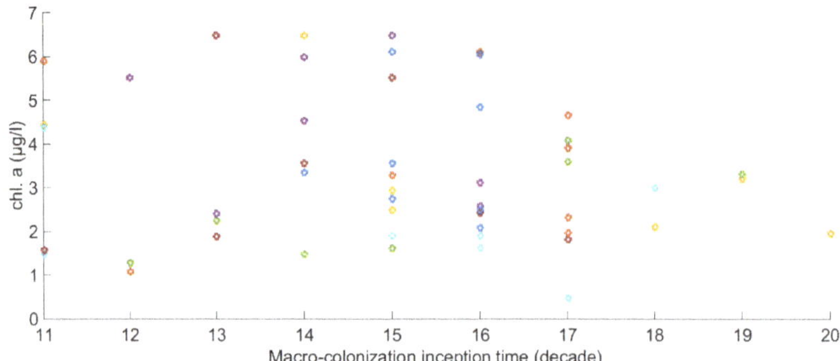

Figure A3. The relationship between macro-colonization starting times (end of the initiation phase) expressed in 10 days; each color represents a year of the 1996–2012 time-series.

References

1. Heaf, N.J. The Effect of Marine Growth on The Performance of Fixed Offshore Platforms in The North Sea. In Proceedings of the Offshore Technology Conference, Houston, TX, USA, 30 April–3 May 1979; p. 14. [CrossRef]
2. Jusoh, I.; Wolfram, J. Effects of marine growth and hydrodynamic loading on offshore structures. *J. Mek.* **1996**, *1*, 77–98.
3. API RP 2A WSD. *Recommended Practice for Planning, Designing, and Constructing Fixed Offshore Platforms*, 21st ed.; American Petroleum Institute: Washington, DC, USA, 2005; Volume 2.
4. DNV. *Recommended Practice Det Norske Veritas*; DNV-RP-C20; DNV: Oslo, Norway, 2010.
5. Faber, M.H.; Hansen, P.F.; Jepsen, F.D.; Moller, H.H. Reliability-Based Management of Marine Fouling. *J. Offshore Mech. Arct. Eng.* **2001**, *123*, 76. [CrossRef]
6. Schoefs, F.; Boukinda, M.L. Sensitivity Approach for Modeling Stochastic Field of Keulegan–Carpenter and Reynolds Numbers Through a Matrix Response Surface. *J. Offshore Mech. Arct. Eng.* **2010**, *132*, 011602. [CrossRef]
7. Boukinda, M.L. Surface de Réponse des Efforts de Houle des Structures Jackets Colonisées par des Bio-salissures. Ph.D. Thesis, Université de Nantes, Nantes, France, 2007.
8. Schoefs, F.; Boukinda, M.L. Modelling of Marine Growth Effect on Offshore Structures Loading Using Kinematics Field of Water Particle. In Proceedings of the Fourteenth International Offshore and Polar Engineering Conference, Toulon, France, 23–28 May 2004; pp. 419–427.
9. Joschko, T.J.; Buck, B.H.; Gutow, L.; Schröder, A. Colonization of an artificial hard substrate by *Mytilus edulis* in the German Bight. *Mar. Biol. Res.* **2008**, *4*, 350–360. [CrossRef]
10. Maar, M.; Bolding, K.; Petersen, J.K.; Hansen, J.L.S.; Timmermann, K. Local effects of blue mussels around turbine foundations in an ecosystem model of Nysted off-shore wind farm, Denmark. *J. Sea Res.* **2009**, *62*, 159–174. [CrossRef]
11. El Hajj, B.; Schoefs, F.; Castanier, B.; Yeung, T. A condition-based deterioration model for the stochastic dependency of corrosion rate and crack propagation in a submerged concrete structure. *Comput. Aided Civ. Infrastruct. Eng.* **2014**, *32*, 18–33. [CrossRef]
12. Ameryoun, H. Probabilistic Modeling of Wave Actions on Jacket Type Offshore Wind Turbines in Presence of Marine Growth. Ph.D. Thesis, Université de Nantes, Nantes, France, 2015.
13. Koojiman, S. *Dynamic Energy and Mass Budgets in Biological Systems*; Cambridge University Press: Cambridge, UK, 2000.
14. Koojiman, S. Dynamic Energy Budget Theory for Metabolic Organization. Cambridge University Press: Cambridge, UK, 2010.

15. Dürr, S.; Thomason, J. *Biofouling*; Wiley-Blackwell: New York, NY, USA, 2009; Available online: http://eu.wiley.com/WileyCDA/WileyTitle/productCd-1405169265.html (accessed on 18 December 2009).

16. Railkin, A.I. *Marine Biofouling: Colonization Processes and Defenses*; CRC Press: Boca Raton, FL, USA, 2003.

17. Liu, Y. Modeling the Time-to-Corrosion Cracking of the Cover Concrete in Chloride Contaminated Reinforced Concrete Structures. Ph.D. Thesis, Virginia Polytechnic Institute and State University, Blacksburg, VA, USA, 1996.

18. Newell, R.I.E. *Species Profiles: Life Histories and Environmental Requirements of Coastal Fishes and Invertebrates (North and Mid-Atlantic)*; Biological Report, 82(11.102) TR EL-82-4 June; US dept of Interior/US Army Corps of Engineers: Baltimore, MD, USA, 1989.

19. Bruijs, M.C.M. *Biological Fouling Survey of Marine Fouling on Turbine Support Structures of the Offshore Windfarm Egmond aan Zee. Report Prepared for Noordzeewind*; 50863511-TOS/PCW 10-4207, OWEZ_R_112_T1_20100226; KEMA Nederland, B.V.: Arnhem, The Netherlands, 2010.

20. Langhamer, O.; Wilhelmsson, D.; Engström, J. Artificial reef effect and fouling impacts on offshore wave power foundations and buoys—A pilot study. Estuarine. *Coast. Shelf Sci.* **2009**, *82*, 426–432. [CrossRef]

21. Gosling, E. *Bivalve Molluscs: Biology, Ecology and Culture*; Wiley-Blackwel: Hoboken, NJ, USA, 2003.

22. Barillé Boyer, A.-L. Contribution à l'étude des potentialités conchylicoles du Pertuis Breton. Ph.D. Thesis, Université d'Aix-Marseille II, Marseille, France, 1996.

23. Garen, P.; Robert, S.; Bougrier, S. Comparison of growth of mussel, *Mytilus edulis*, on longline, pole and bottom culture sites in the Pertuis Breton, France. *Aquaculture* **2004**, *232*, 511–524. [CrossRef]

24. Rosland, R.; Strand, Ø.; Alunno-bruscia, M.; Bacher, C.; Strohmeier, T. Applying Dynamic Energy Budget (DEB) theory to simulate growth and bio-energetics of blue mussels under low seston conditions. *J. Sea Res.* **2009**, *62*, 49–61. [CrossRef]

25. Widdows, J. Physiological ecology of mussel larvae. *Aquaculture* **1991**, *94*, 147–163. [CrossRef]

26. Dutertre, M.; Beninger, P.G.; Barillé, L.; Papin, M.; Rosa, P.; Barillé, A.-L.; Haure, J. Temperature and seston quantity and quality effects on field reproduction of farmed oysters, Crassostrea gigas, in Bourgneuf Bay, France. *Aquat. Living Resour.* **2009**, *22*, 319–329. [CrossRef]

27. Bayne, B.L. Growth and the delay of metamorphosis of the larvae of *Mytilus edulis* (L.). *Ophelia* **1965**, *2*, 1–47. [CrossRef]

28. Bayne, B.L.; Worrall, C.M. Growth and Production of Mussels *Mytilus edulis* from Two Populations. *Mar. Ecol.* **1980**, *3*, 317–328. [CrossRef]

29. Van Harden, R.; Koojiman, S. Application of a Dynamic Energy Budget Model to *Mytilus edulis* (L.). *Neth. J. Sea Res.* **1993**, *31*, 119–133. [CrossRef]

30. Page, H.M.; Hubbard, D.M. Temporal and spatial patterns of growth in mussels Mytihs edulis on an offshore platform: Relationships to water temperature and food availability. *Exp. Mar. Biol. Ecol.* **1987**, *111*, 159–179. [CrossRef]

31. Thomas, Y.; Mazurié, J.; Alunno-Bruscia, M.; Bacher, C.; Bouget, J.-F.; Gohin, F.; Pouvreau, S.; Struski, C. Modelling spatio-temporal variability of *Mytilus edulis* (L.) growth by forcing a dynamic energy budget model with satellite-derived environmental data. *J. Sea Res.* **2011**, *66*, 308–317. [CrossRef]

32. Thompson, R. Production, reproductive effort, reproductive value and reproductive cost in a population of the blue mussel *Mytilus edulis* from a subarctic environment. *Mar. Ecol. Prog. Ser.* **1984**, *16*, 249–257. [CrossRef]

33. REPHY dataset. French Observation and Monitoring program for Phytoplankton and Hydrology in coastal waters. 1987–2016. *Metrop. Data* **2017**. [CrossRef]

34. Hernandez-Farinas, T.; Soudant, D.; Barille, L.; Belin, C.; Lefebvre, A.; Bacher, C. Temporal changes in the phytoplankton community along the French coast of the eastern English Channel and the southern Bight of the North Sea. *Mar. Sci.* **2013**, *70*, 1439–1450. [CrossRef]

35. Barillé, L.; Lerouxel, A.; Dutertre, M.; Haure, J.; Barillé, A.L.; Pouvreau, S.; Alunno-Bruscia, M. Growth of the Pacific oyster (Crassostrea gigas) in a high-turbidity environment: Comparison of model simulations based on scope for growth and dynamic energy budgets. *J. Sea Res.* **2011**, *66*, 392–402. [CrossRef]

36. Handå, A.; Alver, M.; Edvardsen, C.V.; Halstensen, S.; Olsen, A.J.; Øie, G.; Reinertsen, H. Growth of farmed blue mussels (*Mytilus edulis* L.) in a Norwegian coastal area; comparison of food proxies by DEB modeling. *J. Sea Res.* **2011**, *66*, 297–307.

37. Pouvreau, S.; Bourles, Y.; Lefebvre, S.; Gangnery, A.; Alunno-Bruscia, M. Application of a dynamic energy budget model to the Pacific oyster, Crassostrea gigas, reared under various environmental conditions. *J. Sea Res.* **2006**, *56*, 156–167. [CrossRef]

38. Ren, J.S.; Ross, A.H. Environmental influence on mussel growth: A dynamic energy budget model and its application to the greenshell mussel Perna canaliculus. *Ecol. Model.* **2005**, *189*, 347–362. [CrossRef]

39. Barillé, L.; Prou, J.; Héral, M.; Razet, D. Effects of high natural seston concentrations on the feeding, selection, and absorption of the oyster Crassostrea gigas (Thunberg). *J. Exp. Mar. Biol. Ecol.* **1997**, *212*, 149–172. [CrossRef]

40. Bayne, B.L.; Newell, R.C. Physiological energetics of marine molluscs. *Mollusca* **1983**, *4*, 407–515.

41. Van Noortwijk, J.M. A survey of the application of gamma processes in maintenance. *Reliab. Eng. Syst. Saf.* **2009**, *94*, 2–21. [CrossRef]

42. Abdel-Hameed, M. A Gamma Wear Process. *IEEE Trans. Reliab.* **1975**, *R-24*, 152–153. [CrossRef]

43. Cheng, T.; Pandey, M.D.; Van Der Weide, J.A.M. The probability distribution of maintenance cost of a system affected by the gamma process of degradation: Finite time solution. *Reliab. Eng. Syst. Saf.* **2012**, *108*, 65–76. [CrossRef]

44. Van Noortwijk, J.M.; Van der Weide, J.A.M.; Kallen, M.J.; Pandey, M.D. Gamma processes and peaks-over-threshold distributions for time-dependent reliability. *Reliab. Eng. Syst. Saf.* **2007**, *92*, 1651–1658. [CrossRef]

45. Guida, M.; Postiglione, F.; Pulcini, G. A time-discrete extended gamma process for time-dependent degradation phenomena. *Reliab. Eng. Syst. Saf.* **2012**, *105*, 73–79. [CrossRef]

46. Sarpkaya, T. On the Effect of Roughness on Cylinders. *J. Offshore Mech. Arct. Eng.* **1990**, *112*, 334. [CrossRef]

47. Theophanatos, A. Marine Growth and the Hydrodynamic Loading of Offshore Structures. Ph.D. Thesis, University of Srathclyde, Glasgow, UK, 1988.

48. Schoefs, F. Sensitivity approach for modelling the environmental loading of marine structures through a matrix response surface. *Reliab. Eng. Syst. Saf.* **2008**, *93*, 1004–1017. [CrossRef]

49. Morison, J.R.; Johnson, J.W.; Schaaf, S.A. The Force Exerted by Surface Waves on Piles. *J. Pet. Technol.* **1950**, *2*, 149–154. [CrossRef]

50. Stokes, G.G. On the theory of oscillatory waves. *Trans. Camb. Phil. Soc.* **1847**, *8*, 441–455.

51. Wolfram, J.; Jusoh, I.; Sell, D. Uncertainty in the Estimation of Fluid Loading Due to the Effects of Marine Growth, Safety and Reliability Symposium. In Proceedings of the 12th International Conference on Offshore Mechanics and Arctic Engineering (O.M.A.E'93), Glasgow, Scotland, UK, 20–24 June 1993; Volume II, pp. 219–228.

52. Kasahara, Y.; Koterayama, W.; Shimazaki, K. Wave Forces Acting on Rough Circular Cylinders at High Reynolds Numbers. In Proceedings of the Offshore Technology Conference, Houston, TX, USA, 6–9 May 2013; p. 12. [CrossRef]

53. Troesch, A.W.; Kim, S.K. Hydrodynamic forces acting on cylinders oscillating at small amplitudes. *J. Fluids Struct.* **1991**, *5*, 113–126. [CrossRef]

54. O'Byrne, M.; Schoefs, F.; Pakrashi, V.; Ghosh, B. An underwater lighting and turbidity image repository for analysing the performance of image based non-destructive techniques. *Struct. Infrastruct. Eng.* **2018**, *14*, 104–123. [CrossRef]

55. O'Byrne, M.; Schoefs, F.; Pakrashi, V.; Ghosh, B. A Stereo-Matching Technique for Recovering 3D Information from Underwater Inspection Imagery. *Comput. Aided Civ. Infrastruct. Eng.* **2018**, *33*, 193–208. [CrossRef]

56. O'Byrne, M.; Pakrashi, V.; Schoefs, F.; Ghosh, B. Semantic Segmentation of Underwater Imagery Using Deep Networks. *J. Mar. Sci. Eng.* **2018**, *6*, 93. [CrossRef]

57. Zeinoddini, M.; Bakhtiari, A.; Schoefs, F.; Zandi, A.P. Towards an Understanding of the Marine Fouling Effects on VIV of Circular Cylinders: Partial Coverage Issue. *Biofouling* **2017**, *33*, 268–280. [CrossRef]

58. Bakhtiari, A.; Schoefs, F.; Ameryoun, H. Unified Approach for Estimating of The Drag Coefficient In Offshore Structures In Presence Of Bio-Colonization. In Proceedings of the 37th International Conference on Offshore Mechanics and Arctic Engineering (O.M.A.E'18), Madrid, Spain, 17–22 June 2018; p. 78757.

59. Nerzic, R.; Prevosto, M.; Frelin, C.; Quiniou, V. Joint Distributions for Wind/waves/current in West Africa and derivation of Multi Variate Extreme I-FORM Contours. In Proceedings of the 17th International Offshore and Polar Engineering Conference, Lisbon, Portugal, 1–6 July 2007; pp. 81–88.
60. Schoefs, F.; Bakhtiari, A.; Hameryoun, H.; Quillien, N.; Damblans, G.; Reynaud, M.; Berhault, C.; O'Byrne, M. Assessing and modeling the thickness and roughness of marine growth for load computation on mooring lines. In Proceedings of the Floating Offshore Wind Turbine Conference (FOWT 2019), Montpellier, France, 24–26 April 2019.

 Journal of
*Marine Science
and Engineering*

Article

Numerical Analyses of Wave Generation and Vortex Formation under the Action of Viscous Fluid Flows over a Depression

Chih-Hua Chang [1,2]

[1] Department of Information Management, Ling-Tung University, Taichung 408, Taiwan; changbox@teamail.ltu.edu.tw

[2] Natural Science Division in General Education Center, Ling-Tung University, Taichung 408, Taiwan

Received: 21 March 2019; Accepted: 4 May 2019; Published: 12 May 2019

Abstract: Transient free-surface deformations and evolving vortices due to the passage of flows over a submerged cavity are simulated. A two-dimensional stream function–vorticity formulation with a free-surface model is employed. Model results are validated against the limiting case of pure lid-driven cavity flow with comparisons of the vortical flow pattern and velocity profiles. The verification of the free-surface computations are also carried out by comparing results with published potential flow solutions for cases of flows over a depressed bottom topography. The agreements are generally good. Investigations are extended to other viscous flow conditions, where the cavity is set to have the normalized dimension of one by one when scaled by the still water depth. The free-surface elevations and streamline patterns for cases with Froude numbers ranging from 0.5 to 1.1 and different Reynolds numbers (Re = 5000 and 500) are calculated. At the condition of near-critical flow ($Fr \approx 1.0$), the phenomenon of upstream advancing solitons is produced. Viscous effects on the free-surface profile reveal that at a lower value of Re (e.g., Re = 500) larger advancing solitary waves are generated. Vortical flow patterns in the cavity are examined for the cases with $Fr = 1.0$ and various values of Re. When Re = 5000, the vortex pattern includes a primary and a weak, but dominated secondary vortices at the time reaching a nearly quasi-steady motion. For the case of lower Re (e.g., Re = 500), a steady-state vortex pattern can be established with a clockwise primary vortex mostly occupied inside the cavity.

Keywords: solitons; free-surface flow; depressed bottom; viscous fluid flow; vortical flow

---

1. Introduction

In 1834, John Scott Russell discovered what he called a form of a large solitary elevation or later a solitary wave that was emerged in front of a tow-boat after its sudden stop. Since then, describing the solitary wave elevations produced in shallow water either by flow passing through a disturbance or by a moving object has become a fascinating and important research topic. Most disturbances considered in the past were either a surface pressure or a bottom-placed convex object and the use of a bottom cavity, as a submerged object in general is limited. The viscous effect is also ignored. This paper presents the novelty of a numerical study to explore the transient phenomena of wave generation and vortex evolution resulting from a uniform flow passing over a bottom cavity in shallow water.

A cavity-like region is usually observed in a hydraulic channel or a dredged waterway for navigation use. In nature, various materials—such as toxic chemicals, nutrients, planktons, sludge, or sediments—may be deposited and trapped in concaved bottom trenches or in depressed zones of water-covered terrains. Numerical simulation of vortex motions in a cavity can provide critical information to understand the transport mechanism of materials in it. On the other hand, modeling cavity flow is also a classical problem for the study of vortex phenomena in fluid mechanics. A square or rectangular cavity has been commonly used as a test domain for theoretical, numerical, and experimental studies of formed vortices. With the continuous advancement of the research tools, either numerically or experimentally, the results generated from the setting of this simple geometry can demonstrate several important concepts of flow motion. For example, Ryzhov and Koshel [1] and Ryzhov et al. [2] discussed the motion change of point vortices due to boundary current in different circular cavity apertures using the Kirchhoff-Routh stream-function method. It is a simple geometric assumption, but may provide valuable qualitative insight into feasible vortex motion near curved coastlines.

In general, four research topics related to the cavity flow can be defined by the following. Type (1): The studies involve a pure lid-driven cavity flow, i.e., typical flow patterns generated by a lid moving at a constant speed (see Shankar and Deshpande [3]). Type (2): The studies focus on the flows and vortices induced by a uniform flow passing over an open cavity. For example, Chang et al. [4] investigated the differences of the evolved vortices between the conditions of laminar and turbulent flows in a three-dimensional cavity. Zhang and Rona [5] examined the pressure distribution that follows a pressure wave passing through a cavity and Fang et al. [6,7] modeled the process of contaminant removal from a cavity. Type (3): The studies considered cavities of various shapes or rectangular cavities with different aspect ratios, such as the visualization photo of triangular cavity flow shown in the book *An Album of Fluid Motion* by Van Dyke [8]. Later, a similar problem was investigated numerically by Erturk and Gokcol [9]. In addition, Chang and Cheng [10] studied lid-driven air flow within an arc-shape cavity. Yin and Kumar [11] explored the induced flow patterns in a cavity with a flexible boundary and with various aspect ratios. Type (4): The cases include the obstacles in a cavity, which form multiple connected domains. For example, Khanafer and Aithal [12], using a finite element formulation, studied the mixed convective flow and heat-transfer characteristics in a lid-driven cavity that contains a circular cylinder.

The present study investigates the evolving flow patterns and generated vortices as a uniform flow with a free surface passing over a submerged cavity, which is different from the usual lid-driven cavity flow as the so-called rigid lid is replaced by a layer of water that moves with a specified speed on top of the cavity. The formed vortices, instead of confining within the cavity, may expand out of the cavity to interact with the external flow outside of the cavity. The movable free surface may also interact with the flow inside the cavity.

Water waves are naturally generated by external forces, such as wind blowing on the water surface, a landslide moving into water (Grilli et al. [13]), a fish swimming in water (Adkins and Yan [14]), moving vessels (Kara et al. [15]), or the movement of a submerged body (Chang and Wang [16]). In the past, researchers explored the wave patterns of flow passing through a submerged object based on the assumption of steady flow at supercritical or subcritical flows. For example, Hanna et al. [17] applied the Schwartz–Christoffel transformation technique and a series truncation based computational procedure to solve the problem of a steady supercritical flow over a trapezoidal obstacle. Furthermore, a steady turbulent flow model was used by Tzabiras [18] to study the super- and subcritical flows over a hump. In viewing a relative motion, interesting phenomena of generation of solitary waves by an external forcing moving at a transcritical speed has also been studied. The features of the generation of upstream-propagating solitary waves were first investigated numerically by Wu and Wu [19] with solutions that were obtained by solving the generalized Boussinesq equations under the conditions of a surface pressure disturbance moving at a speed close to the transcritical regime. Other external forces in similar studies found the literature included a seabed protrusion or a submerged translating obstacle. The hump-like topography is considered to function as a positive external forcing. On the contrary, a depressed topography is referred to as a negative forcing function (Zhang and Chwang, [20]). The features of wave systems generated by a positive or a negative moving forcing are quite different [20]. To the authors' knowledge, the numerical simulation considering a negative forcing function is more challenging than the cases with a positive one. In other related studies, Grimshaw and Smyth [21], Wu [22], and Camassa and Wu [23] have applied either forced KdV equations or generalized

Boussinesq equations to model solitary waves generated by a moving negative forcing. Grimshaw et al. (2009) [24] investigated flow over a hole using fKdV equations. Their study was inspired by the idea that a semi-infinite positive step generates only an upstream-propagating undular bore (Grimshaw et al. (2007) [25]), and a negative step generates only a downstream-propagating undular bore. Zhang and Chwang [20] also developed a numerical model based on the Euler equations to analyze the roles of the positive and negative forcing functions played on the generated waves. In recent times, Xu and Meng [26] investigated the solitary waves generated by a submerged two-dimensional foil with an angle moving in shallow water at subcritical, super-critical and hypercritical speeds.

In the past, the viscous effect has often been ignored in water wave studies, including those described on wave generation. It appears that the first numerical calculation considering the fluid viscosity for solving the soliton radiation problem was carried out by Chang and Tang [27]. Only a smooth bottom hump was considered, and hence, no vortex wake was found in their simulated results. Zhang and Chwang [28] numerically solved the Navier–Stokes equations to study the transcritical flow passing through a submerged semi-elliptical cylinder. They showed that the vorticities were transported and diffused around the body, but no obvious vortex wake appeared behind the streamlined body. Later, Lo and Yang [29], adopted the vorticity-velocity formulations in calculating the flow passing over a blunt body and indicated that the vortices were clearly observed on the leeside of the body. Employed the stream-function–vorticity formulations with a free surface (SVFS) model—extended from a model developed by Tang and Chang [30]; Chang et al. [31] investigated the vortex patterns generated by a near-critical flow passing over a small square hump at the bottom.

The present study applies the further improved SVFS model with a finer and nonuniform grid system to model the flow motions of the viscous fluid encountering a cavity forcing. To validate the numerical model, the patterns of vortices in the cavity are compared with the results from the limiting cases of pure cavity flow studies. For the free-surface profiles, the present solutions are compared with those from the study of soliton radiation produced by a negative forcing function in potential flow. After the confirmation of model validation, the SVFS is applied to simulate the present study of a viscous uniform flow over a cavity. The cases studied for flows over a cavity include the combinations of two different Reynolds numbers (Re = 500 and 5000) and various Froude numbers (Fr = 0.5 to 1.1). The definitions of Re and Fr are stated in the next section. Free-surface deformations and the evolution of vortices are the focuses of this study and analysis.

2. Mathematical Formulations

The physical problem is sketched in Figure 1, where the free-surface deformations and vortex evolution during the process of a uniform flow interacting with a cavity are investigated. In the study domain, a two-dimensional (2-D) unsteady viscous flow is assumed. The governing equations, along with the initial and boundary conditions, are described below.

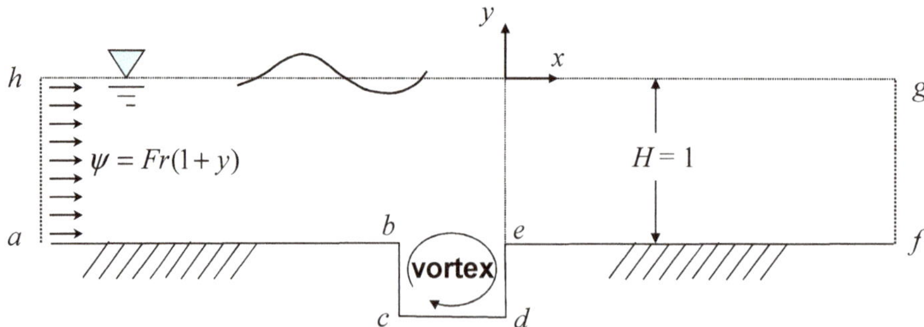

Figure 1. Domain configuration and coordinates.

2.1. Governing Equations

For an incompressible viscous flow problem in two-dimensional domain, it is convenient to reduce the variables by converting the Navier–Stokes equations into the stream-function–vorticity (ψ, ω) formulations to describe the flow fields. All variables are written in dimensionless form, using the length scale of undisturbed mean water depth, H^*, and respectively the velocity and time scales of $\sqrt{gH^*}$ and $\sqrt{H^*/g}$, with g being gravitational acceleration. The variables with superscript "*" represent the dimensional quantities. A transformation from Cartesian coordinates, $(x, y; t)$, to a boundary-fitted coordinate system, $(\xi, \eta; \tau)$, is carried out by introducing the general transformation $\xi = \xi(x, y)$ and $\eta = \eta(x, y)$. The dimensionless governing equations according to the laws of conservations of mass and momentum can be expressed in terms of vorticity ω and stream-function ψ in the defined curvilinear coordinate system as

$$\nabla^2 \omega = Re(\omega_\tau + U\omega_\xi + V\omega_\eta) \tag{1}$$

$$\nabla^2 \psi = -\omega \tag{2}$$

The subscripts denote the partial derivatives. Here, $Re = H * \sqrt{gH^*}/v$ is a representative Reynolds number, in which v is the kinematic viscosity. The symbol ∇^2 is the Laplacian operator, which is defined as

$$\nabla^2 = g^{11}\frac{\partial^2}{\partial \xi^2} + 2g^{12}\frac{\partial}{\partial \xi}\frac{\partial}{\partial \eta} + g^{22}\frac{\partial^2}{\partial^2 \eta} + f^1\frac{\partial}{\partial \xi} + f^2\frac{\partial}{\partial \eta} \tag{3}$$

The variables U and V are the contra-variant components of the relative fluid velocities given as

$$V = (-y_\tau x_\xi + x_\tau y_\xi - \psi_\xi)/J \tag{4a}$$
$$U = (-x_\tau y_\eta + y_\tau x_\eta + \psi_\eta)/J; \quad V = (-y_\tau x_\xi + x_\tau y_\xi - \psi_\xi)/J,$$

$$U = (-x_\tau y_\eta + y_\tau x_\eta + \psi_\eta)/J; \quad V = (-y_\tau x_\xi + x_\tau y_\xi - \psi_\xi)/J, \tag{4b}$$

and

$$g^{11} = (x_\eta^2 + y_\eta^2)/J^2 \tag{5a}$$

$$g^{22} = (x_\xi^2 + y_\xi^2)/J^2 \tag{5b}$$

$$g^{12} = -(x_\xi x_\eta + y_\xi y_\eta)/J^2, \tag{5c}$$

$$f^1 = [(Jg^{11})_\xi + (Jg^{12})_\eta]/J \tag{5d}$$

$$f^2 = [(Jg^{12})_\xi + (Jg^{22})_\eta]/J, \text{ and} \tag{5e}$$

$$J = (x_\xi y_\eta - y_\xi x_\eta) \tag{5f}$$

2.2. Initial and Boundary Conditions

A flow field that is approximately confined in a finite domain as indicated by *abcdefgh* in Figure 1 is assumed. The required initial and boundary conditions are described in the following.

2.2.1. Initial Condition

At the initial state, an approximated potential flow condition is specified across the entire computational domain. Assuming the free surface is undisturbed and a uniform upstream velocity denoted by U^* is provided to enter at the domain inlet. This quantity is a given parameter and is related to the Froude number as $Fr = U^*/\sqrt{gH^*}$.

2.2.2. Boundary Conditions

At the free-surface boundary, defined by $y = \zeta(x,t)$, one of the applicable conditions is the kinematic condition:

$$\psi_\xi + \zeta_\tau x_\xi = x_\tau \zeta_\xi \tag{6}$$

The other is the dynamic free-surface boundary condition:

$$\psi_{\xi\tau}(Jg^{12}) + \psi_{\eta\tau}(Jg^{22}) + \psi_\xi \widetilde{A} + \psi_\eta \widetilde{B} + (u - x_\tau)u_\xi + (v - \zeta_\tau)v_\xi + \zeta_\xi$$
$$+\omega[(u - x_\tau)\zeta_\xi - (v - \zeta_\tau)x_\xi] + \frac{J}{Re}(g^{12}\omega_\xi + g^{22}\omega_\eta) = 0 \tag{7}$$

in which (u, v) denote the velocity components in the (x, y) directions and are defined as

$$u = (x_\xi \psi_\eta - x_\eta \psi_\xi)/J, \; v = (-y_\eta \psi_\xi + y_\xi \psi_\eta)/J \tag{8}$$

Furthermore, the convective coefficients $\widetilde{A}$ and $\widetilde{B}$ are arranged as

$$\widetilde{A} = -\left(\frac{x_\eta}{J}\right)_\tau x_\xi - \left(\frac{\zeta_\eta}{J}\right)_\tau \zeta_\xi; \; \widetilde{B} = \left(\frac{x_\xi}{J}\right)_\tau x_\xi + \left(\frac{\zeta_\xi}{J}\right)_\tau \zeta_\xi \tag{9}$$

According to Tang [32], the vorticity condition at the free surface (ω_f) is given as

$$\omega_f = -2\frac{\partial u_f}{\partial \widetilde{n}} \tag{10}$$

where $\widetilde{n}$ is the unit-normal vector towards the interior fluid domain and u_f is the fluid free-surface particle velocity along the tangential direction. In terms of the stream function, it is written as

$$u_f = \frac{g^{22}\psi_\eta + g^{12}\psi_\xi}{\sqrt{g^{22}}} \tag{11}$$

Then, substitution of Equation (11) into Equation (10) yields

$$\omega_f = -2\frac{g^{22}(u_f)_\eta + g^{12}(u_f)_\xi}{\sqrt{g^{22}}} \tag{12}$$

A uniform flow is assumed to proceed from the upstream boundary (domain *ha*), where

$$\psi = Fr(y + 1), \; \omega = 0 \tag{13}$$

The outlet condition at the downstream boundary (domain *fg*) is approximated by the radiation condition, which is expressed as

$$\vartheta_t + (u_{fg} + \sqrt{1 + \zeta})\vartheta_x = 0 \tag{14}$$

where ϑ is a dummy variable, representing either ψ, ω, or ζ. The u_{fg} denotes the downstream horizontal velocity, which is numerically evaluated using the Neumann condition. The impermeable solid boundary (domain *abcdef*) resembles as a streamline, which can be specified with a constant stream function value, i.e., $\psi = 0$. The no-slip condition along cavity bottom is applied to the derivation of the impermeable surface vorticity. A standard formulation is taken from Nallasamy [33]. On the bottom, the wall vorticity is expressed as

$$\omega = -2\psi_1/\Delta n^2 \tag{15}$$

Here, ψ_1 represents the stream function value at the first grid node adjacent to the wall in the fluid domain and Δn is the normal distance between the wall and the adjacent node.

3. Numerical Method

3.1. Grid Generation and Discretization

In the numerical method, the algebraic grid-generation technique is adopted for curvilinear grids generated to fit the free surface, as it undergoes deformation. The indices of grid points are denoted by $i = 1$ to IM within the computational domain (x_{min}, x_{max}) and $j = 1$ to JM within (y_{min}, y_{max}) in the x- and y-directions, respectively. The size of the mesh in the x direction is fixed, however, distributed nonuniformly in space. As shown in Figure 2, the domain along the x direction is divided into three regions: the first region extends from $i = 1$ to $i = i_c$, the second region covers the cavity from $i = i_c$ to $i = i_{c1}$, and the indices from $i = i_{c1}$ to $i = IM$ are arranged for the third region. However, along the y direction, the domain is divided by the line numbering $j = j_c$ into two parts: the lower region measures from $j = 1$ to $j = j_c$ and the upper one extends from $j = j_c$ to $j = JM$. The grid size in y direction below y_{jc} is fixed while in the upper region (above $y = y_{jc}$) it is transiently evolved with the varying free surface. In order to generate the numerical grid efficiently, the following non-uniform grid system is adopted:

$$x_i = x_{i_c} - \frac{a_0\left(1 - r_L^{i_c - i}\right)}{1 - r_L}, \text{ for } i = 1 \text{ to } i_c \tag{16a}$$

$$x_i = x_{i_c} + \frac{\left(x_{i_{c1}} - x_{i_c}\right)}{i_{c1} - i_c}(i - i_c), \text{ for } i = i_c \text{ to } i_{c1} \tag{16b}$$

$$x_i = x_{i_{c1}} + \frac{a_0\left(1 - r_R^{i - i_{c1}}\right)}{1 - r_R}, \text{ for } i = i_{c1} \text{ to } IM \tag{16c}$$

$$y_j = y_1 + \frac{\left(y_{jc} - y_1\right)}{j_c - 1}(j - 1), \text{ for } j = 1 \text{ to } j_c \tag{16d}$$

$$y_{i,j} = y_{jc} + \frac{\zeta_i - y_{jc}}{JM - j_c}(j - j_c), \text{ for } j = j_c \text{ to } JM \tag{16e}$$

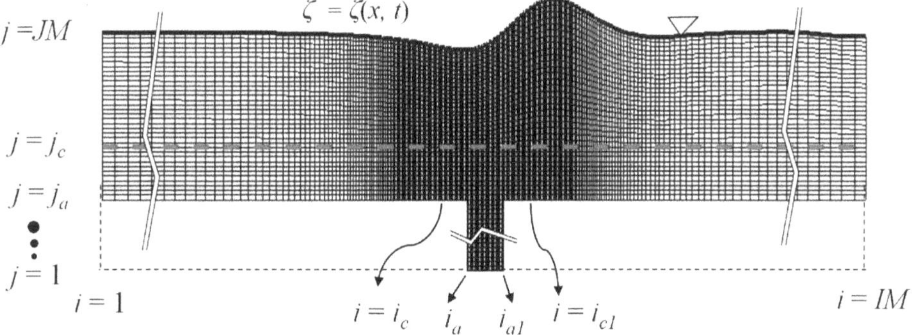

Figure 2. Schematic of grid structure.

It should be noted the grid points including $j = 1$ to $j = j_a$ indices are assigned only inside the cavity with x indices covering from $i = i_a$ to $i = i_{a1}$. An example plot showing a grid system covering the fluid domain, including the free surface and a cavity, is illustrated partially in Figure 2. For this case, the parameter values are set as $a_0 = 0.02$, $r_L = 1.0451$, $r_R = 1.0426$, $i_c = 101$, $i_a = 151$, $i_{a1} = 201$, $i_{c1} = 251$, $j_a = 51$, $j_c = 76$, $x_{i_c} = -2$, $x_{i_a} = -1$, $x_{i_{a1}} = 0$, $x_{i_{c1}} = 1$, $y_1 = -2$, and $y_{j_c} = -0.5$. This gives a grid system with equal spacing along x and y directions within the cavity, where $\Delta = \Delta x_{cavity} = \Delta y_{cavity} = 0.02$. The grid sizes of $\Delta = 0.05$ and $\Delta = 0.01$ have also been selected to test the effect of grid size on the results of the flow condition $Fr = 1.0$ and $Re = 5000$. To analyze the grid-size influence on the evolving vortices, results obtained from using the three different mesh sizes (not shown here) are compared. It is found the vortical flow patterns in a larger scale are generally similar to those from the setup of any of the three different meshes, however, with the differences shown on the corner eddies—where they cannot be revealed in the domain using coarse grid ($\Delta = 0.05$). On the other hand, the simulation in a domain with a finer grid size of $\Delta = 0.01$ costs vast computational efforts to obtain the similar results as those from the case of $\Delta = 0.02$. Therefore, the grid size of $\Delta = 0.02$ within the cavity is utilized to solve all cases in the present study.

The finite analytic method (FAM) developed by Chen and Chen [34] is applied to solve the coupled system of Equations (1) and (2) in the flow domain. Recent applications of the FAM based SVFS model have been addressed in Chang et al. [31] and Chang and Lin [35]. The focus described here is to provide a new numerical scheme applied to solve the boundary conditions. Numerically, it is more challenging in solving the free-surface conditions. An upwind scheme for the free-surface computation is employed. Details of the numerical treatment of this model are described in Appendix A. Once the free-surface variables ψ and ζ are determined, the grid points are regenerated, and the time-marching solutions with increment $\Delta \tau = 0.01$ are obtained by satisfying the governing equation and other associated boundary conditions. The converging criteria for both ψ and ζ are reached when the absolute deviations of the iterated variables are less than 10^{-6}, and ω is less than 10^{-4}.

3.2. Numerical Procedure

The numerical procedure of solving the above-described discretizations for obtaining the converged solutions at each time step is given below.

Initially (at $t = 0$), a still water surface is assumed. The initial values of ζ and ω are zero. Then, the computational grids are generated according to the grid formulations (Equation (16)) and the grid metric coefficients in Equation (5) are determined. Initial values of ψ in the whole flow domain are calculated by solving Equation (2) with the settings of $\omega = 0$, the inlet uniform flow, and other boundary conditions.

With the values obtained for $\psi_{i,j}^n$, $\omega_{i,j}^n$, and $y_{i,j}^n$ (including ζ_i^n) at the time level n, the computation of the unknown variables at the new time level ($t = (n+1)\Delta t$, $n = 0, 1, 2, 3 \ldots$) follows the steps as shown below:

1. Solve the kinematic free-surface boundary condition (Equation (A2)) to obtain ζ_i^{n+1}, and accordingly update $\hat{y}_{i,j}^{n+1}$ from Equation (16e). (Here, the notation of hat "ˆ" represents the provisional solutions).
2. Solve the dynamic free-surface boundary conditions (Equations (A3) and (12)) for the values of $\hat{\psi}_{i,JM}^{n+1}$, and $\hat{\omega}_f^{n+1}$, respectively.
3. Update the wall vorticity from Equation (15).
4. Regenerate the vertical coordinates $\hat{y}_{i,j}^{n+1}$ (Equation (16)) and calculate the grid metric coefficients (Equation (5)).
5. Solve the coupled Equations (1) and (2) to obtain $\hat{\psi}_{i,j}^{n+1}$ and $\hat{\omega}_{i,j}^{n+1}$, respectively, in the flow field.
6. Repeat steps 1–5 until converged solutions are obtained.

The numerical procedure of the above six steps is then carried into the next time step, and the calculation is continued until the final allotted time is reached.

4. Results

4.1. Model Validations

Model validations for the computed free-surface elevations are conducted by comparing the results with published solutions for the cases of a potential uniform flow passing over a bottom-depressed region. In addition, the vortex evolution is validated by the comparisons with the results of a lid-driven cavity flow without a free surface.

4.1.1. Free-Surface Wave Generation Due to a Negative Bottom Forcing Function

The free-surface deformations generated due to a flow passing over a bottom hump-like structure have been frequently studied. For example, Lowery and Liapis [36] studied the wave motions produced by a flow passing over a bottom-mounted, semi-circular cylinder. The hump case is referred to as having a positive forcing effect, and in the presence of a depression, it represents a negative forcing function (Wu, [35]). The free-surface motions resulting from the input of a negative forcing function are very different from those under the condition of a positive forcing function. In general, it is also more difficult to achieve stable solutions in computations. Zhang and Chwang [20] investigated the different wave systems due to the consideration of either a positive or a negative forcing function. They solved for the primitive variables in the Euler equations using the finite difference method. Here, one of their cases is selected for results comparisons. As shown in Figure 3, the bottom shape is expressed by

$$b(x) = \frac{b_m}{2}\left[1 + \cos(\frac{2\pi}{L})\right], \quad -L/2 \le x \le L/2, \tag{17}$$

The selected flow condition is the same as in [20] with the inputted parameters of $Fr = 1.0$, $L = 2.0$, and $b_m = -0.1$. In our numerical calculation, the primary variables (stream function and free-surface elevation) in this potential flow problem are analyzed for their convergence. As shown in Figure 4. The time-step convergence requirement is that the absolute difference between the previous value and the current value must be $< 10^{-6}$. It can be seen from the plot that the convergence of the stream function is slower than that of the free-surface elevation, although convergence is reached eventually.

The results showing the successive free-surface profiles from $t = 0$ to 400 are compared in Figure 5a,b. Comparing the results shown in Figure 5a,b, it is noted the numbers and the temporal variations of the advancing upstream solitons obtained from the present model are in good agreement with those from Zhang and Chwang [20], although with a similar pattern the present model produces slightly stronger trailing dispersive waves downstream of the structure. It is demonstrated through this comparison study that the second-order upwind scheme is suitable to be applied for all cases that follow.

Figure 3. Diagram of the flow over a depression.

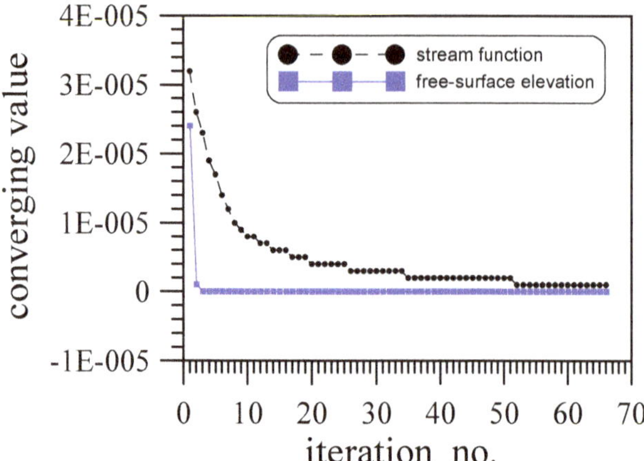

Figure 4. Processes of a typical time step of convergence iterations for the primary variables.

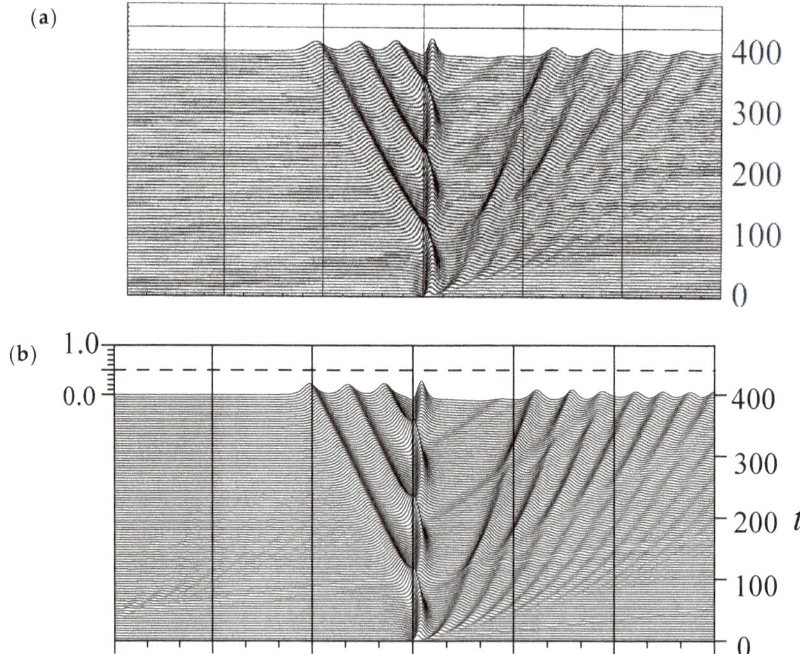

Figure 5. Free-surface evolution over a negative bottom: (a) Zhang and Chwang, adapted from [20], with permission from Copyright © 2001 Cambridge University Press; (b) present results.

4.1.2. Vortex Motion Comparison: Pure Lid-Driven Cavity Flow

The laboratory data or other numerical results for vortex motions induced by uniform flows with a free surface over a cavity are not currently available. Therefore, a lid-driven flow in a closed cavity, as shown in Figure 6, is tested for comparisons with other studies. The flow at the lid is assumed to move with a constant horizontal velocity while the flows at other walls are stationary. The Reynolds number for the cavity flow is defined as $Re_{lid} = u^*_{lid} L^* / v$, in which u^*_{lid} is the velocity at the lid and L^* is the length of the cavity. Hence, the lid-normalized velocity $u = u_{lid} = 1$. Traditionally, the quantitative analyses are made to compare the velocity profiles in the cavity. In case of steady flow, the x-component of velocity (u) and the y-component of velocity (v) are both plotted along the y-axis at x = 0.5 and the x-axis at y = 0.5, respectively. It is compared for five set grids (101 × 101, 121 × 121, 151 × 151, 181 × 181, and 201 × 201) having different numbers of grid nodes for calculating the flow field, as denoted by the velocity profiles presented in Figure 7. In this figure, it can be qualitatively seen that 181×181 has almost reached grid independence. It is conservatively chosen the results of 181 × 181 to compare them with the results of Ghia et al. [37], where a clearer plot of their results was given in Peng et al. [38].

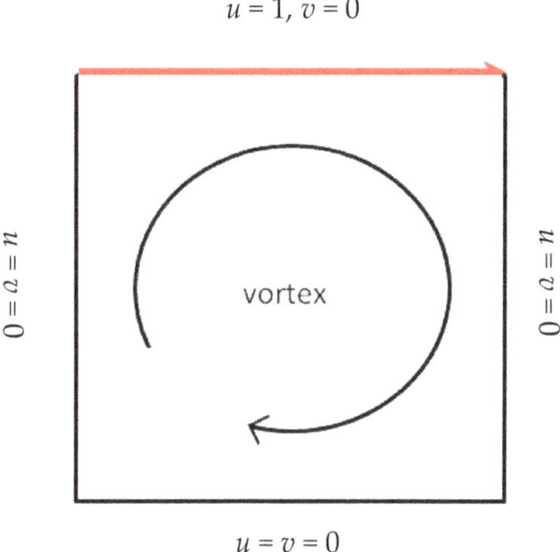

Figure 6. Sketch of a cavity domain with a constant speed at the lid.

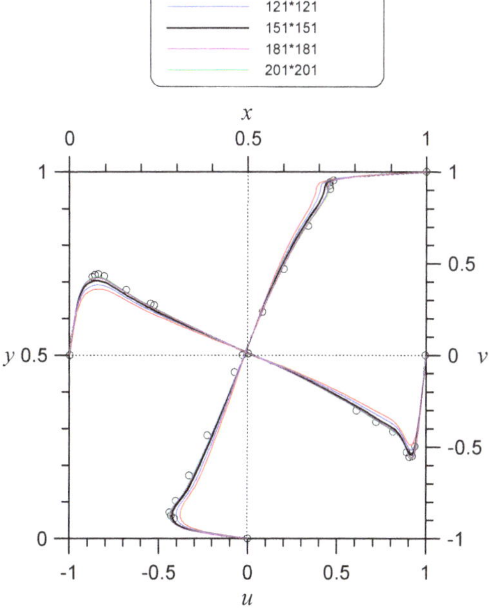

Figure 7. Computed steady velocity profiles for Re_{lid} = 5000.

Figure 8 depicts the cavity flow phenomenon with streamline patterns for a selected flow condition, $Re_{lid} = 5000$. The qualitative comparisons with other studies reveal the achievement of similar patterns. In addition to show the present model results (Figure 8a), plots of the numerical solutions of Ghia et al. [37] and Erturk et al. [39] are presented in Figure 8 for comparisons. The present solutions employ a uniform grid system of 181 × 181, while Ghia et al. [37] used 257 × 257 nodes, and the simulation of Erturk et al. [39] used a 129 × 129 grid system. The solutions from the present investigation are drawn with streamline contours at the interval of 0.01 for the values ranging from −0.11 to 0 and 0.0002 for the values between 0 and 0.048. In this case, the flow cannot reach a steady state until after $t = 40$. The results at $t = 300$ are shown for comparisons to ensure that the steady-state situation has been reached.

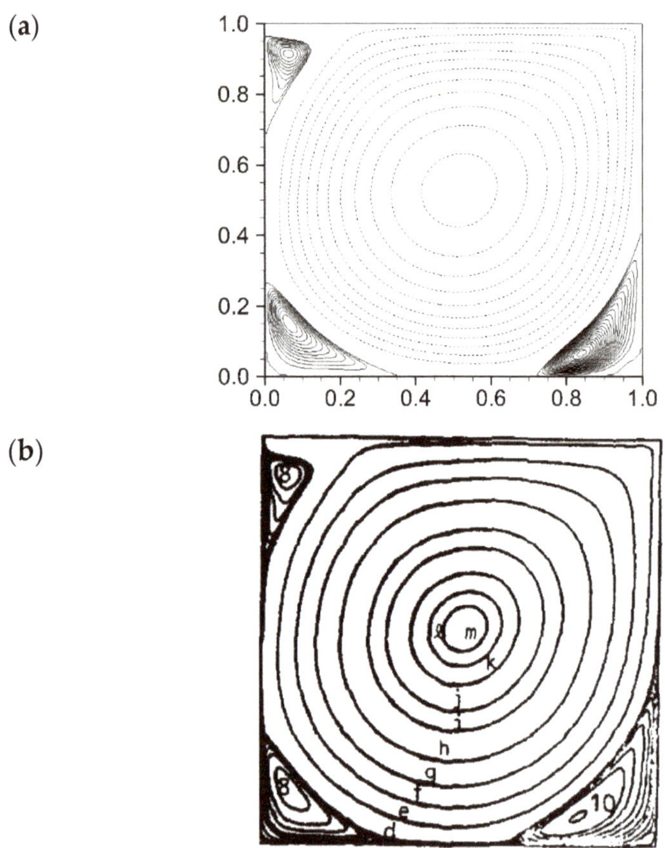

Figure 8. *Cont.*

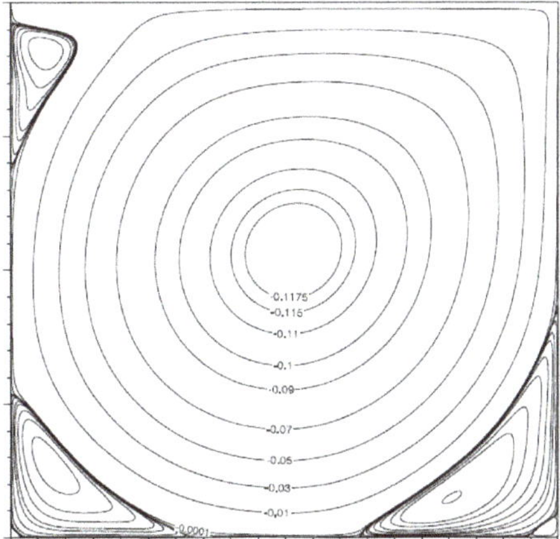

Figure 8. Comparisons of streamline patterns for Re_{lid} = 5000 in a square cavity. (**a**) Present solutions; (**b**) Ghia et al., adapted from [37], with permission from Copyright © 1982 Published by Elsevier Inc.; (**c**) Erturk et al., adapted from [39], with permission from © Mar 11, 2005, John Wiley & Sons Ltd.

4.2. Vortical Flows in a Square Cavity with a Free Surface: Fr = 1.0, Re = 5000 and 500

In this subsection, we discuss the vortical-flow structures at various values of Re. For an inlet flow with a constant speed, Fr = 1.0, the phenomena of generated vortices for flows passing over a 1 × 1 square cavity at two Reynolds numbers, Re = 5000 and 500, are considered.

The case with Re = 5000 is first discussed for the streamline patterns in Figure 9. During the transient process of vortex evolution, it reveals that a shear flow separating the left corner forms a clockwise vortex (Figure 9a) immediately and enlarges to occupy almost the entire cavity space at t = 2 (Figure 9c). As time progresses, a small counterclockwise eddy emerges from the right wall (Figure 9d). This secondary vortex continues to expand to fill nearly half of the space in the lower right part of the cavity (Figure 9e). Later, this bottom counterclockwise vortex transitions to the left wall and extends upwards to reduce the region of the clockwise vortex on top of it (Figure 9j). With the further developments of the vortices under the influence of a near-constant shear-flow velocity passing over the opening of a cavity, as shown in Figure 9p, the streamline patterns show a flattened, oval, clockwise vortex that is visible at the top part of the cavity and appear below it is a large counterclockwise, but weak vortex. It should be noted this counterclockwise vortex is evolved from a secondary vortex. In addition, small eddies, which are too weak to be displayed clearly, appear at both bottom corners. This pattern is different from that of a pure cavity flow at Re_{lid} = 5000, which can eventually reach a steady-state flow condition. In Figure 9, the free surface evolves with time, yielding a movable cavity lid that interacts with the fluid inside and outside of the cavity.

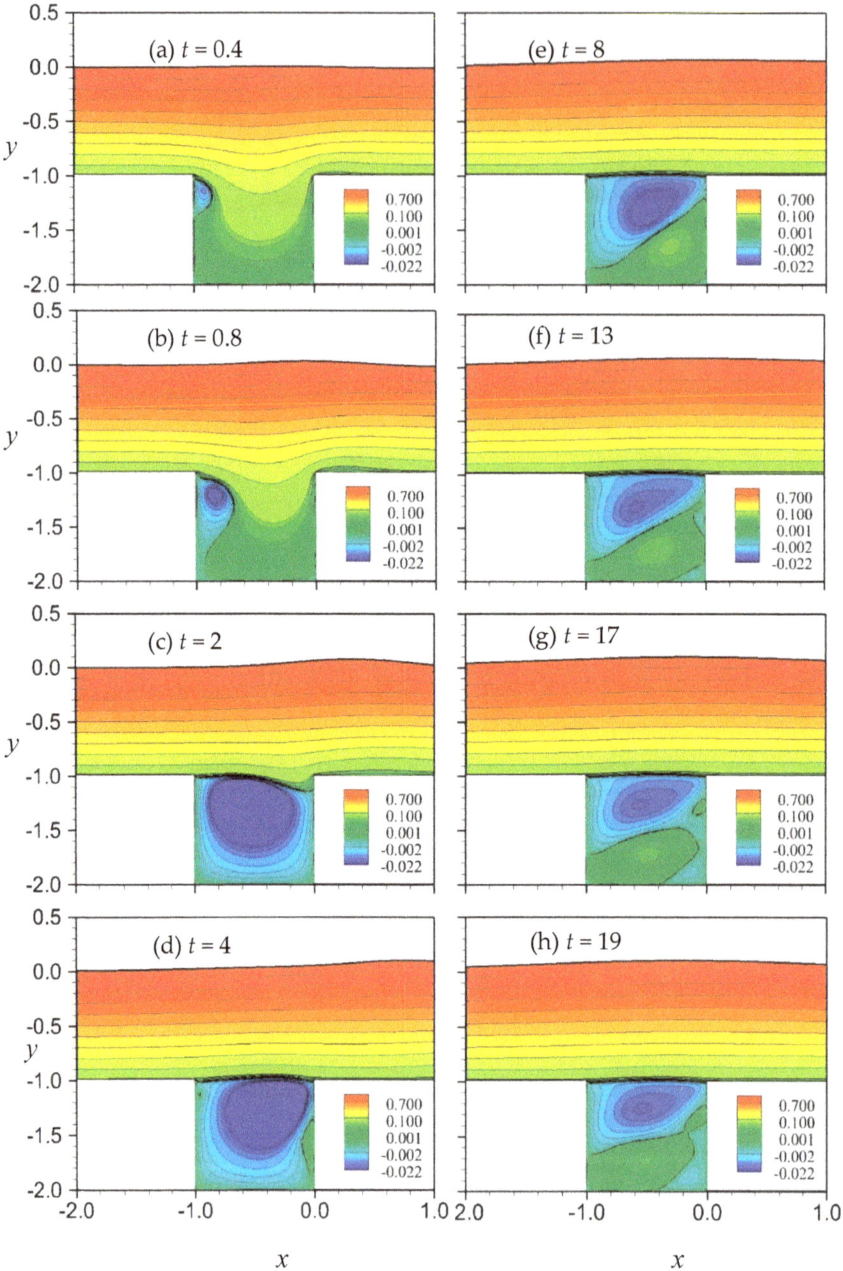

Figure 9. *Cont.*

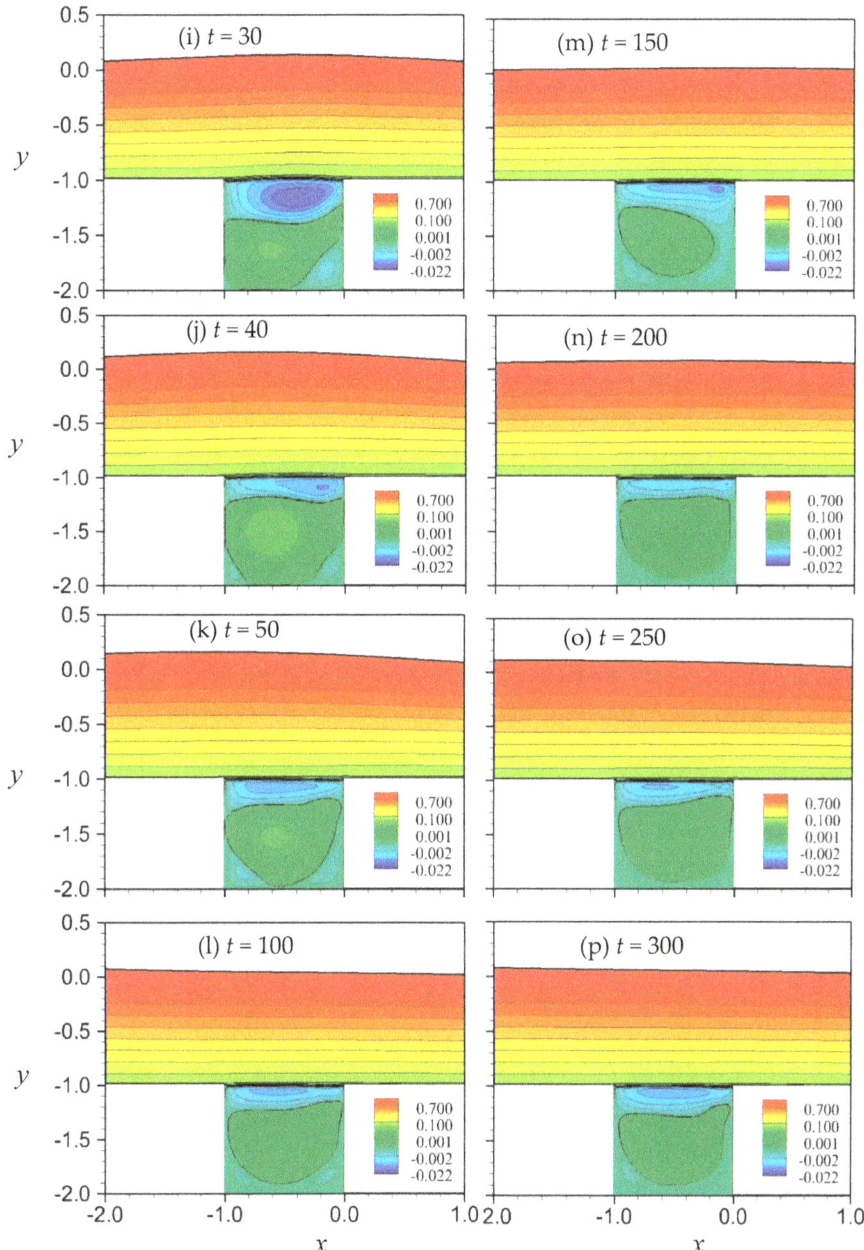

Figure 9. Streamline patterns in a square cavity for $Re = 5000$ and $Fr = 1.0$ at various phases: (**a**) $t = 0.4$, (**b**) $t = 0.8$, (**c**) $t = 2$, (**d**) $t = 4$, (**e**) $t = 8$, (**f**) $t = 13$, (**g**) $t = 17$, (**h**) $t = 19$, (**i**) $t = 30$, (**j**) $t = 40$, (**k**) $t = 50$, (**l**) $t = 100$, (**m**) $t = 150$, (**n**) $t = 200$, (**o**) $t = 250$, and (**p**) $t = 300$.

For the case of the lower Reynolds number, $Re = 500$, Figure 10 illustrates the time variations of the vortex flows in a cavity and the near steady situation is achieved after about $t = 30$ (Figure 10g). The main vortex pattern is clearly shown to be different from that presented in Figure 9. In this case, an obvious eddy occurs at the bottom right corner (Figure 10d). A small but visible eddy appears at the left corner. This feature is similar to the pattern of pure cavity flow for a lower value of Re_{lid} described in Ghia et al. [37]. It is worth to note, although the flow pattern inside the cavity seems to reach a near steady state, the surface waves are still in a transient condition.

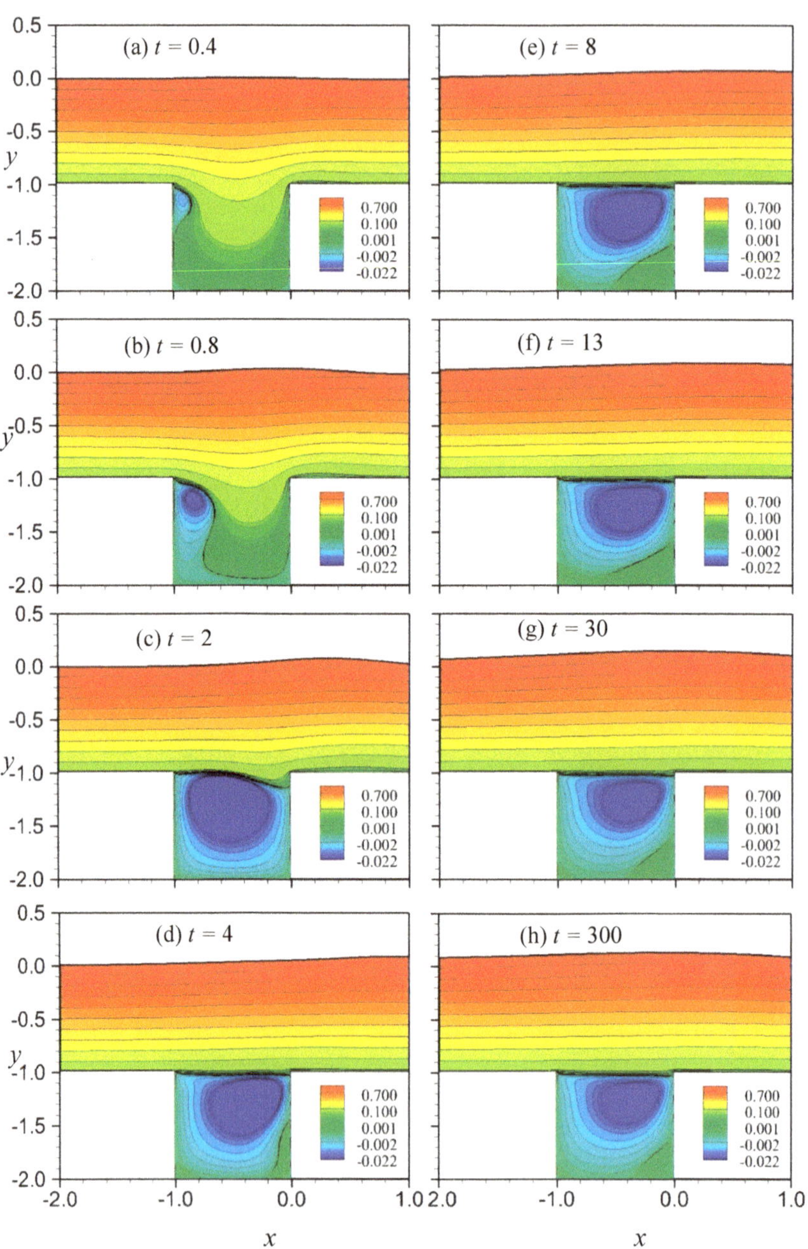

Figure 10. Streamline patterns in a square cavity for $Re = 500$ and $Fr = 1.0$ at various phases: (**a**) $t = 0.4$, (**b**) $t = 0.8$, (**c**) $t = 2$, (**d**) $t = 4$, (**e**) $t = 8$, (**f**) $t = 13$, (**g**) $t = 30$, and (**h**) $t = 300$.

To examine the effect of the Reynolds number on the vortical flow, the viscous flow patterns at $t = 12$ for the cases of $Re = 5000$ and 500 are compared with each other as well as with the results from the potential flow in Figure 11, where the inflow condition is set as $Fr = 1.0$. As can be seen in Figure 11a, the potential flow solutions show no vortices produced in the cavity; which is not realistically reflected in nature. The fluid accelerates in the upward direction downstream of the cavity to force the free surface to rise up violently and eventually reach a condition violating the potential flow theory. For the viscous fluid flow cases, the cavity is almost completely occupied by vortices. For $Re = 5000$, Figure 11b reveals that the flow structure in the cavity is in the developing stage and the space is occupied mainly by a secondary vortex (green part). It finally reaches a quasi-steady like flow (see Figure 9). For $Re = 500$, the primary vortex (blue part, see Figure 11c) is always dominant in the cavity and a steady-state flow can be formed eventually.

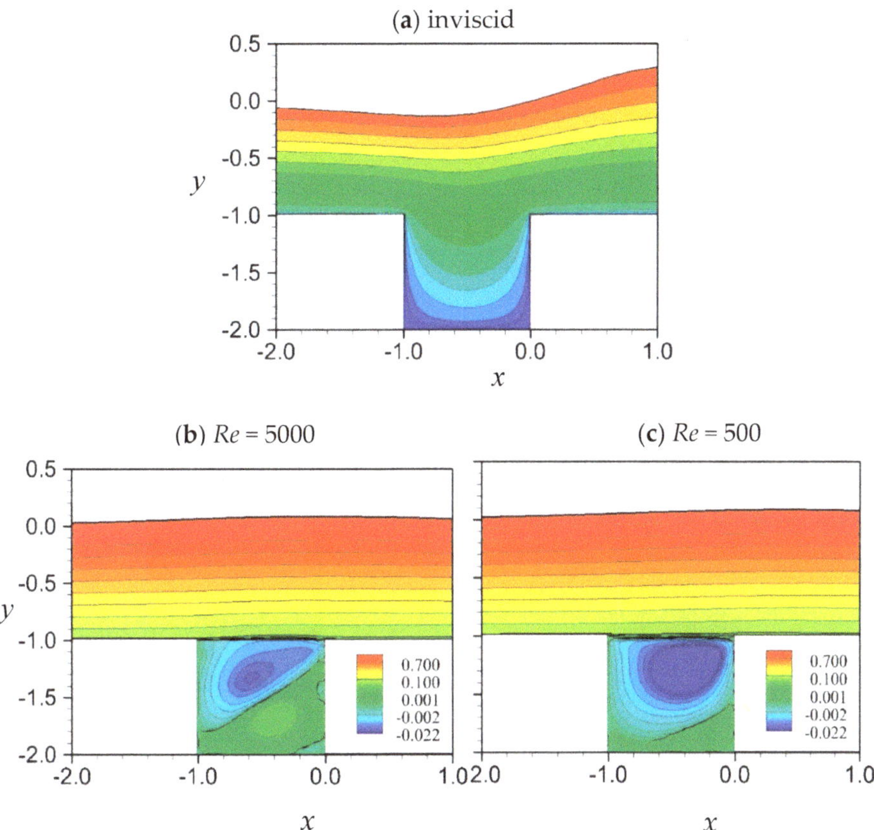

Figure 11. Streamlines for $Fr = 1.0$ at $t = 12$: (**a**) inviscid flow, (**b**) $Re = 5000$, and (**c**) $Re = 500$.

4.3. Free-Surface Elevations for a Square Cavity at Re = 5000 and 500 with Fr = 0.5 to 1.1

To analyze the influence of the Froude number on the free-surface motion, the Reynolds number is fixed at $Re = 5000$ or 500. The water surface elevations at the central cavity position are plotted in time histories ($t = 0$ to 300) for Fr ranging from 0.5 to 1.1 in Figure 12. The flow ranges cover the subcritical, critical, and supercritical regions. The varying displacement of water surface reflects the potential formation of advancing waves for a specified Fr. The tendencies of free-surface pattern are similar between the plots for the cases of $Re = 5000$ and 500 but vary in phase and strength. For a subcritical flow condition (e.g., *Fr* = 0.5), it appears the free surface is reflected with the small undular motions. With an increase of *Fr*, the forms of waves are enhanced and become stronger (see Figure 12). However, as it has not ever been reported for the case of *Fr* ≈ 1.2, this higher supercritical flow may cause unstable situation. This transition regime at 1.1 < *Fr* < 1.2 was ascertained by Lee et al. [40] in analyzing a hump forcing function. They also indicated that for the positive forcing cases the wave breaking was experimentally observed to occur in the advancing solitary waves at lower supercritical speeds (about 1.1 < *Fr* < 1.2).

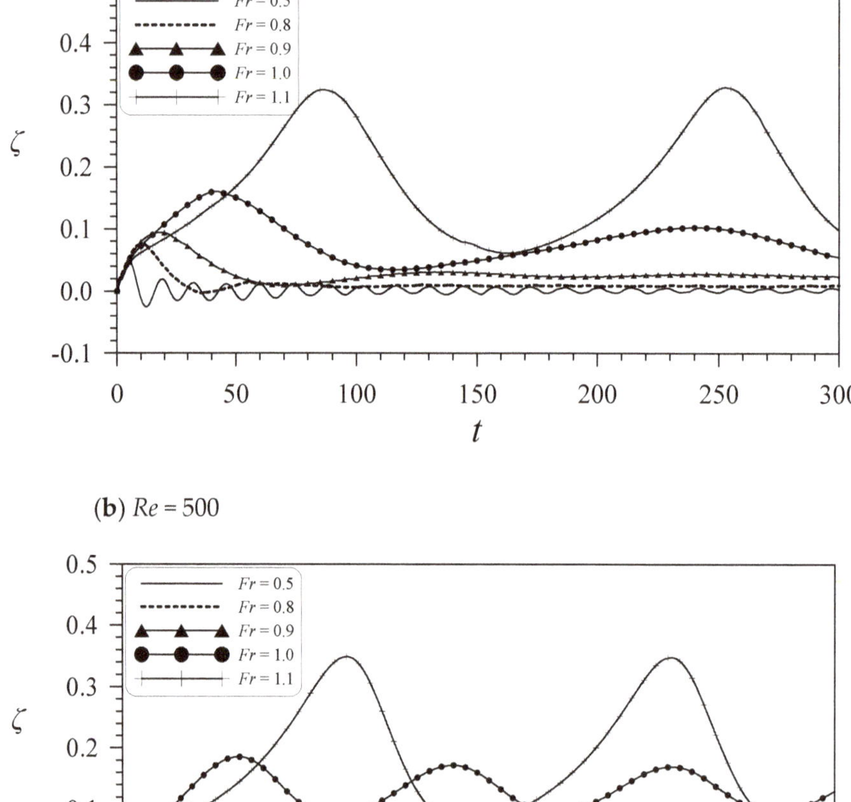

Figure 12. Free-surface elevation at *x* = −0.5 for various *Fr* at (**a**) *Re* = 5000 and (**b**) *Re* = 500.

Figure 13 shows the subplots of the time varying free-surface elevations for various values of *Fr* to reflect the differences of the water surface between the conditions of *Re* = 5000 and 500. For the range of near-critical flow, *Fr* = 0.8, 0.9, 1.0, and 1.1, the waves are large enough to show the obvious viscous influences; including the period between each of the upstream advancing solitons. For the lower value of *Re*, the wave elevation is higher and the period is shorter. By contrast, the viscous effect on the lower subcritical flow (e.g., *Fr* = 0.5) is inconspicuous.

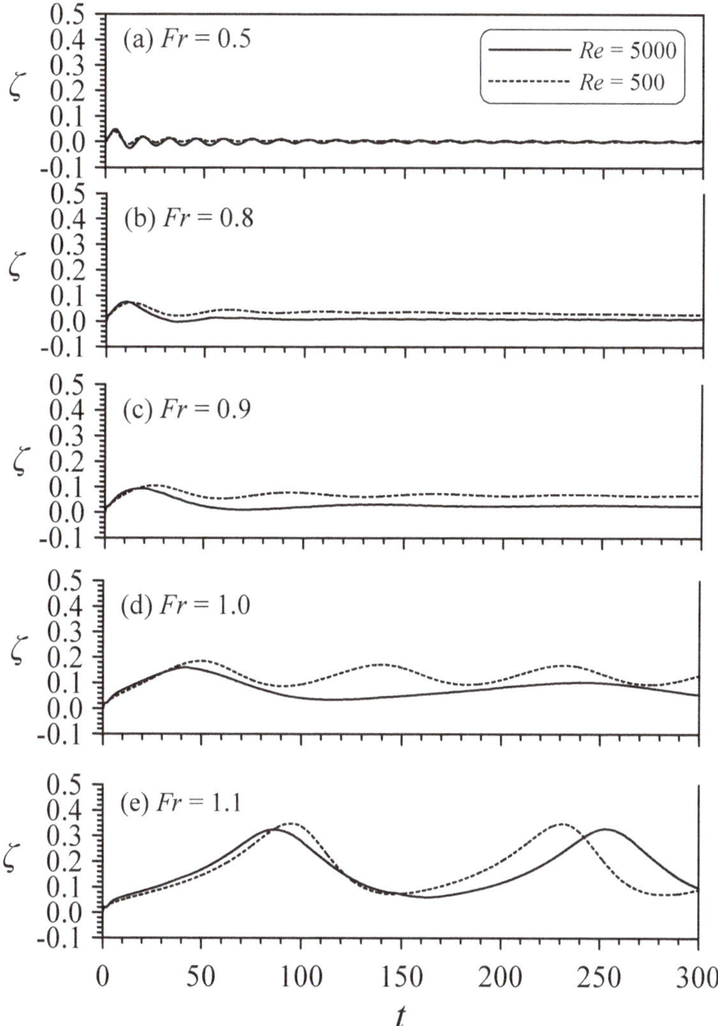

Figure 13. Time varying free-surface elevations at *x* = −0.5 for *Re* = 5000, 500, and at various *Fr*: (a) *Fr* = 0.5, (b) *Fr* = 0.8, (c) *Fr* = 0.9, (d) *Fr* = 1.0, and (e) *Fr* = 1.1.

4.4. Free-Surface Profiles at Various Fr for Re = 5000 and 500

With the same conditions as in Figure 13, the results of the spatial wave profiles at *t* = 300 are plotted in Figure 14. The basic phenomena of the wave pattern produced by transcritical flow over a disturbance have been described explicitly by Lee et al. [40]. Similar undular bores are found in Figure 14d with *Fr* = 1.0 where the indicated three flow regions can be noticed. It is observed that there are three separate solitons generated when *Re* = 500. For the case of higher-subcritical flow (e.g.,

$Fr = 0.9$), the trailing waves are stronger while the amplitude of advancing solitons become larger for lower-supercritical flow (e.g., $Fr = 1.1$). Comparing the advancing solitons generated under the conditions of $Re = 5000$ and $Re = 500$, the wave displacement for $Re = 5000$ is found to be less than that for $Re = 500$. This is potentially a result of the thicker boundary layer from the condition of lower Re allowing the additional strength and the disturbance to be extended to a greater region to produce stronger wave motions. In spite of this, after a long traveling distance, the waves are expected to undergo more damping for a lower value of Re. It is also interesting to note the wave motions are affected by the movement of the lid streamline of the cavity as it interacts with the inlet flow to influence the wave patterns. It was found that when $Re = 500$, the separation streamline lid in the cavity could reach steady state. However, the separation-streamline lid of $Re = 5000$ showed a slightly movable boundary. Overall, the lid of $Re = 500$ was lower than the average of $Re = 5000$. The lower lid of Re = 500 will be shown to be one of the causes of large wave fluctuations.

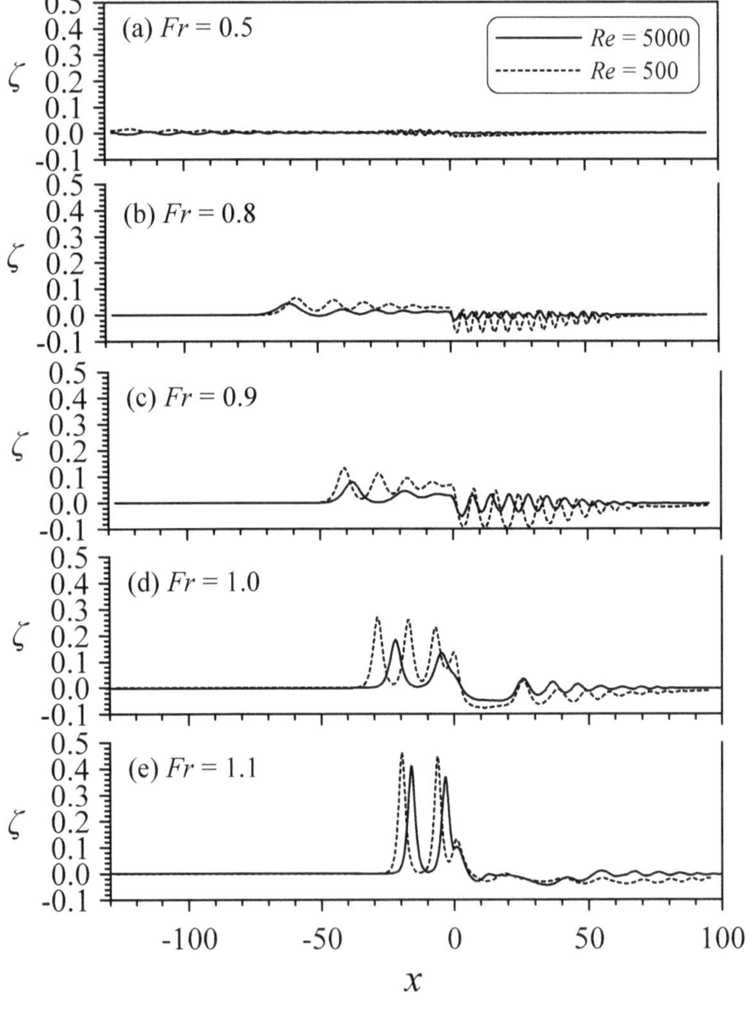

Figure 14. Comparisons of the free-surface profiles at $t = 300$ for $Re = 5000$ and 500 with various Fr: (**a**) $Fr = 0.5$, (**b**) $Fr = 0.8$, (**c**) $Fr = 0.9$, (**d**) $Fr = 1.0$, and (**e**) $Fr = 1.1$.

The near-critical flows create dramatically different wave phenomenon, especially the upstream advancing solitons. Similar phenomena were found in the near-critical flow over a hump in Lee et al. [40]. They addressed both the amplitude of the upstream-advancing waves and their generation period increase very rapidly as Fr approaches the limiting value of about 1.2. This was also one of the incentives behind this article. The flow pasts over a protrusions can produce this phenomenon. So what about the cave and the influence of fluid viscosity? To illustrate the process of the wave formation, the free-surface profiles are distinctly plotted with perspective view plots in the x-t plane, as shown in Figure 15. Only the case of $Re = 500$ is selected to plot the wave elevations under the transcritical flow conditions, $Fr = 0.9$, 1.0, and 1.1. In each subplot, the free-surface profiles are displayed from $t = 0$ to 300 at an interval of three unit times. Similar to the explanations mentioned above, but with more insights, the results show the dramatic differences of upstream and downstream wave patterns among the cases from high subcritical Fr (e.g., $Fr = 0.9$) to low supercritical (e.g., $Fr = 1.1$). The wave trough/peak traces in x-t plane implies the swift movement of the generated waves. The wave patterns of near-critical flows passing over a bottom cavity are similar to those of flows over a hump. The reason is the depressed region will be filled with vortices and a movable cavity lid is formed above the cavity. The movable lid plays a similar role as the hump to produce the surface disturbance for the advancement of a group of solitary waves upstream of the cavity. In terms of the free-surface elevations behind the cavity, the high subcritical flow (e.g., $Fr = 0.9$), when compared to the low supercritical flow, tends to make the downstream oscillatory trailing waves with higher frequency and amplitude resembling a train of modulated cnoidal waves. Also, its upstream undular bores are weaker. In contrast, the low supercritical flow (e.g., $Fr = 1.1$) produces a group of higher and clearly separated solitons but weaker trailing waves with longer periods resembling indeed small-amplitude long waves.

4.5. Influence of Cavity Depth

This section will discuss the effects of the depth of the cavity. Consider $Re = 5000$, as an example. Figure 16 denotes the stream patterns in the cavity with different aspect ratios (shallow half) when $t = 300$. Their vortex motion can reach the quasi-steady state. In addition, both of them also exhibit a flat elliptical eddy in the cavity near the lid. Figure 16a is a shallower-depth cavity. The virtual cavity lid of the shallower one is slightly above -1. Also, to compare their wave elevation at a fix position ($x = -0.5$), Figure 17 shows the shallower cavity makes the shorter wave period and larger waves. The vortex motion in cavity will play a very important role to influence the wave evolutions. The cavities of different depths or shapes will affect the vortices and the wave motions. The more details for the effects of geometry and size of cavity may be discussed in our future work.

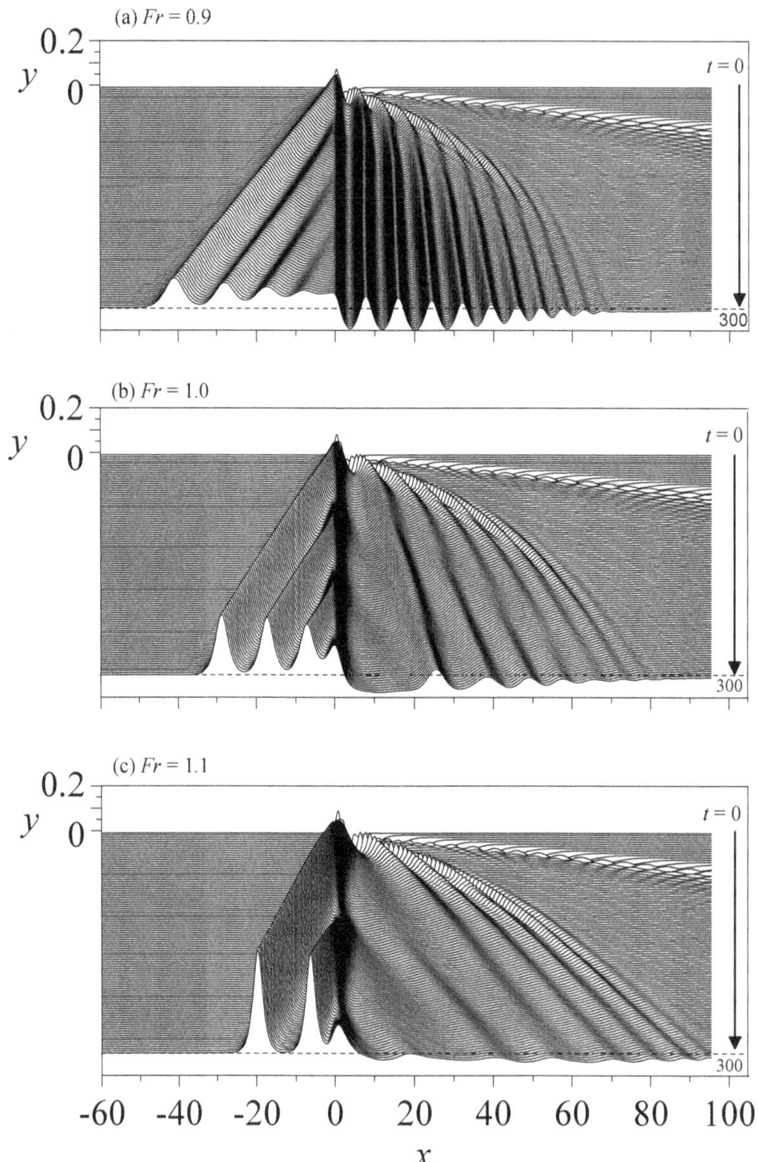

Figure 15. Time series plots of wave elevations at $Re = 500$ for (**a**) $Fr = 0.9$, (**b**) $Fr = 1.0$, and (**c**) $Fr = 1.1$.

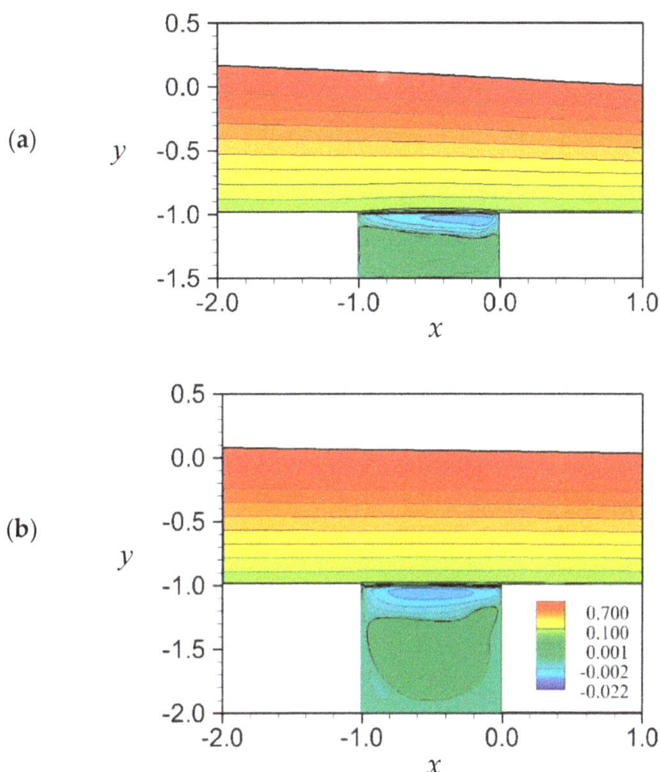

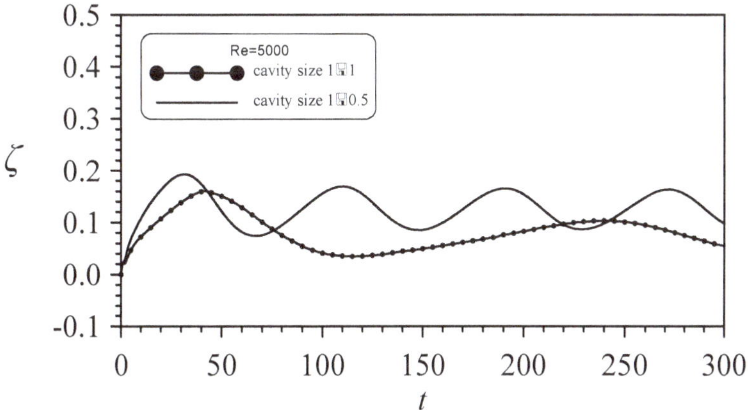

Figure 16. Streamline patterns in a cavity for $Re = 5000$ and $Fr = 1.0$ at $t = 300$ (quasi- steady phase) with cavity size (**a**) 1.0×0.5, and (**b**) 1.0×1.0.

Figure 17. Time varying free-surface elevations for $Re = 5000$ at $Fr = 1$ with different cavity depth.

5. Conclusions

This numerical study presents the phenomena of free-surface deformation with generated upstream advancing solitons and downstream oscillating waves together with the vortical motion due to a uniform flow passing over a bottom cavity. Under the effect of a negative forcing, the present model results are verified with the limiting cases of potential flow solutions of Zhang and Chwang [20]. In addition, the vortical-flow patterns in the cavity are compared with the results from other studies of pure lid-driven cavity flow. The breaking points and important findings are summarized as follows:

1. This paper numerically explores the properties of a viscous free-surface flow over a cavity, which has never been investigated in the past for the vortex motions and the surface waves produced by a negative forcing. Numerical simulations under the consideration of a fixed cavity and various flow conditions ranging Fr from 0.5 to 1.1 at Re =500 and 5000 are carried out with results presented and discussed.

2. Under the condition of $Fr = 1.0$, the vortical flows in the cavity for the cases of lower value of Re (e.g., $Re = 500$) are shown to be similar to the classical closed lid-driven cavity flow pattern, where a steady-state solution can be reached. For the case with a higher Re (e.g., $Re = 5000$), the flow motions, although can establish nearly to a quasi-steady state condition, are more complex and are very different from the patterns of $Re = 500$. Although the cavity flow can reach the steady state, the free surface will remain unsteady. This is a phenomenon worth exploring, one that not been explored by researchers. Therefore, we attempt to provide further possible phenomena in this area of research. As for the viscous influence on the wave development, at a lower Re, stronger advancing solitary waves are generated, possibly because of the increase in the thickness of the boundary layer on the solid walls and the shallower separation-streamline lid of the cavity. This problem is complicated because of the interaction between waves and vortices. Previous researchers have considered inviscid fluid for this problem. This will be very different from the real situation. Therefore, this paper hopes that research in this area can make further progress to consider the effect of viscosity.

3. The wave properties of flows over a cavity is found to resemble those with flows over a hump. The forming of a movable but slightly protruding cavity lid as flows passing over a concavity becomes a forcing mechanism. With the values of Fr ranging from 0.5 to 1.1, a series of cases covering from subcritical to supercritical flow regimes are investigated. The results for the lower subcritical flow (e.g., $Fr = 0.5$) condition indicate that the water surface is disturbed with very weak undular motions. With an increase of Fr (e.g., up to lower supercritical flow conditions), it is noticed the wave height of the upstream advancing waves (solitons) increases. The time required for the development of each emerging solitary wave is also increased.

4. In the literature, the variations in waves caused by flow over depressed terrain have not been widely discussed. When it is, the influence of viscosity is generally missing. Also, this phenomenon is difficult to produce in experiments. That is why the results for cavity flow and free surface were verified separately. We strove to prove the credibility of this model, although indirectly. Therefore, the motivation behind, and purpose of this paper, is to provide some information for investigators who are interested in this issue.

Author Contributions: The author (C.-H.C.) developed the numerical model, performed the computation, and drafted and polished the paper.

Funding: This research was funded by the Ministry of Science and Technology, Taiwan, Republic of China (Grant No. MOST 106-2221-E-275-003) and the APC for publishing this paper was also funded by the Ministry of Science and Technology, Taiwan, Republic of China (Grant No. MOST 106-2221-E-275-003).

Acknowledgments: This work was supported financially by the Ministry of Science and Technology, Taiwan, Republic of China (Serial No. MOST 106-2221-E-275-003). The author is very grateful to Professor Keh-Han Wang of the Department of Civil Engineering at the University of Houston for the improvement of many comments in the article. The authors would like to thank Enago (www.enago.tw) for the English language review.

Conflicts of Interest: The author declares no conflict of interest.

Nomenclature

$\widetilde{A}, \widetilde{B}$	convective coefficients
b	shape of bottom object
b_m	minimum tip of object
Fr	Froude number
g	gravitational acceleration
g^{ij}, f^i	geometric coefficients
H	non-dimensional still-water depth
H^*	dimensional still-water depth
(i, j)	grid-node indices
IM	maximum grid index in x-direction
J	Jacobian
JM	maximum grid index in y-direction
L	length of object
$\widetilde{n}$	unit-normal vector
Re	Reynolds number
Re_{lid}	Re for lid-driven cavity flow
(U, V)	contra-variant fluid velocities
U^*	dimensional inlet velocity
u_f	tangential free-surface fluid particle velocity
(x, y)	Cartesian coordinates
α	solitary-wave height
$\delta_\tau, \delta_\xi, \delta_\eta$	finite-difference operators
Δ	variable increment
Δn	normal distance between the wall and the adjacent node
∇^2	Laplacian operator
ζ	free-surface elevation
ν	kinematic viscosity
(ζ, η)	Curvilinear coordinates
ν	time in transient curvilinear coordinate system
Ψ	Stream function
Ψ_1	stream function at the first grid node to the wall
ω	vorticity

Appendix A. Numerical Method for Free-Surface Calculation

The free-surface elevation (ζ) is calculated from the kinematic free-surface boundary condition (Equation (6)) whereas, the stream function (ψ) on the free surface is evaluated in accordance with the dynamic free-surface condition (Equation (7)).

We define the following operators with a dummy variable f,

$$\delta_\tau f = \frac{f_{i,JM}^{*(n+1)} - f_{i,JM}^n}{\Delta\tau} \tag{A1-a}$$

$$\delta_\xi f = \frac{f_{i+1,JM}^{*(n+1)} - f_{i-1,JM}^{*(n+1)}}{2\Delta\xi} \tag{A1-b}$$

$$\delta_\eta f = \frac{f_{i,JM}^{*(n+1)} - f_{i,JM-1}^{*(n+1)}}{\Delta\eta} \tag{A1-c}$$

where the superscript "$*(n + 1)$" represents the variables with temporary values used in each iteration at the $(n + 1)$ new time level. After a series of calculation tests, a fully implicit second-order upwind scheme is utilized on the free-surface boundaries. In one of authors' early studies [31] on modeling flows over a bottom-mounted square cylinder, a mixed explicit–implicit scheme was used and was found to be stable and accurate for the flow in the context of a positive-forcing problem. However, in this study, using a fully implicit second-order upwind scheme is shown to be a better approach in solving the problem of flows interacting with a negative forcing like body. Therefore, Equations (6) and (7) can be discretized and arranged as

$$\zeta_i^{n+1} = \zeta_i^n - \Delta\tau\left[\frac{\delta_\xi\psi}{\delta_\xi x}\right] \tag{A2}$$

$$\psi_{i,JM}^{n+1} = \left[-\frac{C_1}{2}\psi_{i+1,JM}^{*(n+1)} + \frac{C_1}{2}\psi_{i-1,JM}^{*(n+1)} + C_2\psi_{i,JM-1}^{*(n+1)} + C_3 + C_4\right]/C_2 \tag{A3}$$

where

$$C_1 = Jg^{12} + \widetilde{A} = Jg^{12} + \Delta\tau\left[-\delta_\tau(\frac{x_\eta}{J})\delta_\xi x - \delta_\tau(\frac{\zeta_\eta}{J})\delta_\xi\zeta\right]$$

$$C_2 = Jg^{22} + \widetilde{B} = Jg^{22} + \Delta\tau\left[\delta_\tau(\frac{x_\xi}{J})\delta_\xi x + \delta_\tau(\frac{\zeta_\xi}{J})\delta_\xi\zeta\right]$$

$$C_3 = -\Delta\tau\left[u^{*(n+1)}u_\xi + v^{*(n+1)}v_\xi + \delta_\xi\zeta\right]$$

$$C_4 = Jg^{12}\delta_\xi\psi + Jg^{22}\delta_\eta\psi$$

In coefficient C_3, u_ξ and v_ξ (generalized as φ_ξ in expressions shown below) are modelled with a second-order upwind scheme as

$$\varphi_\xi = \begin{cases} (\varphi_{i-2,JM}^{*(n+1)} - 4\varphi_{i-1,JM}^{*(n+1)} + 3\varphi_{i,JM}^{*(n+1)})/2, \; if \; \varphi^{*(n+1)} > 0, \\\\ (-\varphi_{i+2,JM}^{*(n+1)} + 4\varphi_{i+1,JM}^{*(n+1)} - 3\varphi_{i,JM}^{*(n+1)})/2, \; if \; \varphi^{*(n+1)} < 0. \end{cases} \tag{A4}$$

In addition, the finite difference method is utilized on the downstream boundary conditions, which are discretized backward in space using the first-order scheme and with the first-order temporal difference in this calculation.

References

1. Ryzhov, E.A.; Koshel, K.V. Steady and perturbed motion of a point vortex along a boundary with a circular cavity. *Phys. Lett. A* **2016**, *380*, 896–902. [CrossRef]
2. Ryzhov, E.A.; Koshel, K.V.; Sokolovskiy, M.A.; Carton, X. Interaction of an along-shore propagating vortex with a vortex enclosed in a circular bay. *Phys. Fluids* **2018**, *30*, 016602. [CrossRef]
3. Shankar, P.N.; Deshpande, M.D. Fluid mechanics in the driven cavity. *Annu. Rev. Fluid Mech.* **2000**, *32*, 93–136. [CrossRef]
4. Chang, K.; Constantinescu, G.; Park, S.O. Analysis of the flow and mass transfer processes for the incompressible flow past an open cavity with a laminar and a fully turbulent incoming boundary layer. *J. Fluid Mech.* **2006**, *561*, 13–145. [CrossRef]
5. Zhang, X.; Rona, A. An observation of pressure waves around a shallow cavity. *J. Sound Vib.* **1998**, *214*, 771–778. [CrossRef]
6. Fang, L.C.; Nicolaou, D.; Cleaver, J.W. Transient removal of a contaminated fluid from a cavity. *Int. J. Heat Fluid Flow* **1999**, *20*, 505–613. [CrossRef]
7. Fang, L.C.; Nicolaou, D.; Cleaver, J.W. Numerical simulation of time-dependent hydrodynamic removal of a contaminated fluid from a cavity. *Int. J. Numer. Meth. Fluids* **2003**, *42*, 1087–1103. [CrossRef]
8. Van Dyke, M. *An Album of Fluid Motion*; Parabolic Press: Stanford, CA, USA, 1982.
9. Erturk, E.; Gokcol, O. Fine grid numerical solutions of triangular cavity flow. The European Physical. *J. Appl. Phys.* **2007**, *38*, 97–105.
10. Chang, M.H.; Cheng, C.H. Predictions of lid-driven flow and heat convection in an arc-shape cavity. *Int. Commun. Heat Mass Transf.* **1999**, *26*, 829–838. [CrossRef]

11. Yin, X.; Kumar, S. Two-dimensional simulations of flow near a cavity and a flexible solid boundary. *Phys. Fluids* **2006**, *18*, 063103. [CrossRef]

12. Khanafer, K.; Aithal, S.M. Laminar mixed convection flow and heat transfer characteristics in a lid driven cavity with a circular cylinder. *Int. J. Heat Mass Transf.* **2013**, *66*, 200–209. [CrossRef]

13. Grilli, S.T.; Vogelmann, S.; Watts, P. Development of a 3D numerical wave tank for modeling tsunami generation by underwater landslide. *Eng. Anal. Bound. Elem.* **2002**, *26*, 301–313. [CrossRef]

14. Adkins, D.; Yan, Y.Y. CFD simulation of fish-like body moving in viscous liquid. *J. Bionic Eng.* **2006**, *3*, 147–153. [CrossRef]

15. Kara, F.; Tang, C.Q.; Vassalos, D. Time domain three-dimensional fully nonlinear computations of steady body–wave interaction problem. *Ocean Eng.* **2007**, *34*, 776–789. [CrossRef]

16. Chang, C.H.; Wang, K.H. Generation of three-dimensional fully nonlinear water waves by a submerged moving object. *J. Eng. Mech.* **2011**, *137*, 101–112. [CrossRef]

17. Hanna, S.N.; Abdel-Malek, M.N.; Abd-el-Malek, M.B. Super-critical free-surface flow over a trapezoidal obstacle. *J. Comput. Appl. Math.* **1996**, *66*, 279–291. [CrossRef]

18. Tzabiras, G.D. A numerical investigation of 2D, steady free surface flows. *Int. J. Numer. Methods Fluids* **1997**, *25*, 267–598. [CrossRef]

19. Wu, D.M.; Wu, T.Y. Three dimensional nonlinear long waves due to moving surface pressure. In Proceedings of the 14th Symposium of Naval Hydrodynamics, Ann Arbor, MI, USA, 1 January 1982; National Academy Press: Washington, DC, USA; pp. 103–129.

20. Zhang, D.; Chwang, A.T. Generation of solitary waves by forward- and backward-step bottom forcing. *J. Fluid Mech.* **2001**, *432*, 341–350.

21. Grimshaw, R.H.J.; Smith, N. Resonant flow of a stratified fluid over topography. *J. Fluid Mech.* **1986**, *169*, 429–464. [CrossRef]

22. Wu, T.Y. Generation of upstream advancing solitons by moving disturbances. *J. Fluid Mech.* **1987**, *184*, 75–99. [CrossRef]

23. Camassa, R.; Wu, T.Y. Stability of forced steady solitary waves. *Philos. Trans. R. Soc. Lond. A* **1991**, *337*, 429–466.

24. Grimshaw, R.H.J.; Zhang, D.H.; Chow, K.W. Transcritical flow over a hole. *Stud. Appl. Math.* **2009**, *122*, 235–248. [CrossRef]

25. Grimshaw, R.H.J.; Zhang, D.H.; Chow, K.W. Generation of solitary waves by transcritical flow over a step. *J. Fluid Mech.* **2007**, *587*, 235–254. [CrossRef]

26. Xu, G.D.; Meng, Q. Waves induced by a two-dimensional foil advancing in shallow water. *Eng. Anal. Bound. Elem.* **2016**, *64*, 150–157. [CrossRef]

27. Chang, C.H.; Tang, C.J. Viscous effects on nonlinear water waves generated by a submerged body in critical motion. In Proceedings of the 17th National Conference on Theoretical and Applied Mechanics, Taipei, Taiwan, 10–11 December 1993; pp. 35–42.

28. Zhang, D.; Chwang, A.T. Numerical study of nonlinear shallow water waves produced by a submerged moving disturbance in viscous flow. *Phys. Fluids* **1996**, *8*, 147–155. [CrossRef]

29. Lo, D.C.; Young, D.L. Numerical simulation of solitary waves using Velocity–Vorticity Formulation of Navier–Stokes Equations. *J. Eng. Mech.* **2006**, *132*, 211–219. [CrossRef]

30. Tang, C.J.; Chang, C.H. Flow separation during a solitary wave passing over a submerged obstacle. *J. Hydraul. Eng.* **1998**, *124*, 732–749. [CrossRef]

31. Chang, C.H.; Tang, C.J.; Wang, K.H. Vortex pattern and wave motion produced by a bottom blunt body moving at a critical speed. *Comput. Fluids* **2011**, *44*, 267–278. [CrossRef]

32. Tang, C.J. Free Surface Flow Phenomena ahead of a Two-Dimensional Body in a Viscous Fluid. Ph.D. Thesis, The University of Iowa, Iowa City, Iowa, 1987; pp. 14–15.

33. Nallasamy, M. Numerical solution of the separation flow due to an obstruction. *Comput. Fluids* **1986**, *14*, 59–68. [CrossRef]

34. Chen, C.J.; Chen, H.C. Finite analytic numerical method for unsteady two-dimensional Navier-Stokes equations. *J. Comput. Phys.* **1984**, *53*, 209–226. [CrossRef]

35. Chang, C.H.; Lin, C. Effect of solitary wave on viscous-fluid flow in bottom cavity. *Environ. Fluid Mech.* **2015**, *15*, 1135–1161. [CrossRef]

36. Lowery, K.; Liapis, S. Free-surface flow over a semi-circular obstruction. *Int. J. Numer. Meth. Fluids* **1999**, *30*, 43–63. [CrossRef]
37. Ghia, U.; Ghia, K.N.; Shin, C.T. High-Re solutions for incompressible flow using the Navier-Stokes equations and a multigrid method. *J. Comput. Phys.* **1982**, *48*, 387–411. [CrossRef]
38. Peng, Y.F.; Shiau, Y.H.; Hwang, R.R. Transition in a 2-D lid-driven cavity flow. *Comput. Fluids* **2003**, *32*, 337–352. [CrossRef]
39. Erturk, E.; Corke, T.C.; Gorke, C. Numerical solutions of 2-D steady incompressible driven cavity flow at high Reynolds numbers. *Int. J. Numer. Meth. Fluids* **2005**, *48*, 747–774. [CrossRef]
40. Lee, S.J.; Yates, G.T.; Wu, T.Y. Experiments and analyses of upstream-advancing solitary waves generated by moving disturbances. *J. Fluid Mech.* **1989**, *199*, 569–593. [CrossRef]

MDPI

St. Alban-Anlage 66

4052 Basel

Switzerland

Tel. +41 61 683 77 34

Fax +41 61 302 89 18

www.mdpi.com

Journal of Marine Science and Engineering Editorial Office

E-mail: jmse@mdpi.com

www.mdpi.com/journal/jmse

9 783039 281824